Environmental Division 2016

Core Programming Area at the 2016 AIChE Annual Meeting

San Francisco, California, USA
13 - 18 November 2016

ISBN: 978-1-5108-3435-4

Printed from e-media with permission by:

Curran Associates, Inc.
57 Morehouse Lane
Red Hook, NY 12571

Printed by Curran Associates, Inc. (2017)

For permission requests, please contact AIChE
at the address below.

AIChE
120 Wall Street, FL 23
New York, NY 10005-4020

Phone: (800) 242-4363
Fax: (203) 775-5177

www.aiche.org

Additional copies of this publication are available from:

Curran Associates, Inc.
57 Morehouse Lane
Red Hook, NY 12571 USA
Phone: 845-758-0400
Fax: 845-758-2633
Email: curran@proceedings.com
Web: www.proceedings.com

TABLE OF CONTENTS

475551 Toward Understanding the Atmospheric Chromium Chemistry

Abstract:

Research Interests:

Trivalent chromium (Cr(III)) and hexavalent chromium (Cr(VI)) are the two common oxidation states of chromium in the environment. Inhalation exposure to Cr (VI) may lead to cancer, nasal damage, asthma and bronchitis. In contrast, Cr(III) is a trace element essential for the proper function of living organisms. Therefore, it is important to accurately discriminate between these two species in atmospheric particulate matter (PM) samples. In the United States, Cr(VI) is one of the criteria pollutants which has been monitored through the National Air Toxics Trends Station (NATTS) Network. Thus, measurement of Cr(VI), is one of the focal points for hazardous air pollutants (HAPs) monitoring in the US.

The focus of my PhD research at Clarkson University was centered around understanding the atmospheric chromium chemistry and improving the sampling of atmospheric Cr(VI). Chromium is available as soluble or insoluble in the atmosphere. When soluble chromium is dissolved in water, the transformation of Cr(VI) is expected to take place in the water associated with particles (aqueous chemistry of chromium). Airborne particles that include soluble salts like ammonium sulfate or ammonium nitrate can spontaneously take up water (deliquescence) from the air and form a solution on the surface of the particle. This liquid layer allows chemistry to take place. I have shown that keeping the sampling particulate matter below the point at which it can take up water provides significant improvements in the amount of Cr(VI) retained in the sample. I have found that Cr(III) is converted to Cr(VI) in the presence of gaseous species like ozone and reactive oxygen species (ROS) at dry condition (gas-solids chemistry of chromium) implying that Cr(III) conversion can produce false positive measurements for Cr(VI) sampling. I have demonstrated that conversion decreases at lower temperature values. Thus, based on my work, it is likely that a sampler that keeps the sample dry and cool will improve the recovery of Cr(VI) that is present in the sample. I built a prototype sampler and deployed 4 of them in a field study in NJ, side-by-side with 4 of the samplers currently used by USEPA in their HAPs monitoring network. These tests showed that the improved samplers did provide better preservation of Cr(VI). Presently, I aim to develop a dynamic model for the simulation of aqueous chromium chemistry in the atmosphere ($1<pH<5$) and during the sampling period (24 hours- pH 9).

Recent studies showed that insoluble Cr(III) and insoluble Cr(VI) dominate atmospheric chromium mass. Regardless of few data in my PhD thesis, no information is available in the literature on the reaction kinetics of insoluble chromium speciation in airborne particles. Therefore my future research will concentrate on insoluble chromium chemistry (gas-solids chemistry of chromium) in the atmosphere (within next five years). My long term research plan (beyond next five years) is to investigate the chemistry/toxicity of soluble and insoluble chromium compounds in human lungs and their impact on lung surfactants. I have proposed three projects on atmospheric chromium. The first project investigates the reaction kinetics of insoluble Cr(III) and insoluble Cr(VI) with reactive oxygen species and volatile organic compounds, respectively (gas-solids reactions). The second project is aimed at designing and field testing a new Cr(VI) sampler to preserve both Cr(VI) and Cr(III) during sampling. The third project focuses on modeling Cr reactions (soluble and insoluble reactions) in the atmosphere and during the sampling. The details of my future research plan and potential funding agencies, current research, previous teaching and undergraduate research mentoring experiences are available during the AIChEÂ â€œMeet the Faculty Candidatesâ€ poster session.

Teaching Interests:

I have the past experiences of teaching the following courses:

Capstone Design Projects in Chemical Engineering, Process Simulation using Aspen Plus, Transport Phenomena, Safety Engineering and Risk Management, Process Calculations (mass and energy balance)

475988 Membranes at the Water-Energy-Food Nexus: Experimental and Modeling Approaches

Abstract:

Research Interests:

This poster will provide an overview of my computational and experimental research about the design and fabrications of different types of membrane (microfiltration, ultrafiltration, reverse osmosis and forward osmosis) at the water-Energy-Food Nexus. My Ph.D. dissertation focused on the application of polymeric nano-composite membranes (Mixed Matrix Membrane) to water purification. During the first part of my research, I worked on understanding the mechanism of natural organic matter (NOM) fouling on an ultrafiltration membrane by relating the operational viewpoint for fouling (macroscopic) to the microscopic viewpoint. Then, in the second part of my research, I used the power of nanoparticles to increase the antifouling properties of the polysulfone membrane. The mixture of titanium dioxide nanoparticles and multiwall carbon nanotube with the optimum mass ratio was embedded with the novel and simple method into the polymeric membrane and interestingly increased both the membrane permeability and rejection two times compared to the pure polysulfide membrane. Also, the nanocomposite membranes showed the higher antifouling behavior compared to the pure membrane at different operational conditions and flaunt physicochemical properties. In my post-doctoral research, I have continued to work on the membrane application for water treatment and energy production. Forward osmosis (FO) (also, pressure retarded osmosis) is a prime membrane technology of interest due to its potential for low energy use in water – energy nexus. The main goal of my post-doctoral project was investigation of effective parameters on transportation of water and solute molecules in forward osmosis membrane to improve FO membrane design and process efficiency. This project combines computational fluid dynamics (CFD) modeling with experimental fabrication of FO membranes to examine the assumptions imbedded in current transport equations and identify sources of inaccuracies in water flux predictions. This effort can lead us to the development of transport equations that can be used by all to support FO research, enabling the

deployment of FO for real-world problem solving with economically competitive performance. Based on my knowledge of membranes, instrumentation design and componential fluid dynamic (CFD) modeling, my future research will focus on both experimental and modeling aspects of membrane technology issues for the water – energy –food nexus. In details, I plan to mainly focus my academic career on parallel research area of the computational modeling and experimental synthesis, modification and fabrication of different types of novel mixed matrix membranes for using at different connected interdisciplinary areas at the water – energy – food nexuses.

Some research areas that I plan to pursue in my first five years as an Assistant Professor:

Membrane fouling mitigation** Forward osmosis and pressure retarded osmosis design, surface modification and characterization**Mixed matrix ultrafiltration/microfiltration**Nanomaterials environmental safety: transport and fate of nanomaterials

Teaching Interests:

My first preference would be to teach the chemical engineering core courses, such as Transport Phenomena and Chemical Engineering Thermodynamics. However, since my background and previous experience is diverse, I am flexible to handle other courses within the University Chemical Engineering Curriculum. Additionally I would like (provided the opportunity is given) to develop new courses, as the special topics, including topics of water / wastewater treatment, membrane technology and environmental science.

476060 Removal of Contaminants from Water and Wastewater: (Bio)Sorption, Membrane Filtration, Advanced Oxidation

Abstract:

Overcoming the emerging global water shortages and pollution are grand challenges faced by humanity today. This poster will provide an overview of my research upon the development and optimizations of different biological, chemical and physical methods for removal of toxic and carcinogenic contaminants (both organic and inorganic) from water and wastewater. Summary of my top achievements include: During my Master, I worked on optimization of biosorption process which is an eco-friendly, effective and extremely low-cost technology for removal of synthetic dyes and hexavalent chromium which are toxic and carcinogenic contaminants, from water by using available biosorbents/adsorbents (such as algae, aquatic plants, etc.). Finally, potent organic/inorganic-binding biomass types for removal of pollutants form aqueous solution were discovered. My current PhD research is on Optimization of advanced oxidation process (pulsed corona electrical discharge, UV/photocatalysts) as a highly efficient method, for removal of public health dangers contaminants (synthetic dyes, humic substances) from water/wastewater without using any chemicals (eco-friendly) and optimization of a unique hybrid advanced oxidation technology /nanocomposite ultrafiltration membrane for removal of synthetic and natural organic public health dangers contaminants form water.

Research Interests:

- Using low cost biosorbents/adsorbents for wastewater treatment
- Membrane-based hybrid processes for treating impaired quality waters
- Providing important insights into fundamental mechanisms of membrane fouling under a wide range of operating conditions.
- Development of novel approaches to enhance performance of advanced oxidation processes (electrical discharge, UV photocatalysts, ozone) for water and waste water treatment

Future Direction: My research experiences and objectives give me confidence that I will be able to obtain funding and add significant improvements in areas related to the water and wastewater treatment. I will be able to establish a strong collaboration with industry and academia to make my research objectives achievable.

Teaching Interests: Â As a faculty member, I would like to continue my research in the water and wastewater treatment. I am interested to teach graduate courses in the field of Advanced Topics in Environmental Chemistry, Chemical Principles of Environmental Systems, and the graduate laboratory courses on Environmental Instrumentation. I also would like to teach theÂ undergraduate and core courses on kinetic, transport phenomena, environmental material balances and coordinate in the undergraduate Capstone Course.

Selected publications:

- **Negin Koutahzadeh**, Milad Rabbani Esfahani, Pedro E Arce, Removal of Acid Black 1 from water by the pulsed corona discharge advanced oxidation method, *Journal of Water Process Engineering, Volume 10, April 2016, Pages 1â€"8.*
- **Negin Koutahzadeh**, Milad Rabbani Esfahani, Pedro E Arce, A sequential use of UV/H_2O_2 â€" (PSF/TiO_2/MWCNT) mixed matrix membranes for dye removal in water purification: Membrane permeation, fouling, rejection and decolorization, *Journal of Environmental Engineering Science, 2016* (Accepted, *In press).*
- **Negin Koutahzadeh**, E. Daneshvar, M.Kousha, M.S.Sohrabi, A.Bhatnagar, Biosorption of hexavalent chromium from aqueous solutions by six brown macro algae biomasses: Equilibrium and kinetics studies, *Journal of Desalination and Water Treatment*, Volume 51, Issue 31-33, 2013.

476108 Application of Shrinking Core Model Applied for Gas Hydrate-Based CO_2 Capture in Presence of Porous Hydrogels

Abstract:

Research Interests: CO_2 Capture, Gas Hydrate, Process Modelling, Reservoir Simulation, Flow Assurance, Asphaltene Deposition and Precipitation, Enhanced Oil Recovery

Teaching Interests: Thermodynamics, Transport Phenomena, Chemical Engineering Simulation Software, Heat Transfer, Kinetics and Reactor Design

A salient portion of the global pollutant gases emission is contributed to the carbon dioxide (CO_2) released from industrial sectors. Currently, stiff environmental requirements do not allow industries to proceeds beyond the standard limitations for such a harmful exhaust. Hence, deduction of CO_2 released from industrial processes has been the subject of researches for many years seeking novel, efficient and cost-effective technologies for worldwide applications. Among the various methodologies, the hydrate-based CO_2 capture (HBCC) process is considered as a promising approach to achieve this goal. This process, however, is in its infancy and must be examined through a vast range of experiments and modelling studies prior to being well commercialized. Advancements in both the process intensification and the materials and technology development associated with this method can be found in a few recent reports. While most works focus on the development of new materials to enhance the kinetic of this process, a very small number of studies have been focused on the mechanism of the hydrate-based CO_2 capture process in the presence of porous materials which will lead to design and improvement of new materials for industrial application. In this article, porous hydrogel (HYD) microspheres are investigated for application in CO_2 clathrate hydrates. A new variation of so-called shrinking core model is developed and analyzed for this case study in which the intrinsic hydrate formation rate is combined with transport phenomena to comprehensively capture the role of controlling mechanisms including the diffusivity of the CO_2 gas molecules, the water distribution and conversion rate in various phases of the medium and operational parameters involved. The proposed model allows the prediction and explanation of hydrate formation and CO_2 capture rates in hydrate structure under different pressures and temperatures. The model is tested against experimental data which have been obtained in our research group. A range of sensitivity analysis is carried out to investigate the influence of the extracted parameters.

456510 Size-Dependent Chemical Compositions in Particulate Matters from Major Outdoor Sources in a Megacity and Corresponding Inhalation Exposure Assessment

Abstract:

A variety of existing studies have characterized the chemical compositions in particulate matters (PM) for various megacities. However, these studies rarely accounted for the size-dependence of chemical constituents in PM, hindering the application of the data for inhalation exposure assessment and thus epidemiological studies. In this work, a series of field measurements were conducted to characterize the major emission sources of Singapore, i.e. traffic and industry emissions (coal combustion is rare in Singapore and thus disregarded). The particle size distribution and corresponding size-dependent chemical compositions were identified. Existing respiratory deposition models were employed to analyze the deposition distributions of particles and chemical compositions in the human respiratory system, based on the data of particle size and chemical compositions. Finally, potential protection suggestions were proposed to mitigate the PM exposure for susceptible populations. The data in this work will be also representative to other megacities whose particulate air pollution is mainly derived from traffic and industry emissions.

Research Interests:

Dr. Siming Youâ€™s research interests fall under the umbrella of the interactions between environment and energy systems. He is trying to explore the potential environmental problems underlying existing energy systems, and understand and solve environmental problems from an energy point of view. Graduated from Nanyang Technological University as the top PhD student in Thermal and Fluids Engineering in 2014, he accumulated significant interdisciplinary research experiences by studying human exposure to harmful particulate matters (PM), especially, pathogens, and thus infection transmission from a physicochemical point of view. He then pursued his postdoctoral research at the Department of Civil & Environmental Engineering, Massachusetts Institute of Technology, and gained a deeper understanding of the transport dynamics of harmful PM in environments and inhalation exposure-related health implications. He is now a research fellow at Professor Chi-Hwa Wangâ€™s group at the National University of Singapore, where he is investigating the environmental impacts and challenges of sustainable energy systems and proposing solutions to the potential environmental problems facing sustainable energy systems.

Teaching Interests:

I would like to teach Fluid Mechanics for undergraduate students, and Aerosol Technology and Environmental Health for both undergraduate and graduate students.

Fluid Mechanics is one of the most actively developing fields of physics and one of the basic core courses of engineering. Significant parts of my PhD studies and postdoctoral study at MIT have devoted to understanding how harmful particles transport in environments under the effects of aerodynamic forces. Specifically, I have accumulated a significant amount of knowledge about turbulence, such as the characteristics of turbulent bursts and the methods of quantifying turbulence parameters. These past research experiences and knowledge did lay a solid foundation to enable me to participate in this course on Fluid Mechanics.

Aerosol is ubiquitous throughout the world and the related technology has been widely employed in various fields such as microelectronic industry, pharmaceutical industry, environmental sciences, etc. The course of Aerosol Technology would cover the fundamental properties and behaviors of aerosols, aerosol sampling methods, and the application of aerosol technology in various fields such as industry and environmental health sciences. This course would equip students with the ability to sample, analyze, and apply aerosols for their future requirements in either industry or academics.

Environmental Health would cover the knowledge about how natural and built environments may affect human health. It would present an overview of methods used to understand environmental health hazards, ranging from global issues such as climate change to regional issues such as air, water, and energy to local issues such as food safety, and occupational health, and to conduct risk assessment and risk management for environmental exposure-related health implications. This course would not only equip students with necessary knowledge and skills to address the potential concerns related to environmental health, but also encourage them to make positive changes in their behavior and attitude to protect the environment.

470274 Environmental Applications of Field Portable Low Temperature Porous Layer Open Tubular Cryoadsorption Headspace Sampling and Analysis

Abstract:

Building on the successful application in the laboratory of PLOT-cryoadsorption as a means of collecting vapor (or headspace) samples for chromatographic analysis, we now introduce a field- portable apparatus. This device fits inside of a briefcase (aluminum tool carrier), and can be easily transported by vehicle or by air. The portable apparatus functions entirely on compressed air, making it suitable for use in locations lacking electrical power, and for use in flammable and explosive environments. The apparatus consists of four aspects: a field capable PLOT-capillary platform, the supporting equipment platform, the service interface between the PLOT-capillary and the supporting equipment, and the necessary peripherals. Vapor sampling can be done with either a hand piece (containing the PLOT capillary) or with a custom fabricated standoff module. Both the hand piece and the standoff module can be heated and cooled to facilitate vapor collection and subsequent vapor sample removal. The service interface between the support platform and the sampling units makes use of a unique countercurrent approach that minimizes loss of cooling and heating due to heat transfer with the surroundings (recuperative thermostatting). We will also describe the use of this apparatus applied to several analytes of interest. We conducted tests with coumarin and 2,4,6-trinitrotoluene (two solutes that were used in initial development of PLOT-cryo technology), naphthalene, aviation turbine kerosene, and diesel fuel, on a variety of matrices and test beds. We demonstrated that these analytes can be easily detected and reliably identified using the portable unit for analyte collection. By leveraging efficiency-boosting temperature control and the high flow rate multiple capillary wafer, very short collection times (as low as 3 s) yielded accurate detection. For diesel fuel spiked on glass beads, we determined a method detection limit below 1 ppm. We observed greater variability among separate samples analyzed with the portable unit than previously documented in work using the laboratory-based PLOT-cryo technology. We identify three likely sources that may help explain the additional variation: the use of a compressed air source to generate suction, matrix geometry, and variability in the local vapor concentration around the sampling probe as solute depletion occurs both locally around the probe and in the test bed as a whole. This field-portable adaptation of the PLOT-cryo approach has numerous and diverse potential applications.

470549 Effect of Beaded Activated Carbon Fluidization on Adsorption of Volatile Organic Compounds

Abstract:

Adsorption on activated carbon is a widely used technique for controlling emissions of volatile organic compounds (VOCs); however, one challenge in removing VOCs from gas streams is irreversible adsorption or heel formation, which occurs during successive adsorption/desorption cycles. For adsorbents to be most effective, heel buildup should be minimized. This research investigates the effect of adsorbent bed configuration on VOC adsorption with beaded activated carbon (BAC). Five-cycle adsorption/desorption tests using a single VOC (1,2,4-trimethylbenzene) and a mixture of nine VOCs intended to simulate industrial painting emissions were completed using fixed and fluidized bed adsorber configurations. Adsorption tests were completed with full loading of the adsorbent. All regeneration cycles were completed in the fixed bed arrangement to standardize the regeneration method and focus results on the impact of bed-type during adsorption. Results show that the adsorption capacity of the BAC is not affected by the adsorption bedâ€™s configuration in case of reaching full breakthrough. For the VOC mixture, however, 30% less heel buildup was observed for the fluidized bed configuration. Thermogravimetric analysis of the regenerated samples and gas chromatography â€“ mass spectrometry during adsorption were also completed to assess the heel buildup and adsorption kinetics, respectively. Higher accumulation of heavy adsorbates was found in fixed bed as opposed to fluidized bed, justifying the higher heel buildup in the former. On the other hand, improved mass transfer was found across the entire fluidized bed as a result of better gas-particle contact. These results show that, besides the expected engineering advantages of a fluidized bed adsorption system (e.g., lower pressure drop), decreased heel buildup is an additional advantage when using fluidized bed adsorption in industrial settings.

462578 Modeling of Mercuric Chloride Removal Using Raw Activated Carbon

Abstract:

The physical adsorption of mercuric chloride ($HgCl_2$) onto raw activated carbon was studied in an entrained flow and fabric filter based on adsorption kinetics and equilibrium data obtained from fixed-bed system. An operation study was conducted where high $HgCl_2$ removal was achieved by injecting raw activated carbon sorbent at a high loading for a short time followed by no sorbent injection until the end of the cleaning cycle. This injection method helped in increasing sorbent utilization for the same amount of $HgCl_2$ removed as compared to continuous injection. The pressure drop across the filter cake was estimated using Darcyâ€™s law. The pressure drop across the filter cake was found to be within the operating limits for a typical fabric filter. The capture performance for mercury in an entrained flow was found to be insignificant with a removal efficiency of $<1\%$. It was found that 90% $HgCl_2$ removal efficiency was achieved within the first 9 min of filter cake residence time using the injection method. The role of internal and external mass transfer was studied in detail. It was found that while the external mass transfer played an insignificant role, the role of internal mass transfer was pronounced. The result shows that fundamental adsorption kinetics and equilibrium can be used to predict $HgCl_2$ capture.

459249 Improved Performance of Beaded Activated Carbon for Volatile Organic Components Capture Via Chemical Surface Modification

Abstract:

Volatile organic compounds (VOCs) associated with painting processes are typically captured on activated carbon so that clean air is released to atmosphere. The VOCs consist of aromatic and aliphatic hydrocarbons, ketones, esters and alcohols. In the present study, adsorption of organic vapours onto microporous beaded activated carbon (BAC) and regeneration of the saturated BAC are carried out. Further, the surface chemistry of the BAC was modified by various chemical treatment methods and the effects of the resultant changes in porosity and surface chemistry on adsorption characteristics of the BAC samples were studied. Characterizations conducted with elemental analysis, physical adsorption and scanning electron microscope demonstrate that both the porosity and surface chemistry of the carbon beads were tuned by the chemical treatment methods without any alteration of the integrity of the desirable overall spherical morphology. Adsorption evaluation with thermogravimetric analysis under various conditions indicate that chemically treated BACs have relatively high VOC adsorption capacity compared to untreated BAC with fast adsorption/desorption kinetics, superior durability, and good selectivity. Both the unique spherical form (mean diameter 0.72 mm) and the superior adsorption performance render the modified BAC a promising candidate for VOC capture especially in automotive industries.

468435 Effect of Adsorbate Properties on Heel Formation during Regeneration of Activated Carbon

Abstract:

Regeneration conditions such as high heating rate, low purge gas flow rate, and other factors can lead to heel formation and adsorption capacity loss in activated carbon. This research investigates the effects of adsorbate properties on heel formation in activated carbon, under high heating rates and low purge gas flow. For this purpose, five-cycle adsorption/desorption tests were completed using eight alkyl benzenes including toluene, 1,2,4 trimethylbenzene (TMB), 1,3,5 TMB, o-xylene, p-xylene, m-xylene, isopropylbenzene, and neopentylbenzene as adsorbates. Heel formation showed differences for these alkyl benzene adsorbates with cumulative heel formation ranging from 8.9% (toluene) to 17.0% (neopentylbenzene). Thermogravimetric analysis identified adsorbate decomposition as the main heel formation mechanism in all cases, showing a highly stable heel with low removal (less than 31.5%) at temperatures as high as 800 °C. Higher heel formation was observed for adsorbates with high boiling point and low thermal stability (i.e., longer branches with Î^{2} and Î^{3}-carbons), which provided favourable conditions for adsorbate decomposition. The results from this study contribute to a better understanding of heel formation mechanisms during activated carbon regeneration.

469399 A Theoretical Investigation of Sulfur Adsorption on Calcium Oxide

Abstract:

Sulfur capture on coal fly ash plays a critical role in limiting SO_x emissions in power plants. The retention on the fly ash is heavily dependent on the alkali and alkaline earth metal (AAEM) species in coal such as Ca. In oxy-combustion, this mechanism becomes particularly important as the sulfur concentration (SO_2 + SO_3) in the flue gas can be several times greater than traditional air-combustion due to the unique staging required of oxy-fuel systems. While higher rates of sulfur capture on fly ash have been observed for oxy-combustion compared to air-combustion, this increased retention can create concerns in utilization of the fly ash for cement and concrete production.

Although there have been previous experimental studies, the surface mechanism of sulfur retention on Ca-based materials is not well understood. In this study, computational modeling was employed to examine the binding of SO_x species to the most prominent AAEM species, calcium oxide. Density functional theory (DFT) calculations were performed using Vienna *ab initio* Simulation Package (VASP) to investigate the binding mechanism for SO_x on a CaO(100) surface. Calculations were also run with various other co-adsorbates, including CO_2, H_2O and SO_x, that are pre-adsorbed to the surface to examine the effect of their presence on any subsequent SO_x reactions on the surface. The energetic, geometric and electronic properties of the simulated surfaces were analyzed to discern any possible trends in the SO_x binding mechanism with respect to the specific pre-adsorbed molecule. The results were then compared to experimental data where CaO was exposed to a variety of flue gas compositions. In this way, a more complete picture of the SO_x binding mechanism on calcium oxide was determined allowing for a better understanding of the initial stages of sulfur capture in combustion systems.

472770 A Microfluidic Study of Liquid-Liquid Extraction Mediated By Carbon Dioxide

Abstract:

Liquid-liquid extraction is an important separation and purification method, however it faces a challenge in reducing the energy consumption and the environmental impact of solvent (extractant) recovery. The reversible chemical reactions of switchable solvents (nitrogenous bases) with carbon dioxide (CO_2) can be implemented in reactive liquid-liquid extraction to reduce significantly the cost and energy requirements of solvent recovery. The development of new effective switchable solvents reacting with CO_2 and the optimization of extraction conditions rely of the ability to evaluate and screen the performance of switchable solvents in extraction processes. We report a microfluidic strategy for time- and labour-efficient studies of CO_2-mediated solvent extraction. The platform utilizes a liquid segment containing an aqueous extractant droplet and a droplet of a solution of a switchable solvent in a non-polar liquid, with gaseous CO_2 supplied to the segment from both sides. Following the reaction of the switchable solvent with CO_2, the solvent becomes hydrophilic and transfers from the non-polar solvent to the aqueous droplet. By monitoring the time-dependent variation in droplet volumes, we determined the efficiency and extraction time for the CO_2-mediated extraction of different nitrogenous bases in a broad experimental parameter space. The platform enables a significant reduction in the amount of switchable solvents used in these studies; provides accurate temporal characterization of the liquid-liquid extraction process, and offers the capability of high-throughput screening of switchable solvents.

465291 Modeling of Water Quality Downgradient of Mulch Biowall with Nearby Surface Water Receptor

Abstract:

Groundwater treatment biowalls are commonly placed at a property boundary to prevent contaminant transport off-site or adjacent to a surface water body to protect from contaminant discharges to the surface water. Groundwater contaminants passing through the biowall are treated within the biowall or immediately downgradient of the biowall. A biowall typically contains bark mulch or other agricultural waste material that serves as an electron donor to the degradation process. One of the results of biowall treatment activity is the generation of anaerobic groundwater, which may result in increased sulfide levels if influent groundwater contains moderate to high sulfate concentrations. Sulfide represents a degraded water quality of the groundwater exiting the biowall, which is of significant concern if the biowall is located in close proximity to a groundwater discharge to surface water. A time-dependent one-dimensional fate and transport model was developed to evaluate concentrations of sulfide that may be generated in such a situation. The model was developed by applying mass governing equations over a control volume of the aquifer between a biowall and a surface water body receptor. COMSOL Multiphysics was used to solve the family of equations for a timescale extending to approximately 4 years, using appropriate parameters and initial / boundary condition values. The mass species glucose (surrogate for carbon source upon degradation of solid carbonaceous material), lactic acid (surrogate for fatty acids), and electron donor sulfate were modeled with the partial differential equations. The model results provide insight into the advance of sulfide levels in groundwater downgradient of a biowall, and in close proximity of a surface water receptor. The carbon source, sulfate and sulfide data is compared to actual site data obtained from literature or other sources.

Effects of Entrapped nZVI in Alginate Polymer on BTEX Removal

Maha M. El-Shafei[1], Ahmed S. Mahmoud[2], Mohamed K. Mostafa[3] and Robert W. Peters[4]*

[1]Professor of Chemical Engineering, [2]Researcher of Inorganic Chemistry, [3]Doctor of Environmental Engineering and [4]Professor of Environmental Engineering

[1,2]Sanitary and Environmental institute (SEI), Housing and Building National Research Center (HBRC), Egypt

[3]Faculty of Environmental Engineering, Zewail City of Science and Technology, Egypt

[4]Department of Civil, Construction and Environmental Engineering, University of Alabama at Birmingham, Birmingham, AL

Abstract

Nanoparticle especially Nano Zero Valent Iron (nZVI) and entrapped nZVI showed highly reactivity towards several series of poly-aromatic hydrocarbons (PAH) contaminants in aqueous solutions. The primary purpose of this study was to use encapsulated nZVI in alginate biopolymer to reduce mono-aromatic hydrocarbons (BTEX) contaminants in aqueous solutions without use of surfactants. The nZVI entrapped in alginate biopolymer and the standard BTEX samples were prepared in the laboratory. The entrapped nZVI was characterized by using x-ray diffraction (XRD) and UV spectrophotometry and the experimental results are in agreement with previous research investigations. The optimum condition was determined by studying the effect of pH, adsorbent dose, contact time, temperature, and initial BTEX mixture concentrations. The results showed that the optimum pH value for the removal of BTEX was for pH in the range of 6.5 and 7.5. The minimum effective stirring rate for removal was 100 rpm, and the effective temperature was 50°C. The most effective time for removal was between 30 min and 45 min. The results also showed that the removal efficiency increased with decreasing BTEX sample concentration. For the above mentioned parameters, the effective dose for removing BTEX compounds was 30 gm/L and the removal efficiency was about 95%.

Introduction and Background

Volatile organic compounds (VOCs) are liquid or solid organic compounds that can easily evaporate (volatilize) into the gaseous phase. VOCs may exist in soil in four different forms: dissolved in water, sorbed on solid particles, comprising the soil gas and the individual liquid phase known as NAPL (non-aqueous phase liquid) due to their limited solubility. In 1997, the Agency for Toxic Substances and Disease Registry (ATSDR) characterized the VOC drinking water pollutants as a past public health hazard and focused its study on VOCs related diseases and caused death for humans, such as cancer related disease, damage to the nervous and immune system. BTEX (benzene, toluene, ethyl benzene, and xylene) compounds are VOCs according to EPA -SW-846 Method 8260 and are classified as human genotoxic and carcinogen substances in Canada and the European Union (Organization, 2006). BTEX compounds are compounds contained in crude petroleum, coal tar, and other petroleum products. Gasoline consists of BTEX (approximately 18 percent by weight) and other hydrocarbons (approximately 82 percent by weight). BTEX compounds include benzene, toluene, ethyl benzene, and xylene, and cause water contamination because its solubility in water.

Nanocatalysts technology, such as zero-valent iron, has been widely used in wastewater and water treatment due to their higher reactivity and surface area (El Saliby *et al.,* 2008). Calcium alginate beads are porous material that allows target contaminants to come in contact with the entrapped material (Bezbaruah *et al.*, 2014). Hence, it has many industrial applications such as in pharmaceutical industries, food industries, as well as in water treatment (Brachkova *et al.*, 2010, Hill and Khan, 2008, Roy *et al.*, 1987). Alginate beads has been used for phenanthrene and other organic contaminants removal using various techniques (Shi *et al.*, 2008). The nZVI entrapped in Ca–alginate is a modern technology that has been used to remove trichloroethylene and nitrate from contaminated water (Bezbaruah *et al.*, 2009, Bezbaruahl., 2011). Encapsulation of nZVI in Ca–alginate has many advantages including higher efficiency of retention for the contaminants spatially when entrapping nZVI. As a result, it has been widely used in water and wastewater treatment (Mohamed and Borik, 2013).

Materials and Methods

2.1 Preparation of encapsulated nZVI

Synthesis of nano-scale Zero Valent Iron (nZVI); 2.1624 g $FeCl_3.6H_2O$ was dissolved in a 4/1 (v/v) ethanol/water mixture (96 mL ethanol + 24 mL deionized water) and agitated well. Also, 0.1 molar from sodium borohydride ($NaBH_4$) solution was prepared by adding 1.5128 g $NaBH_4$ in 400 mL of deionized water. Excess borohydride is needed to ensure better formation of nZVI. The $NaBH_4$ solution was poured in a burette and added at a rate of one drop per second into the $FeCl_3 \bullet 6H_2O$ solution placed in a magnetic stirrer. Sodium alginate (0.2 g) was dissolved in 100 mL deionized water (DW) at room temperature (22 ± 2 °C). The alginate–water mixture was stirred for 20 minutes using magnetic stirrer at 600 rpm until complete dissolution was achieved. The alginate–water mixture was then left for another 30 minutes at room temperature to ensure that the alginate did not float in the solution and to allow air/gas bubbles that are generated during the mixing and stirring processes to escape. The alginate solution containing 2.0 grams/100 mL distilled water (2% w/v) was mixed with 0.3 g of nZVI and mixed using a magnetic stirrer at 500 rpm until the entire mixture turned black. The mixture was promptly dropped into a 2 gram $CaCl_2$ solution/100 mL distilled water (2% w/v) $CaCl_2$ solution at room temperature using a syringe (drop by drop without any bubbles). The Ca-alginate gel beads started to form once the sodium alginate/nZVI mixture came in contact with the $CaCl_2$ solution as shown Figure 1.

Figure 1. Encapsulated nZVI in alginate polymer.

2.2 Effect of operating parameters

2.2.1 Contact time

About 30 g/L of encapsulated nZVI dosage was added to total BTEX mixture concentration 12 μg/L at different contact times, nominally 15, 30, 45, and 60 min, at a temperature of 40 ± 3 °C, pH 7.0 ± 0.2, and stirring rate 100 rpm.

2.2.2 Adsorbent dose

Various encapsulated nZVI dosages (10, 20, 30, and 40 g/L) were added to total BTEX mixture with concentration 12 μg/L for 45 min at pH 7.0 ± 0.2, temperature 40 ± 3 °C, and stirring rate 100 rpm.

2.2.3 Effect of pH

Approximately 30 g/L of encapsulated nZVI was added to the total BTEX mixture containing 12 μg/L for 45 minutes at different pH values (4, 5, 6, 7, 8 and 9) at a temperature 40 ± 3 °C, and stirring rate 100 rpm.

2.2.4 Effect of stirring rate

Approximately 30 g/L of encapsulated nZVI dosage was added to the total BTEX mixture concentration 12 μg/L for 45 minutes at different stirring rate (100, 200 and 300 rpm) using nZVI and encapsulated nZVI at a temperature 40 ± 3°C, and pH 7 ± 0.2.

2.2.5 Effect of the initial concentration:

Approximately 30 g/L of encapsulated nZVI dosage was added to different total BTEX mixture concentrations (36, 24, and 12 μg/L) for 45 min at a temperature 40 ± 3 °C, pH 7.0 ± 0.2, and stirring rate 100 rpm.

2.3 Batch Adsorption Studies

The adsorption of BTEX compounds onto encapsulated nZVI was studied by the batch adsorption technique. A known weight of adsorbent dosage (nominally 1, 2, 3 and 4 g) was equilibrated with 100 mL of the aqueous BTEX solution of known concentration (36, 24, and 12 μg/L) in 250 mL of Erlenmeyer flasks and was agitated at 100 rpm at temperature of 40 ± 2 °C for a known period of time (ranging from 15 to 60 min). After equilibration, the suspension of the adsorbent was separated from solution by filtration using Whatman No. 42 filter paper and the concentration of BTEX compounds remaining in solution was measured using Master DANI GC (fast gas chromatography) using capillary column DN5 non polar column with diameter 60 m x 0.32 mm x 0.25 μm, wool linear 2 mm, split/splitless injector and FID detector according to ASTM 2005 stander method for water and wastewater (Federation and Association, 2005). The removal percentages of BTEX compounds were calculated using Equation 1. The amount of BTEX sorbed by weight of encapsulated nZVI in alginate was calculated using Equation 2.

$$\text{Sorption [\%]} = [C_o - C_e/C_o] \times 100 \quad (1)$$

where C_o is the initial concentration of BTEX compounds in solution, μg/l
C_e is the equilibrium concentration of BTEX compounds in solution, μg/l

$$q_e\ [\text{mg/g}] = [[C_o - C_e]V]/\ m \quad (2)$$

where qe is the equilibrium adsorption capacity, μg/g
V is the volume of aqueous solution, L
m is the dry weight of the adsorbent, g

2.4 Isotherm studies

Adsorption isotherm study used to predict and evaluate the sorption capacity of a sorbent (Chiang, 1993). This equilibrium study is described by expressing the amount of solute adsorbed per unit weight of adsorbent (q) and the concentration of solute remaining in solution C (Nkansah et al., 2012). The most widely used isotherms for water and wastewater treatment applications are the Freundlich and the Langmuir equations (Weber, 1972). Langmuir and Freundlich isotherm models were used to describe the equilibrium sorption data. Langmuir assumes monolayer coverage of adsorbate over a homogeneous adsorbent surface. The linearized form of the Langmuir equation is given by Equation 3. The plot of C_e/q_e versus C_e is employed to generate the values of q_{max} and K_L.

$$C_e/q_e = 1/K_L\, q_{max} + C_e/\, q_{max} \tag{3}$$

where q_e = the amount of BTEX adsorbed by the biomass at the equilibrium, μg/g
C_e = the BTEX concentration in the solution at the equilibrium, μg/L
q_{max} = the maximum monolayer adsorption capacity, mg/g
K_L = the Langmuir constant related to the affinity of the binding sites and adsorption energy, l/mg

The Freundlich isotherm is an empirical equation employed to describe heterogeneous adsorption surface and is given Equation 4. K_f and n can be evaluated by plotting ln (q_e) and ln(C_e).

$$\ln (q_e) = \ln (K_f) + (1/n) \ln (C_e) \tag{4}$$

where K_f = Freundlich constant related to adsorption capacity, (μg/g) $(\mu g/L)^{-1/n}$
n = Freundlich constant related to adsorption intensity, dimensionless

Results and Discussion

The BTEX compounds have been removed at different contact times (15, 30, 45 and 60 min) with encapsulated nZVI. As shown in Figure 2, the percentage removal of benzene, toluene, ethyl benzene, *o*-xylene, *m*-xylene, and *p*-xylene was 70, 70, 70, 70, 70, and 69 %, respectively, at 15 min; 87, 88, 88, 82, 86, and 89 %, respectively, at 30 min; 96, 95, 96, 95, 96, and 96 %, respectively, at 45 min; and 97, 97, 96, 97, 100, and 97 %, respectively, at 60 min.

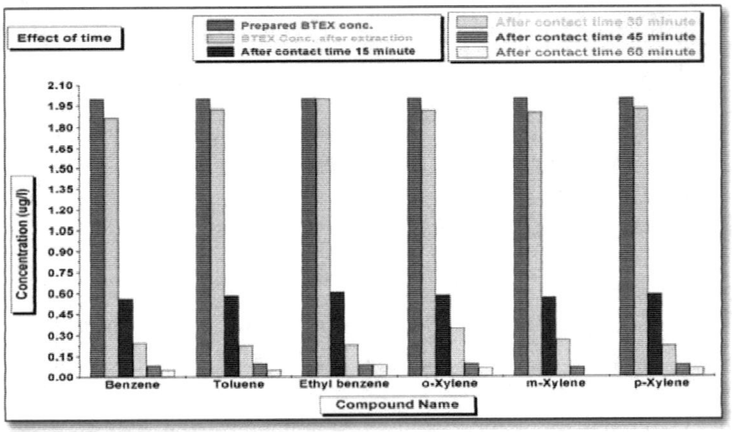

Figure 2. Effect of encapsulated nZVI contact time on BTEX removal.

As shown in Figure 3a, the percentage removal of benzene, toluene, ethyl benzene, *o*-xylene, *m*-xylene, and *p*-xylene was 83, 84, 83, 83, 83, and 83 %, respectively, for a dosage of 10 g/L; 89, 86, 86, 86, 87, and 89 %, respectively, for a dosage of 20 g/L; 96, 95, 96, 95, 96, and 96 %, respectively, for a dosage of 30 g/L; and 100, 100, 100, 100, 100 and 100 %, respectively, for a dosage of 40 g/L. The most effective dose was approximately 30 g/L, as shown in Figure 3b.

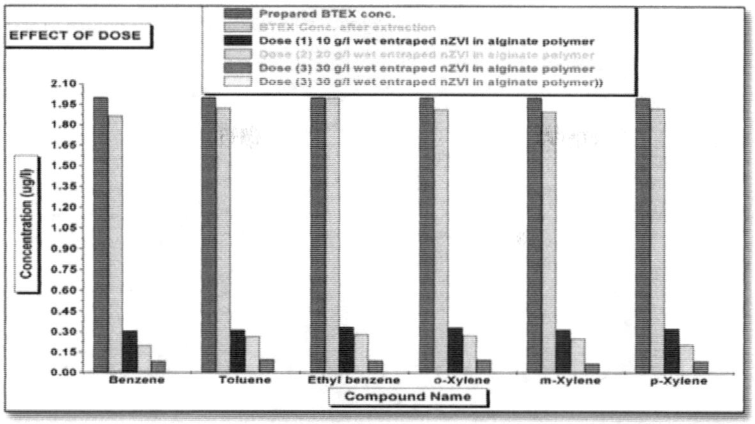

Figure 3a. Effect of encapsulated nZVI dose on BTEX removal.

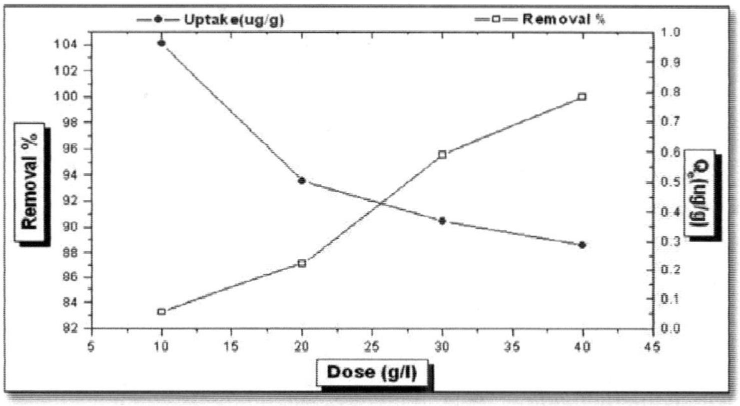

Figure 3b. The minimum effective dose of encapsulated nZVI on BTEX removal.

As shown in Figure 4, the percentage removal of benzene, toluene, ethyl benzene, *o*-xylene, *m*-xylene, and *p*-xylene, was 51, 53, 53, 51, 52, and 52 %, respectively, at pH 4; 57, 60, 60, 58, 58, and 59 %, respectively, at pH 5; 74, 76, 77, 75, 76, and 75%, respectively, at pH 6; 87, 88, 88, 87, 88, and 88 %, respectively, at pH 7; 75, 76, 77, 76, 76, and 76 %, respectively, at pH 8; and 58, 59, 60, 57, 59, and 59 %, respectively, at pH 9.

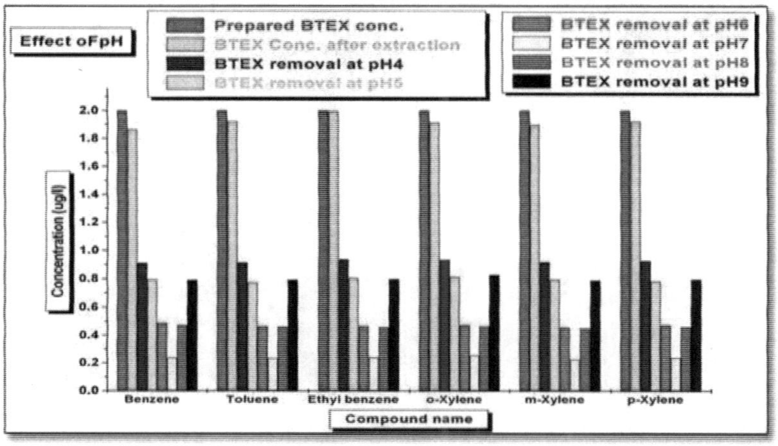

Figure 4. Effect of pH on BTEX removal.

As shown in Figure 5, the percentage removal of benzene, toluene, ethyl benzene, *o*-xylene, *m*-xylene, and *p*-xylene was 94, 94, 95, 95, 96 and 95 %, respectively, for 100 rpm; 96, 95, 96, 95, 96, and 96 %, respectively, for 200 rpm; and 96, 97, 97, 96, 97, and 96 %), respectively, for 300 rpm.

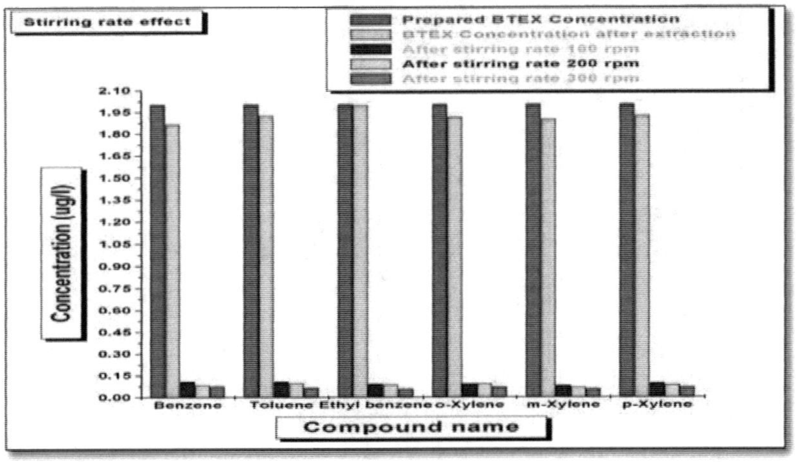

Figure 5. Effect of stirring rate on BTEX removal.

The removal efficiency was also investigated at different initial total BTEX concentrations. As shown in Figures 6, the percentage removal of benzene, toluene, ethyl benzene, *o*-xylene, *m*-xylene, and *p*-xylene, was 86, 86, 86, 86, 86 and, 86 %, respectively, for a mixture containing 34.096 μg/L; 88, 87, 88, 87, 88, and 88 %, respectively, for a mixture containing 22.839 μg/L; and 94, 94, 95, 95, 96, and 95 %, respectively, for a mixture containing 11.521μg/L.

The Freundlich isotherm model had a correlation coefficient (R^2) around 0.999, K_f (μg/g) .$(\mu g/L)^{-1/n}$ was around 25.003, and *n* around 1.107 showed a better fit of adsorption data than the Langmuir isotherm model with (R^2) around 0.899, q_{max} (μg/g) around 6.667 and K_l around 0.005. These obtained results describe the mechanism of adsorption process of BTEX compound into encapsulated nZVI surface. The adsorption process was heterogeneous adsorption process and this result agree with previous researches which study sorption of organic compounds from aqueous solutions (Nkansah et al., 2012). This non-linear model is based on the assumption that the sorption system involves heterogeneous surfaces, limited sorption sites and potential energy interactions (Ahn et al., 2005).

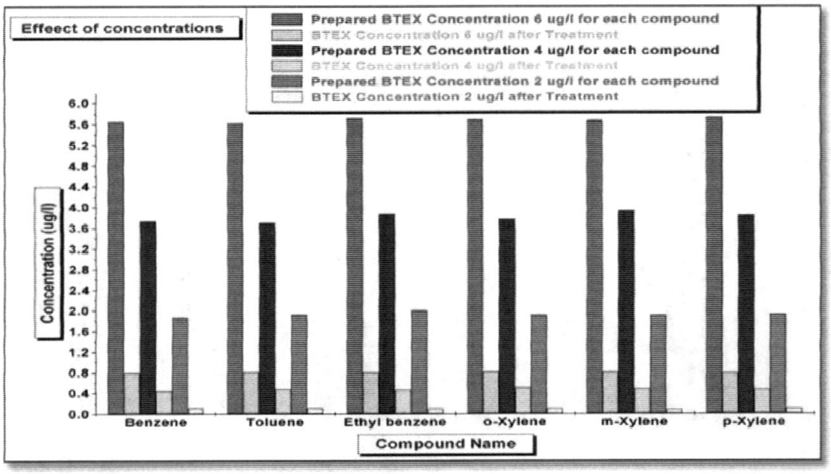

Figure 6. Effect of BTEX compounds initial concentration on the removal efficiency.

Finally, the removal efficiency of BTEX compounds was investigated at different initial temperatures. As shown in Figure 7, the percentage removal of benzene, toluene, ethyl benzene, *o*-xylene, *m*-xylene, and *p*-xylene was 94, 94, 94, 94, 94, and 94 %, respectively, at 30°C; 96, 95, 96, 95, 96, and 96 %, respectively, at 40°C; 100, 99, 98, 100, 99, and 97 %, respectively, at 50°C; and 94, 94, 95, 96, 95 and 95 %, respectively, at 60°C. The results showed that higher temperatures have a negatively effect on the alginate beads; the optimal removal percentages occurred for the most suitable temperature for BTEX removal of 50°C.

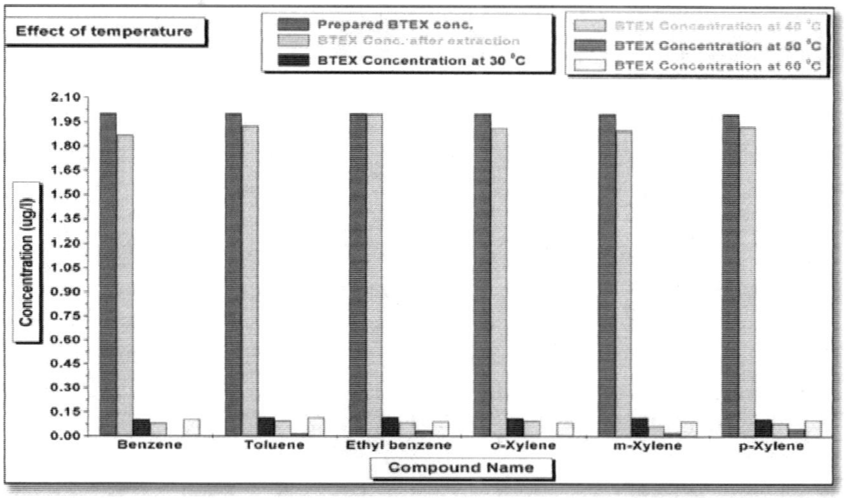

Figure 7. Effect of temperature on BTEX removal.

Summary and Conclusions

Results from the batch studies conducted during this research successfully indicated that encapsulated nZVI in alginate have specific characters on BTEX removal from aqueous solution. By studying the effect of time, dose, pH, stirring rate, concentration and temperature, the alginate biopolymer increase efficiency of removal of BTEX from aqueous solution was investigated. The effect of time was studied for various treatment times (15, 30, 45 and 60 min) for encapsulated nZVI; the resulting removal percentage was 70, 87, 96 and 97 %, respectively. The effect of dosage of encapsulated nZVI was studied (10, 20, 30 and 40 g/L) for encapsulated nZVI; the resulting removal percentage was 83, 87, 96, and 100%, respectively. The effect of pH was studied at pH of 4, 5, 6, 7, 8, and 9, for encapsulated nZVI in alginate and the resulting removal percentages was 52, 59, 76, 88, 76 and 58%, respectively. The stirring rate effect was studied (100, 200 and 300 rpm) and the resulting removal percentage was 95, 96, and 97% respectively. The concentration effect was studied using different BTEX mixtures (36, 24 and 12 µg/L), these values decreased after extraction, with a resulting purification of 34.096, 22.836, 11.521 µg/L, and the removal percentage was 86, 88, and 95 %, respectively.

References

AHN, S., WERNER, D., KARAPANAGIOTI, H.K., MCGLOTHLIN, D.R., ZARE, R.N., LUTHY, R.G. (2005) Phenanthrene and pyrene sorption and intraparticle diffusion in polyoxymethylene, coke, and activated carbon. ***Environmental science and technology 39***, 6516-6526.

BEZBARUAH, A., ALMEELBI, T. B., QUAMME, M., and KHAN, E. 2014. Calcium-Alginate Entrapped Nanoscale Zero-Valent Iron (NVZI). Google Patents.

BEZBARUAH, A. N., KRAJANGPAN, S., CHISHOLM, B. J., KHAN, E., and BERMUDEZ, J. J. E., 2009. "Entrapment of Iron Nanoparticles in Calcium Alginate Beads for Groundwater Remediation Applications", ***Journal of Hazardous Materials***, ***166***: 1339–1343.

BEZBARUAH, A. N., SHANBHOGUE, S. S., SIMSEK, S. & KHAN, E., 2011. "Encapsulation of Iron Nanoparticles in Alginate Biopolymer for Trichloroethylene Remediation", ***Journal of Nanoparticle Research***, ***13***: 6673-6681.

BRACHKOVA, M. I., DUARTE, M. A., and PINTO, J. F., 2010. "Preservation of Viability and Antibacterial Activity of *Lactobacillus spp.* in Calcium Alginate Beads", ***European Journal of Pharmaceutical Sciences***, ***41***: 589–596.

EL SALIBY, I., SHON, H., KANDASAMY, J., and VIGNESWARAN, S. 2008. "Nanotechnology for Wastewater Treatment: in Brief", ***Encyclopedia of Life Support System (EOLSS)***.

HILL, C. B., and KHAN, E., 2008. "A Comparative Study of Immobilized Nitrifying and co-Immobilized Nitrifying and Denitrifying Bacteria for Ammonia Removal from Sludge Digester Supernatant", ***Water, Air, and Soil Pollution***, ***195***: 23–33.

MOHAMED, S. F., and BORIK, R. M., (2013) "Modern Trends in Using Marine Algae for Treatment of Aquatic Pollution", ***International Journal of ChemTech Research***, Vol.5, No.6, pp 2863-2889.

Nkansah, M.A., Christy, A.A., Barth, T., Francis, G.W. (2012) The use of lightweight expanded clay aggregate (LECA) as sorbent for PAHs removal from water. Journal of Hazardous Materials 217, 360-365.

ORGANIZATION, W. H., 2006. *Protecting Ground Water for Health-Managing the Quality of Drinking-water Sources*, World Health Organization. London, UK.

ROY, D., GOULET, J., and LE DUY, A., 1987. "Continuous Production of Lactic Acid from Whey Permeate by Free and Calcium Alginate Entrapped *Lactobacillus helveticus*", ***Journal of Dairy Science***, ***70***: 506–513.

SHI, L., HARMS, H., and WICK, L. Y., 2008. "Electroosmotic Flow Stimulates the Release of Alginate-Bound Phenanthrene", ***Environ. Sci. Technol.***, ***42***: 2105–2110.

WATER ENVIRONMENT FEDERATION, AND AMERICAN PUBLIC HEALTH ASSOCIATION, 2005. ***Standard Methods for the Examination of Water and Wastewater***, American Public Health Association (APHA): Washington, DC, USA.

472386 The Effect of Advanced Oxidation on Physicochemical Properties of Natural Organic Matter (NOM) in Water

Abstract:

Natural organic matters (NOM) present in all surface, ground waters and terrestrial ecosystems. Due to their complex nature, microbial degradation of NOM is not sufficient to completely mineralize them under natural conditions. The presence of NOM in water effect on water quality by causing color, taste and odor problems. NOM are the precursor for formation of disinfection by product (DBPs) which are suspected as the carcinogenic compounds. Therefore, NOMs need to be removed from drinking water. The major limiting factors for using membrane technology for removal of NOM from water is that NOM fouls the membrane surface and decreases process efficiency and membrane lifetime. UV-based advanced oxidation processes (UV-AOPs) have been shown as effective methods for the removal of organic materials in water. In this work, a batch photoreactor with a 450-W UV lamp with high-pressure mercury vapor was used to evaluate the UV/H_2O_2 oxidation process for the removal of humic acids from water. We found 88% decrease in total organic carbon (TOC) also 95% decrease in concentration of humic acid as representative NOM from solution by using UV/H_2O_2. Also, UV/H_2O_2 treatment changed the physicochemical properties of humic acid (HA) aggregates by reducing the super-micrometer size range (>1Î¼m) aggregates to a bimodal distribution of two sub-micrometer size ranges <1Î¼m (10â€"100 nm and 100â€"1000 nm) and increased the zeta potential of HA solution due to the production of new functional groups. Furthermore, sequential using of AOPs / membrane filtration enhanced membrane filtration performance. We found membrane flux decline at 120 min improved from 21% up to 80% and increased the Flux Recovery Ratio (FRR%) from 58% to 80% for the cross flow polysulfone (PSF) ultrafiltration membrane. At the end, the removal of humic acid (HA) was compared by other photocatalytic processes including UV/TiO_2.

Fenton Oxidation Process for the Removal of Microcystin-LR In Nakdong River, Korea

Jeong-Ann Park [1], Boram Yang[1], Sanghyup Lee [1, 2*], Chanhyuk Park[1]

[1]Center for Water Resource Cycle, Korea Institute of Science and Technology, Seoul 02792, Republic of Korea, [2]Green School, Korea University, 145 Anam-ro, Seongbuk-gu, Seoul 136-701, Republic of Korea

*Corresponding author: yisanghyup@kist.re.kr, *Tel: +82-2-958-5911*

Introduction

Harmful algal bloom is caused by excessive algal growth that leads producing of harmful toxins. Microcystin-LR is one of the most commonly observed and studied hapatotoxins in surface water. The conventional water treatment processes, including coagulation, flocculation, and sedimentation, are effective for intracellular MC-LR; however, they are not sufficient for removing extracellular MC-LR. Thus, additional chemical processes are required. Advanced oxidation processes (AOPs) are generally more effective for the reduction of MC-LR than conventional oxidation processes. The Fenton oxidation process is an attractive method in AOPs as H_2O_2 is an environmentally friendly oxidant because it decomposes into hydrogen and oxygen and because iron is highly abundant and non-toxic. However, limited studies have been conducted using the Fenton process to remove MC-LR in distilled water. Thus, the aim of this study is the Fenton ($Fe(II)/H_2O_2$) process was evaluated to investigate the removal of MC-LR in distilled water, and Nak-dong river water from drinking water treatment plant.

Results and discussions

The optimal concentrations of Fe(II) and H_2O_2 for the removal of MC-LR were determined as 5 mg/L Fe(II) and 5 mg/L H_2O_2, respectively. The Fenton process is a fast reaction, mostly complete within 5 min (Fe(II), H_2O_2 = 0.2 – 10 mg/L); the process had a $t_{1/2}$ of less than 5 min at Fe(II) and H_2O_2 concentrations of 5 mg/L and k value of 0.146 min^{-1}. The degradation intermediates of MC-LR were observed as m/z 1029.5, 1011.5, 835.5, 795.4, and 783.4 using kinetic analysis. The major intermediates were m/z 1029.5 and 1011.5 caused by double hydroxyl addition on the conjugated diene bond of Adda, and single hydroxyl substitution. The second oxidation route tends to remove the Adda chain followed by diene bond cleavage through ·OH attack. The toxicity of microcystin-LR is associated with the diene bond in the Adda chain; thus, cleavage of this bond would be anticipated to alleviate toxicity.

The effects of the initial MC-LR concentration (2 – 200 μg/L) and solution pH (3 - 11) were also examined that a decreasing initial MC-LR concentration and solution pH increased the degradation efficiency of MC-LR. The efficiency of MC-LR degradation was 92.06 ± 0.89% after a 30-min reaction with an initial MC-LR concentration of 2 μg/L. The removal efficiency was highest (77.27%) at pH 3, according to the highest $[\cdot OH]_{ss}$ (4.2×10^{-13} M). In a strong basic solution (pH 9–11), the removal efficiency was less than 10% because of a decrease in the Fe(II) concentration due to the formation of $Fe(OH)_3^-$, $Fe(OH)^+$, $Fe(OH)_2^{2+}$ and iron hydroxide precipitation

The Nak-dong river water from a drinking water treatment process was also tested. The water sample had the following characteristics for application of Fenton process: DOC = 4.06 mg/L, SUVA value = 2.64 L/mg-m, turbidity = 0.78 NTU, color = 7 Pt-Co unit, Fe (III) = 0.23 mg/L, pH = 7.64, electrical conductivity (EC) = 475 μs/L, geosmin = 23.10 ng/L, and 2-MIB = 25.60

ng/L. The degradation efficiency was 33.13% for a 5-min reaction in raw water from drinking, while the degradation efficiency was lower than that of DI water because of the competing effect (NOM, geosmin, and 2-MIB) and the higher pH. Nevertheless degradation efficiency was lower than that of DI water in every initial MC-LR concentration by the adverse effect of water quality parameters, the degradation efficiency was over 52% with an initial MC-LR concentration of 2 μg/L within 5 min. Therefore, the final concentration of MC-LR was below the WTO drinking water guideline value (1 μg/L) when the initial MC-LR concentration of 2 μg/L in the river water. This study provided knowledge of the Fenton process for use as an alternative cost-effective method to remove MC-LR in drinking water source.

454505 Electro-Fenton Degradation of Organic Contaminants Using a Carbon-Based Electrode

Abstract:

Nowadays, the wide application of dyes and pharmaceutical personal cares products arouses increasing concerns about their potential adverse impacts to both environment and human beings. In this study, a new carbon-based electrode was developed by graphene oxide and carbon nanotube and used in Electro-fenton reaction for the effective degradation of Methyl Blue (MB) and Chloramphenicol (CP). Several ecxperimental parameters including oxygen flow rate, solution pH and current density were systematically investihated in terms of the production of hydrogen peroxide (H2O2). Results showed that the electrode can work well in a wide pH range of 5- 9, which is favorable for the practice application. The conversion rate of hydroxyl radical from the generated H2O2 and the relative degradation efficiency were investigated under the different current density. The role of different components in the developed electrode was also compared to understand the degradation mechanism. The removal efficiency of MB and CP can respectively reach abount 80% and 95% in the filtration mode.

459925 Pyrolysis of Soils Contaminated with Heavy Hydrocarbons: Reaction Mechanisms and Morphology of the Produced Char

Abstract:

Soil contamination by petroleum and other heavy hydrocarbons is a major global environmental problem. Although offshore oil rigs and tankers are responsible for occasional large-volume spills, 98% of oil spills occur on land, with an average of 70 spills per day (more than 25,000 per year) reported to the U.S. EPA. Without adequate response, the effects of major spills could last decades. Whereas numerous remediation approaches have been proposed, most existing technologies have severe drawbacks. For example, *in situ* bioremediation of sites impacted by petroleum release can take years, particularly when recalcitrant species such as high molecular weight hydrocarbons are involved. Thermal technologies, on the other hand, can remediate sites quickly and efficiently (hours to months), often removing over 99% of a wide range of hydrocarbon fractions. However, current thermal technologies often destroy key soil constituents and severely degrade soil properties such as organic carbon content, water retention, stability and microbial activity. Thus, treated soils become unsuitable for reuse since they cannot support vegetative growth or even provide erosional stability. We have recently reported[1] that pyrolytic treatment at 400-450°C of soils contaminated with heavy petroleum crudes reduced total petroleum hydrocarbons (TPH) to levels well below regulatory standards (<0.1% by weight). A fraction of the original hydrocarbons remained in the treated soil in the form of a highly inert carbonaceous material (char). Standard EPA tests revealed that the post-pyrolysis levels of toxic polycyclic aromatic hydrocarbons (PAHs) were well below applicable standards. More importantly, plant growth studies showed significantly higher (by 80-900%) biomass production of *Arabidopsis thaliana* and *Lactuca sativa* (Simpson black-seeded lettuce) in pyrolyzed soils than in either contaminated or incinerated soils. And because it is carried out at lower temperatures, pyrolysis requires less energy than other thermal methods (like incineration) that can achieve similar TPH reduction levels. Overall, these results suggest that soil pyrolysis could be a viable thermal treatment to quickly remediate soils impacted by weathered oil while improving soil fertility, potentially enhancing revegetation.

We present here results from a follow-up study aimed at elucidating the fundamental mechanisms of the pyrolysis process and the morphology of the produced char.

The dispersion of the char in the pyrolytically treated soils will determine to a large extent its ultimate fate in the environment and modulate its interactions with water and the regenerated soil microbiome. Optical microscopy images of soils treated with pyrolysis and incineration led us to hypothesize that the produced char is distributed as a very thin film coating the soil particles. To test our "char film" hypothesis, we carried out controlled oxidation of pyrolyzed soils followed by surface analysis using XPS. We exposed treated soil to oxygen and reacted it in the TGA for one hour at temperatures between 350 and 450°C in order to progressively burn off the char film. Weight loss measurements and the XPS results confirmed our char film theory. As we burnt the film away, the percentage of C-C bonds on the surface of soil particles decreased and the amount of Si increased (our soils had high quartz content). On the other hand, the relative amounts of C and Si detected remained constant when uncontaminated "background" soil was exposed to the same treatment of controlled oxidation. Surface analysis by XPS detected the same relative amounts of C and Si when uncontaminated soil and pyrolyzed soil were oxidized at 450°C. This treatment "burnt away" the char film that covered the particles of pyrolyzed soil.

To further differentiate our process from existing thermal desorption technologies, we employed thermogravimetry with online evolved gas analysis (TG-MS) and differential scanning calorimetry (DSC) to gain a better understanding of the fundamental pyrolysis mechanisms. Our results show that as mixtures of soil and petroleum crudes are heated in an anoxic atmosphere, we first observe desorption of the lighter hydrocarbons at temperatures between 100 and 350°C. At temperatures between 350 and 450°C, however, thermal cracking and condensation reactions dominate and convert the heavier hydrocarbons (resins and asphaltenes) into char (C:H atomic ratio approximately equal to 1) and light volatiles. If the temperature rises above 550°C, however, key soil constituents (like carbonates) start to decompose and release CO_2 that becomes the dominant peak in the time-resolved MS spectra. Carbonate decomposition can explain the significant increase of the pH and the accompanying loss of fertility we observed when we incinerated contaminated soils at 650°C. MS data also show that H_2O evolves between 300 and 500°C, indicating dehydroxylation of clay constituents of soil. All these observations have significant implications for secondary reactions that may occur when pyrolytic treatment of contaminated soils is carried in commercial reactors.

Knowing the temperatures at which decomposition of soil constituents occurs will allow us to determine the least damaging thermal treatment to reach desired cleanup levels. The clearest lesson that can be gleaned from our studies is that the lowest effective treatment temperature should be used to minimize soil decomposition. While allowing for effective treatment of nearly all hydrocarbons, the high temperatures utilized by incineration and other thermal remediation technologies can cause extensive soil damage, such as decomposition of clays, carbonates, and organic content. This will physically alter the soil and change its geochemical, biological, and fertility properties.

---

[1] Vidonish, J.E., et al., "Pyrolytic treatment and fertility enhancement of soils contaminated with heavy hydrocarbons." *Environmental Science & Technology*, **50** (5), 2498–2506 (2016).

468482 Spatial and Temporal Features of the Catalyzed Hydrocarbon Trap

Abstract:

Spatial and Temporal Features of the Catalyzed Hydrocarbon Trap

Po-Yu Peng, Michael P. Harold, Dan Luss***

Department of Chemical and Biomolecular Engineering, University of Houston, Houston, TX 77204

*mharold@uh.edu, **dluss@uh.edu

The diesel oxidation catalyst (DOC) system plays an important role in promoting CO and hydrocarbons (HCs) oxidation by the precious group metals Platinum (Pt) and Palladium (Pd). The DOC is designed to operate at very high HC conversion (>90%) after a sufficiently operating temperature has been reached. However, during the "cold start", most of the engine-out hydrocarbons remain unreacted and are emitted to the environment.[1, 2] A large-pore zeolite such as beta-zeolite (BEA) has been added to the DOC to trap hydrocarbons during the cold start. In particular, large molecular weight hydrocarbons are especially prone to trapping in BEA-modified DOCs. Although extensive research has been conducted on the BEA storage capacity, there is a need to gain insight on how BEA affects the performance of HC light-off (LO) and of the HC oxidation transient behaviors. [3-7]

DOC and DOC/BEA (0.5 g/in^3 BEA) monolith catalysts were used to study the transient behavior after the HCs (Dodecane, $C_{12}H_{26}$) has been pre-stored on the catalyst, The impact of BEA on the LO performance and HCs trapping were studied by flowing a mixture of 10% O_2 in Ar during temperature-programed oxidation (TPO). The transient behavior of the hydrocarbon trap was measured by both coherent optical frequency domain reflectometer (c-OFDR) and capillary inlet mass spectrometer (Spaci-MS) enabling the collection of transient temperature and concentration profiles the complete spatio-temporal data during the TPO. .

Figure 1 reports the feed HC concentration (150 ppm Dodecane, C12) and C12 effluent concentration. The figure reveals the completion of the HC pre-store and the removal of weakly-bonded HC by the lean gas purge (10% O_2 balanced by Ar). The pre-stored HC amount of 14.0 mg C12/g sample was determined by integrating the area between the feed concentration and effluent concentration. During the TPO, the effluent CO_2 concentration reveals the existence of multiple hydrocarbon LO points where the CO_2 formation occurs sequentially. Specifically, there are three major peaks that form during the TPO, indicating there are three LO stages within the monolith catalyst.

Figure 2 reports and the 3-D temperature dependence on position and time. The plot reveals the temperature rise at different locations and different times in the monolith catalyst. The three temperature hot zones occur due to the local HC combustion. Once the local HC combustion is complete the temperature drops until the next local LO. This indicates that the pre-stored HC is not release at once but occurs sequentially. We conclude by analyzing the HC LO locations and timing that the multiple HC LO phenomena is caused by a front of released HC which is trapped in the downstream and lights-off later when the local temperature exceeds the LO value.

The multiple hydrocarbon LO phenomena is strongly related to a tug-of-war between HC trapping and oxidation, thus, the steady-state HC oxidation has been investigated. Figure 3 shows the spatially-resolved C12, C_3H_6 and CO_2 concentrations at steady-state for a furnace temperature of 220 ^oC conditions for which most of HCs are fully oxidized. The CO_2 concentration fluctuated with a feed of 150 ppm C12 but was stationary under a feed of 0.2% C_3H_6. The effluent CO_2 concentration exhibited periodic fluctuations. These results suggest the existence of a "tug-of-war" between HC trapping and oxidation. Therefore, this phenomena provided a good opportunity to design more efficient catalyst.

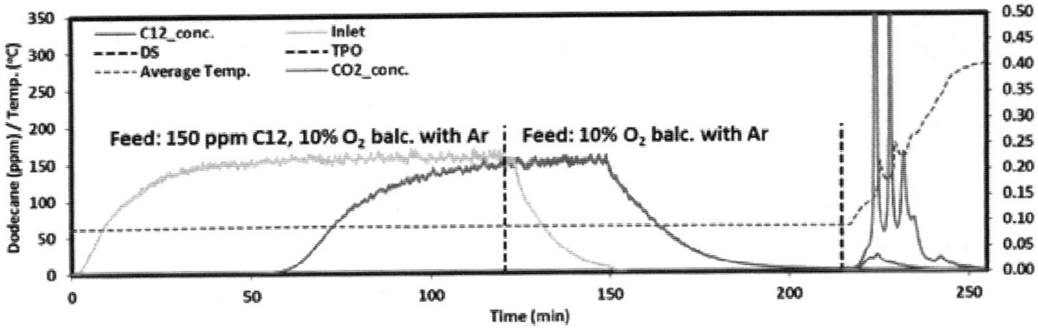

Figure 1. Dodecane and CO_2 concentration measured during adsorption and desorption followed by TPO with BEA-zeolite added Pt/Pd/Al_2O_3 monolith catalyst. Condition during saturation: 150 ppm dodecane and 10% O_2 balanced with Ar. Condition during desorption and TPO: 10% O_2 balanced with Ar.

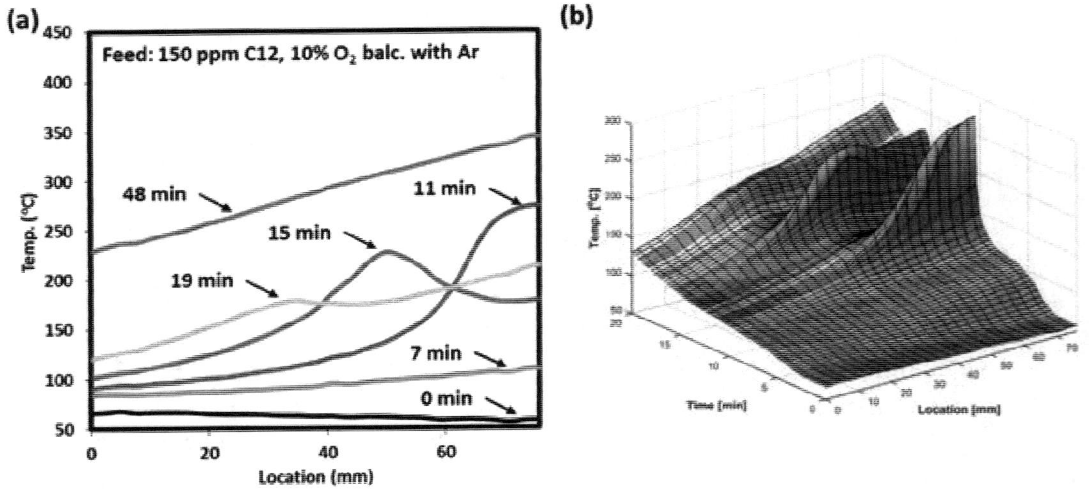

Figure 2. (a) Spatially-resolved temperature profile measured at different time in the C12 pre-stored catalyst and (b) the 3-D plot of the spatially-resolved temperature versus time in feed of 10% O_2 balanced with Ar. (Catalyst: D-05)

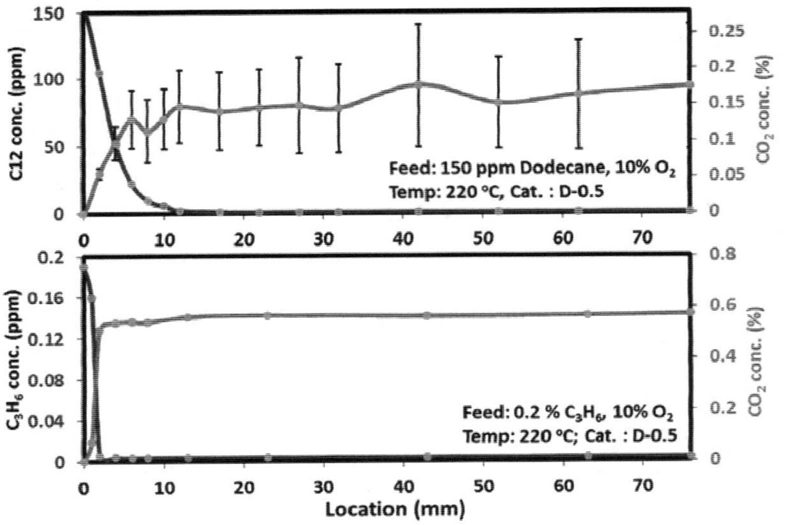

Figure 3. Spatially-resolved C12 and CO_2 concentration at steady-state at 220 °C with the feed of (a) 150 ppm C12 and 10% O_2 balanced with Ar, and (c) 0.2% C_3H_6 and O_2 balanced with Ar. Spatially-resolved C12 and CO2

conversion and selectivity at steady-state at 220 ^{o}C with the feed of (b) 150 ppm C12 and 10% O_2 balanced with Ar, and (d) 0.2% C_3H_6 and O_2 balanced with Ar.

References

1. Mukai, K., et al., *Adsorption and desorption characteristics of the adsorber to control the HC emission from a gasoline engine*. 2004, SAE Technical Paper.

2. AL Harbi, M., et al., *Competitive no, co and hydrocarbon oxidation reactions over a diesel oxidation catalyst.* The Canadian Journal of Chemical Engineering, 2012. **90**(6): p. 1527-1538.

3. Li, H.-X., et al., *Application of zeolites as hydrocarbon traps in automotive emission controls.* Studies in Surface Science and Catalysis, 2005. **158**: p. 1375-1382.

4. Chang, H., et al., *Gasoline Cold Start Concept (gCSC™) Technology for Low Temperature Emission Control.* SAE Int. J. Fuels Lubr, 2014. **7**(2): p. 480-488.

5. Murakami, K., et al., *Development of a high performance catalyzed hydrocarbon trap using Ag-zeolite*. 2004, SAE Technical Paper.

6. Elangovan, S., et al., *A comparative study of zeolites SSZ-33 and MCM-68 for hydrocarbon trap applications.* Microporous and mesoporous materials, 2006. **96**(1): p. 210-215.

7. Elangovan, S., et al., *Silicoaluminophosphate molecular sieves as a hydrocarbon trap.* Applied Catalysis B: Environmental, 2005. **57**(1): p. 31-36.

463188 Formation of Nitrosamines in the Desorber of Tertiary Amine-Based Carbon Dioxide Capture Systems

Abstract:

Post-combustion CO_2 capture is one of the most developed and commonly employed technologies for power plants to reduce greenhouse gas emissions. Amine-based CO_2 absorption technology can be easily retrofitted into existing power plants and especially suitable for the flue gases containing low concentration of CO_2. Tertiary amines are considered as alternatives to the benchmark solvent monoethanolamine for CO_2 absorption, because they provide the benefits of higher absorption capacity and lower heat of reaction. However, public concerns have been raised for amine-based CO_2 capture due to the formation of nitrosamines, which are toxic to human health and ecological systems.

In this work, the influences of amine structural characteristics and desorber conditions on the formation of nitrosamines were investigated. Generally, tertiary amines were less reactive towards nitrosamines comparing with secondary amines. The formation of total nitrosamine from tertiary amine was first-order dependent on nitrite concentration and CO_2 loading, but was not dependent on amine concentration. For tertiary amines with the same number of 2-hydroxyethyl groups, less hindered amines showed faster nitrosamine formation and higher nitrosamine yields based on nitrite consumption. The distribution of specific nitrosamine formation from tertiary alkanolamines implied the preferential cleavage of 2-hydroxyethyl group over methyl group, but comparable to ethyl group, and therefore shed light on the nitrosation mechanisms.

458015 Natural Attenuation and Biostimulation of Petroleum-Hydrocarbon Contaminated Soil in a Simulated Column

Abstract:

Various physiochemical and biological remediation methods can be used to remediate petroleum-hydrocarbon contaminations. Using microorganisms in biological remediation methods is considered as a cost-effective and environmental friendly solution. In the current research, two bioremediation-type processes, natural attenuation (NA) and biostimulation (BS) were applied to simulate biodegradation of petroleum-hydrocarbon contaminated soil collected from a refinery. During 60 days of operation, 88% and 65% of initial petroleum-hydrocarbon contamination biodegraded in BS and NA modes, respectively. Moreover, population of indigenous microorganisms in the soil column increased noticeably. Both bioremediation bodes showed that they are compatible with first-order reaction kinetic. In the NA mode, reaction rate constant was calculated 0.02 d^{-1} while it was 0.039 d^{-1} for the BS mode.

462581 Study of Reaction Kinetics for Elemental Mercury Vapor Oxidation over $CuCl_2$ for Mercury

Abstract:

The reaction kinetics for elemental mercury (Hg(0)) vapor oxidation over $CuCl_2$/Î±-Al_2O_3 has been studied. The reaction kinetic data were obtained with respect to the reaction temperature and inlet Hg(0) concentration. The objective of this study is to determine the intrinsic reaction rate expression for the reaction of Hg(0) with $CuCl_2$. Based on the grain model, a description has been developed to represent the gas-solid reaction taking place within the $CuCl_2$ grains. During the reaction, the $CuCl_2$ grains were found to agglomerate, and thus a sintering description was added to the model in order to take into account the incomplete conversion of $CuCl_2$.Â The reaction rate constants and activation energy were determined by comparing the experimental data with the mathematical model. The detailed experimental and simulation results will be presented.

453270 Advances in the Use of a Bio-Physicochemical Model to Characterize, Optimize the Chemical Absorption-Biological Reduction Integrated System for NO Removal

Abstract:

One promising technology for NO_x removal from flue gas is the chemical absorption-biological reduction (CABR) integrated process, in which NO is absorbed by aqueous solution of Fe(II)EDTA and the absorption product, Fe(II)EDTA-NO, is reduced biologically. At the same time, Fe(II)EDTA, oxidized by O_2 in flue gas, could be regenerated in a function of iron-reducing bacteria.

A kinetic model based on mass transfer coupled with chemical reaction and bioreduction was developed to simulate and predict the NO_x removal by the CABR system in a bio-trickling filter. The developed model was validated by the experimental results (the average relative deviation kept within 5.72%) under different operating conditions, e.g. NO/O_2 concentration, gas/liquid flow rate, and subsequently was used to predict the system performance. NO distribution at different inlet concentrations in the gas phase along the bio-trickling filter was also modeled and predicted.

Furthermore, the liquid flow rate and total iron concentration were optimized for >90% NO removal efficiency, which were 38.8 L h^{-1} and 10 mM under the typical operating condition (2 L min^{-1} gas stream containing 3% (v/v) oxygen and 500 ppm NO) respectively. Sensitivity analysis found that the performance of the CABR process was controlled by the bioreduction activity of Fe(III)EDTA, and the least activity has been achieved. This work could help us with the scale-up design and operation of the CABR process in the industrial application.

Graphic abstract

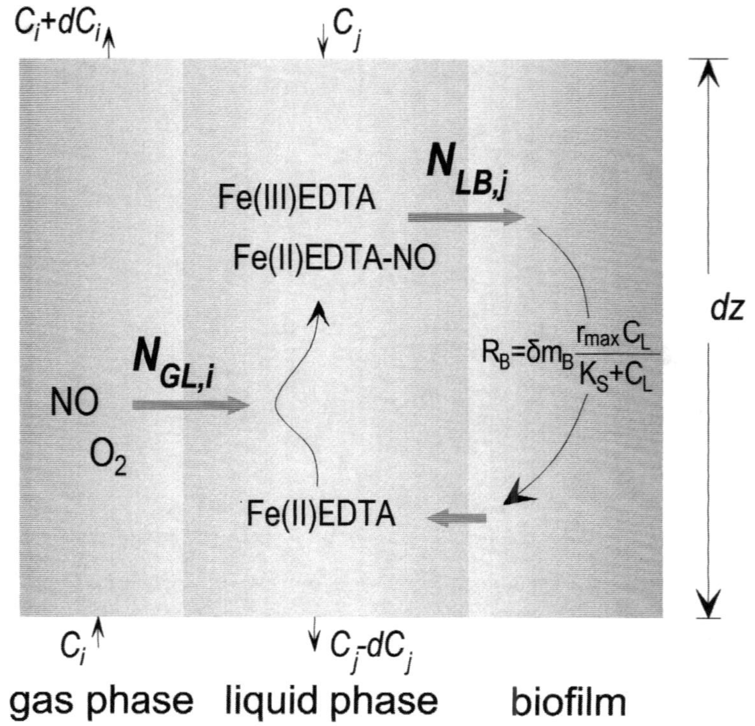

474701 Session Keynote - Integration, Control, and Optimization of Hydrogen Energy Storage in Utility Grid Networks

Abstract:

Global warming, air pollution, and energy security concerns are evolving the way energy is delivered and converted over the next century with huge increases in the levels of clean and renewable energy while reducing energy demand in all economic sectors. Renewable power is the first and foremost option for decarbonizing electricity generation since it is abundant, sustainable, has lesser emissions and environmental impacts, and is widely available globally. It has been shown that multiple combinations of renewable resources including solar, wind, geothermal, hydropower, and biopower are capable of meeting 80% of total U.S. electricity demand in 2050 while achieving deep reductions in electric sector emissions. However, greater levels of renewable power use pose new challenges to electric grids due to geographical dependency, transmission constraints and most importantly power output variability. Therefore, increasing penetrations of renewable power must be accompanied by higher flexibility in the electricity grid through a portfolio of supply- and demand- side technologies, including energy storage, transmission and distribution infrastructure expansion, more highly responsive loads, and enhanced power system operation.

A compressed hydrogen energy storage system, comprised of an electrolyzer that produces hydrogen from water by using otherwise curtailed renewable electricity, and a fuel cell that utilizes the hydrogen to provide power to the grid, has been considered less favorable than batteries due to lower round-trip efficiency and relatively high cost. However, for integration with large-scale wind energy, accomplishing massive energy storage, and for storing over long periods of time (e.g., seasonal), large energy capacity and low self-discharge become more important than round trip efficiency; therefore such hydrogen energy storage systems become more attractive.

A dynamic model to simulate the integration of hydrogen energy storage systems in utility grid networks has been developed to evaluate the potential of hydrogen energy storage in mitigating the impacts of intermittent renewable power on the grid. A reduction in the renewable energy curtailed and the ramping events has been observed under the application of different hydrogen production, storage and power supply strategies. Data from the state of California has been used as a reference for energy consumption and power generation, assuming the evolution of the power market towards the fulfillment of goals set by the different policies and laws of the state. It was found that both diurnal and seasonal load shifting are possible with gaseous storage facilities that are already available and in use for natural gas storage in the state. Moreover, with the optimal management of the storage facilities, it is viable to dispatch the compressed hydrogen, supplying a significant amount of fuel for transportation applications and averting the need for installing additional electrical transmission and distribution equipment.

463243 Frost Control Using Osmotic Membrane Dehumidification

Abstract:

FROST CONTROL USING OSMOTIC MEMBRANE DEHUMIDIFICATION

Frost buildup is a major concern for heat pumps and refrigerator/freezers. Osmotic membrane dehumidification represents an opportunity for controlling frost simply and effectively. The unique application of this technique to heat pumps provides the advantage of independent dehumidification in the cooling mode and frost control in the heating mode.

An osmotic membrane dehumidifier can use a semi-permeable membrane to facilitate capillary condensation of water vapor and the transport of condensed water through the membrane into a salt solution by osmosis. Here a humid gas stream is brought into contact with a semi-permeable membrane, which separates the gas stream from an osmotic (e.g., salt) solution. Some of the pores of the membrane are small enough to permit capillary condensation. Liquid formed within these pores can connect with liquid formed in adjacent pores, collectively forming continuous paths of liquid. These liquid bridges extend across the thickness of the semi-permeable membrane and provide paths by which water can travel across the membrane. Because the membrane is so thin, water concentration gradients across the membrane can be large. This can provide a large driving force for water transport between the humid air and the osmotic fluid. An illustration of this two-step spontaneous process is given below in Fig. 1.

The effectiveness of the membrane, measured as the rate of water removal per unit area of membrane, is key to the compactness and longevity of the dehumidification system. Automotive air conditioning systems have significant constraints on size, while residential, commercial and industrial air conditioning systems are expected to last for generations. High performance membranes can have a major impact on all of these systems, with no significant energy requirements for dehumidification and with reject heat from the vapor compression cooling process used to re-concentrate the osmotic solution, as shown in Fig. 2.

Typically, capillary condensation occurs in the pores of hydrophilic membranes where surface tension attracts and draws condensed vapor along the sides of the pores. **We have now confirmed that hydrophobic pores in a limited size range will enhance both capillary condensation and the transport of condensate out of the pores.** Test results using membranes containing these pores demonstrate uniquely high fluxes of water.

Beckstein and Sansom, in an article in Proceedings of the National Academy of Science demonstrate under condensing conditions that hydrophobic pores will remain void of condensate if they are under a diameter of 0.8 mm and will fill with liquid if over 1.4 nm (Ref1); that is, capillary condensation can happen away from the surface in a large pore. The hydrophobic surface will drive water vapor away from the surface and cause a density gradient that will result in condensation and cluster formation between water molecules. That is typical of biological pores that selectively transport water, aquaporins, which are lined with hydrophobic residues. Kaneko and Iiyama show the importance of cluster-like water assemblies in hydrophobic nanopores using small angle x-ray scattering to elucidate clusters (Ref 2). *Permeation through nanometer pores is rapid because of the reduced friction resulting from repulsion of water from the hydrophobic surfaces.*

Fig. 1

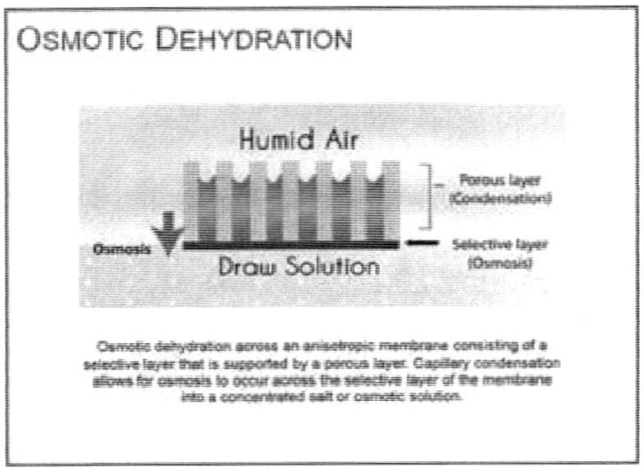

Fig. 2

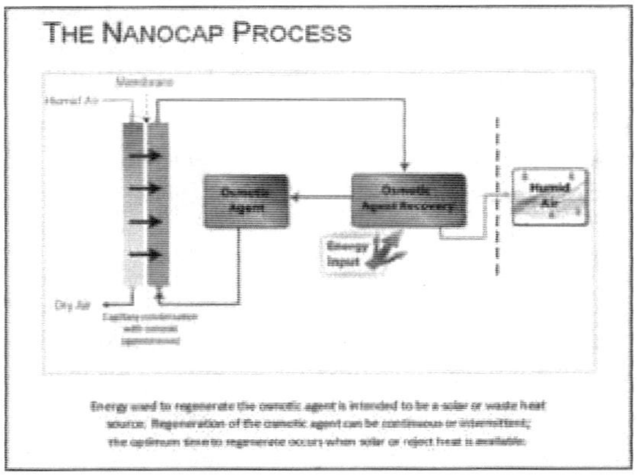

Osmotic dehumidification has been demonstrated in a laboratory scale device using hydrophobic membranes fabricated by Aquaporin Inside, with magnesium chloride used as an osmotic fluid. Water fluxes obtained were more than twice those obtained with available hydrophilic polymeric membranes.

Enhanced performance of the hydrophobic membrane system extends the potential for effective application of osmotic dehumidification to areas like <u>frost control</u> on surfaces containing refrigerant. A heat pump uses a vapor compression system to cool the interior space in warm weather and, by reversing refrigerant flow, to heat the interior space in cold weather. In the cooling mode, independent dehumidification leads to substantial energy savings; in the heating mode, dehumidifying some outside air can inhibit and substantially prevent the formation of frost.

Frost is generally caused by water vapor in very humid air at temperatures somewhat <u>above</u> the freezing point of water freezing on surfaces containing refrigerant at temperature well <u>below</u> the freezing point. Inhibiting the formation of frost involves reducing the humidity of the outside air in contact with cold surfaces and/or raising the temperature of the surfaces in contact with the outdoor air. An osmotic dehumidifier located outdoors in very effective in accomplishing both objectives simultaneously.

As shown in the attached schematic diagram (Nanocap Heat Exchanger.pdf), cold refrigerant exiting the expansion process to the outside is warmed in a liquid-liquid heat exchanger by the osmotic fluid used in the dehumidification process. Outside air that has been dehumidified is transported in a duct to another heat exchanger to warm the refrigerant further. The air is dry enough to eliminate the potential for frost formation, even with refrigerant temperatures below the normal freezing point. The efficiency of the process is enhanced as the temperature is reduced because capillary condensation occurs more readily at low temperature.

Fig. 3

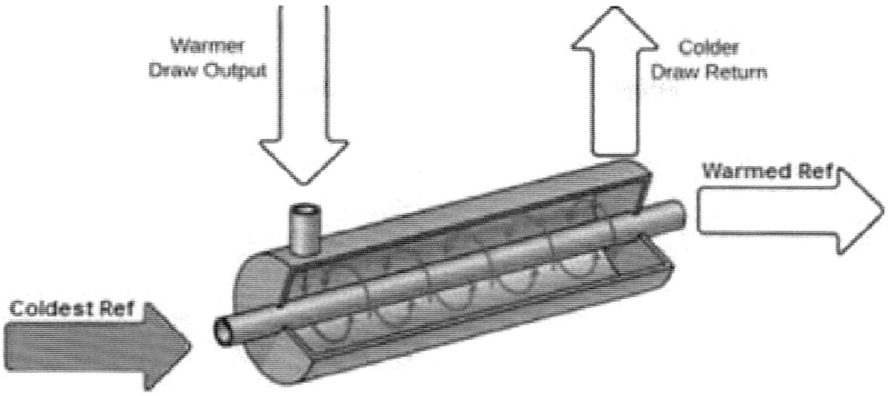

References:

1. Beckstein and Sansom, Liquid-Vapor Oscillation of Water in Hydrophobic Nanopores, Proceedings of the National Academy of Sciences Vol 100, No 12(2003) 7063-7068

2. Kaneko and Iiyama, Structural Understanding of Water Confined in Hydrophobic Nanopores, WATSURF, Les Houches, France Apr. 2013

464623 Optimal Control of a LiBr/Water Absorption Chiller

Abstract:

A LiBr/water absorption refrigeration system driven by solar energy is studied to find its optimal control profile subject to disturbances of the temperature and atmospheric conditions during the day. This work is developed in two phases, first a computational mathematical model was developed to accomplish the optimal control of the chiller, which was then used in a second phase to build a prototype.

A single-stage LiBr/water absorption chiller is made up of four main components: a generator, an absorber, an evaporator and a condenser, which can be modeled as a heat exchanger. The circulation of the solution is assured by a solution pump, and a solution heat exchanger is used to internally recover thermal energy.

Recent works [1-3] had presented dynamic models to evaluate the long-term behavior of a single-effect LiBr/water absorption chiller and also to follow its daily operational profile with good precision. These models may be used to study the interaction of the absorption unit with the other components of a solar-assisted cooling plant, in aim of optimizing both the layout of the system and the control logic of the chiller.

The use of dynamic optimization for analyzing the control possibilities allows to identify the optimal process behavior and select the best control strategy based on quantitative criteria. This implies finding the variable profiles that minimize the objective function in a system subject to disturbances [4]. From the mathematical model or system restrictions a total of 254 variables define the process and a total of 245 equations model the single stage absorption cooling cycle.

The process is modeled by the Differential-Algebraic Equation (DAE) system that represents de dynamics behavior of the system. Then, the optimal control problem is solved by discretizing the time domain, which converts it into a Non-Linear Programming (NLP) problem.

Work [5, 6] results shows that the optimal control strategy is able to optimize the overall performance of the thermal system, achieving a reduction in energy consumption between 0.73 and 2.55% compared with traditional control strategies. With this work we seek to reduce energy consumption for an absorption cycle by applying an optimal control strategy as it has been accomplished for other thermal systems.

Considering that the most inefficient energy consumption systems, identified in countries like Colombia, come from using equipment for cooling and space conditioning [7] and, besides, that during the summer season the energy generation is insufficient to satisfy the demand, the absorption refrigeration system becomes as an opportunity to solve this problem in a more definitive way.

The Non-Linear Programming (NLP) problem is implemented in Gams® and solved with the solver CONOPT.

Reference

[1] G. Evola, N. Pierres, F. Boudehenn, P. Papillon.

[2] P. Kohlenbach, F. Ziegler, "A dynamic simulation model for transient absorption chiller performanc. Part I: The model" International Journal of Refrigeration, pp. 217-225. 2008

[3] P. Kohlenbach, F. Ziegler, "A dynamic simulation model for transient absorption chiller performanc. Part II: Numerical results and experimental verification" International Journal of Refrigeration, pp. 217-225. 2008

[4] J. Cuthrell, L. Biegler, "On the optimization of differential-algebraic process system" Annual AIChE Meeting, Miami, Fl, 1986.

[5] Z. Ma, S. Wang, "Supervisory and optimal control of central chiller plants using simplified adaptive models and genetic algorithm" Applied Energy, pp. 198-211, 2011

[6] T. Chow, G. Zhang, Z. Lin, C. Song, "Gobal optimization of absorption chiller system by genetic algorithm and neural network" Energy and Buildings, pp. 103-109, 2002.

[7] "Plan Energetico Nacional Colombia: Ideario Energético 2050", Unidad de Planeación Minero Energetica, 2015

464689 B-C and B-Fe Core-Shell Nanoparticles for the Enhancement of Biofuel Combustion

Abstract:

Liquid biofuels such as ethanol are good candidates to partially replace fossil fuels. However, the adoption of biofuels is inhibited by low energy densities and incomplete combustion. These characteristics are problematic for applications with a high energy demand, such as ramjets. Boron additives are a promising solution because they are among the most energy-dense materials. Boron can increase the overall heat release of the fuel and reduce its ignition temperature. However, boron gradually oxidizes when stored under ambient conditions, may not completely burn, and the used metal particles lack recyclability and are subject to environmental regulations. These problems are less limiting in ramjets and other specialty combustion applications.

In this study, we synthesized core-shell Boron-Carbon (B-C) and Boron-Iron (B-Fe) nanoparticles for the combustion of ethanol. A variety of templated, non-templated, and wet-chemical techniques were used to produce elemental boron (zero valent) with either a carbon or iron shell. The composites were characterized by N_2 Physisorption, TGA/DSC, ICP/AES, XRD, SEM/EDX, TEM, and SAXS. The composites had surface areas from 20-150 m^2/g, pore sizes from 5-40 nm, and shell thicknesses between 5-20 nm. Boron was well-distributed and mostly amorphous, indicating that reactions with iron or carbon were minimal. The combustion properties with ethanol were determined through batch combustion in a calorimeter. The B-C and B-Fe composites lowered ignition temperatures and raised the rate of fuel conversion in almost all cases. The best B-C composites were synthesized with polymer templates and had relatively large shell thicknesses and pore sizes that positively affected oxygen transport. The best B-Fe composites could be broken down into smaller particles, which is beneficial when fluidized in a ramjet combustor.

474452 Platinum-Nickel Nanowires As Electrocatalysts in Alkaline Hydrogen Oxidation and Evolution

Abstract:

Anion exchange membrane (AEM) fuel cells and electrolyzers have advanced over recent years to become competitive with proton exchange membrane (PEM) devices. The primary benefit of AEM systems is the potential ability to use less expensive, non-platinum (Pt) group metal (PGM) catalysts, and the non-PGM stability that the alkaline environment allows. While these systems have had polymer limitations, recent improvements in the chemical stability of alkaline membranes suggest that catalysts will become the limiting factor in the near future.[1, 2]

Catalyst development in AEM systems typically focuses on oxygen reduction and evolution, since the reactions are kinetically orders of magnitude slower than hydrogen oxidation and evolution. Alternatives, however, exist to PGMs in these reactions: silver in oxygen reduction; and nickel and cobalt in oxygen evolution.[3, 4] While hydrogen oxidation and evolution are kinetically faster reactions, they are roughly two orders of magnitude slower on Pt in base compared to acidic environments. Non-PGM catalyst options are also less clear, and generally struggle to justify the AEM cost benefit, producing activities orders of magnitude lower than PGMs at higher overpotentials.[5]

Recently, advanced Pt electrocatalysts have been developed in an effort to thrift the amount of PGMs in AEM fuel cells and electrolyzers. Pt-nickel (Ni) nanowires, previously developed for acidic oxygen reduction, were studied for their activity in hydrogen oxidation and evolution.[6] These materials were formed by spontaneous galvanic displacement, a process that occurs when a metal template contacts a nobler metal cation. At low levels of displacement, small amounts of Pt were deposited to produce high electrochemical surface areas. Subsequent post-synthesis processing was used to integrate the Pt-rich and Ni-rich zones, compressing the Pt lattice and improving its activity for hydrogen oxidation and evolution. Compared to carbon-supported Pt nanoparticles (Pt/HSC), Pt-Ni nanowires produced hydrogen oxidation/evolution exchange current densities 9 times greater.

[1] B. Pivovar, Alkaline Membrane Fuel Cell Workshop Final Report, in: U.S. Department of Energy (Ed.), http://www1.eere.energy.gov/hydrogenandfuelcells/pdfs/amfc_may2011workshop_report.pdf, 2011.

[2] K.J.T. Noonan, K.M. Hugar, H.A. Kostalik, E.B. Lobkovsky, H.D. Abruña, G.W. Coates, Journal of the American Chemical Society, 134 (2012) 18161-18164.

[3] R. Subbaraman, D. Tripkovic, K.-C. Chang, D. Strmcnik, A.P. Paulikas, P. Hirunsit, M. Chan, J. Greeley, V. Stamenkovic, N.M. Markovic, Nat Mater, 11 (2012) 550-557.

[4] J.S. Spendelow, A. Wieckowski, Physical Chemistry Chemical Physics, 9 (2007) 2654-2675.

[5] J.K. Nørskov, T. Bligaard, A. Logadottir, J.R. Kitchin, J.G. Chen, S. Pandelov, U. Stimming, Journal of The Electrochemical Society, 152 (2005) J23-J26.

[6] S.M. Alia, B.A. Larsen, S. Pylypenko, D.A. Cullen, D.R. Diercks, K.C. Neyerlin, S.S. Kocha, B.S. Pivovar, ACS Catalysis, 4 (2014) 1114-1119.

474645 Characterization of Multicomponent Transport in Membranes for Solar Fuels Devices

Abstract:

Ion conducting membranes are of interest for various energy applications including fuel cells and solar fuel generators. Within the context of artificial photosynthesis towards solar fuel production, membranes are required to facilitate the ion transport necessary to promote electrochemical water splitting and CO2 reduction while meeting additional selectivity and permeability demands. Generally, membranes that limit the crossover of CO2 reduction products are desired, as high rates of product permeation through the membrane typically reduces device efficiency. Herein, we demonstrate the use of in-situ ATR FT-IR spectroscopy to quantitatively resolve the concentration of single component and multicomponent mixtures of select CO2 reduction products; methanol, formate and acetate. We then apply this methodology to the in-situ monitoring of single component and multicomponent permeation across commercially available membranes such as Nafion. Emergent transport phenomena, where the transport of a small molecule solute is affected by the presence of another solute, are explored via multicomponent permeation measurements. Membrane permeabilities and selectivities derived from single component permeation experiments are found to deviate, in some cases significantly, from permeabilities and selectivities measured in multicomponent permeation experiments.

474783 The Evaluation of Ionomers in Alkaline Membrane Electrode Assemblies for Water Electrolysis

Abstract:

The cost of water electrolysis can be reduced by transitioning from an acid to an alkaline environment. The chemistry and electrochemistry of alkaline electrolyte-based electrolyzers allows for less expensive materials-of-construction and catalysts.

The ionomers used to construct the three-dimensional electrodes on the anion conducting membrane (collectively called the membrane electrode assembly (MEA)) contribute to the overall performance of the electrolyzer. First, the ionomer provides the ion pathway between the anion conducting membrane and catalyst particle. Second, the ionomer must solubilize and transport the chemical reactants to the catalyst surface and facilitate the removal of the reaction products. Third, an electrochemical double layer is formed within the ionomer at the catalyst surface. The nature of the double layer affects the overvoltage at the electrode which contributes to the overall cell potential. Fourth, the ionomer provides mechanical stability and adhesion of the particles in the three-dimensional electrode. Thus, the ionomer should have high chemical and mechanical stability so that it does not degrade by the electrochemical or have excess swelling and adversely affect the mass and ion transport to the catalyst.

In this project, a number of ionomers were studied in an alkaline electrolysis cell. The overvoltage and chemical stability of the ionomers was compared. The ionomers tested in this project include those with an all-hydrocarbon backbone, ionomers with partial fluorination in the backbone and ionomers with a block copolymer backbone.

476338 Review and Perspective on Nexus System Analysis

Abstract:

As the global population grows and increasingly lives in cities with lifestyles based on greater material consumption, more attention is being given to the integrated system that supplies our food, energy and water, the Nexus. The interaction between these three essential commodities can be seen in many ways. For example, production of sufficient food requires use of man-made fertilizer, to an extent that food and energy prices have become linked. Agriculture now accounts for 70% of water withdrawn from aquifers, streams and lakes. In many places fresh water is critically constrained, which affects our ability simultaneously to produce more food, grow crops for biofuels, and maintain natural ecosystems. And there is mounting concern about the effects on the Nexus of climate change, itself largely the result of fossil fuel consumption and land-use change (for agriculture).

Evidence for this growing interest in the Nexus is illustrated by a review of published literature (1). Over the 5-year period 2011-2016 the rate of publication of studies on the Food-Water-Energy Nexus in the scientific literature grew by a factor about ten. These publications underline the importance of Nexus concerns to policy-makers, transnational and regional institutions, and business/commerce. The study purposes of this body of literature can be summarized as

- reviews and Nexus issue discussion papers;
- quality-of-life studies to understand Nexus trade-offs or develop responses to achieve sustainability objectives (e.g., human well-being, healthy ecosystems);
- product studies to understand Nexus implications or develop responses related to providing products (e.g., a renewable energy technology, a service such as clean water); and
- studies to develop integrated system modelling tools.

Case studies tend to be of two types: those related to regional development, and those related to specific sectors. So Hoff's (2) background paper for the Bonn2011 conference outlines ten Nexus case studies, including Jordan and Ethiopia, the arid Ningxia region of China, and Gujarat. The Brazilian Cerrado savanna area offers a study in the competition between biofuel and food crops for land and water resources; the Nile Basin offers the same trade-off, with hydropower also playing a role (2). Responses by several companies to Nexus challenges in their own business areas are outlined in a report by Conservation International (3). These responses cover a wide range of policies and activities, but there are several common themes, including

- developing greater resource efficiency
- conducting resilience planning for price volatility and climate change
- systems thinking to design synergistic solutions (e.g., business decisions that also yield multiple environmental and social benefits)
- gathering new data to understand current and future trends
- stakeholder engagement to better understand the Nexus
- ecosystem valuation to help decision-making and balance trade-offs.

Particularly important in view of the trend of more people to live in cities are the (relatively few) studies of the "Urban Nexus", dealing with what may be termed the "metabolism" of the city. It is concluded that this metabolism, a hybrid of natural and artificial structures presents an opportunity for greater efficiency through better design and regulation. At present it is poorly understood, and studies are hampered by lack of data.

Case studies show the vital role played by stakeholder involvement in clarifying issues, priorities, and values. They also demonstrate the importance of an integrated systems view of the complex interrelationships of the Nexus when planning effective remedies. Assessments conclude that transformative social and political change is needed to create new structures, markets, and governance to deal with the Nexus if we are to meet agreed-upon sustainable development goals.

Needs for the future of Nexus studies include

- the publication of more case studies, particularly including stakeholder involvement;
- the creation of new frameworks and methodologies of analysis, including some that will integrate societal, environmental, and technical dimensions;
- the development of modelling tools accessible to a wider range of investigators (at present these tend to be highly specialized);
- the promotion of best practice in Nexus evaluation and effective use of study results;
- improved accessibility of Nexus data;
- metabolism modelling and engineering for the urban Nexus; and
- promotion and support from funding agencies and institutions for multidisciplinary and multinational studies of Nexus analysis, issues, and applications.

(1) Keairns DL, Darton RC, Irabien A 2016 ***The Energy-Water-Food Nexus*** Annu. Rev. Chem. Biomol. Eng. 7:239–62

(2) Hoff H. 2011. ***Understanding the Nexus. Background Paper for the Bonn2011 Conference: The Water, Energy and Food Security Nexus.*** Stockholm: Stockholm Environ. Inst.

(3) Gerholdt J, Pandya S. ***Resources: The Energy, Water, Food Nexus***. Arlington, VA: Conserv. Int. Bus. Sustain. Counc.

476892 Overview of U.S. DOE Environmental and Waste Cleanup Programs

Abstract:

The Department of Energy has a liability for the safe cleanup of the environmental legacy brought about from decades of nuclear weapons development and government-sponsored nuclear energy research. The Departmentâ€™s Office of Environmental Management (EM) is responsible for this cleanup function and over the 25-plus years of its existence have safely cleaned up 91 sites, leaving 16 sites, with these remaining 16 sites posing the greatest technical and programmatic challenges. Four of the remaining 16 sites are tank waste sites where wastes resulting from reprocessing of spent fuel are stored in over 200 large underground tanks, containing approximately 100 gallons of highly radioactive and chemically toxic wastes. EM has a nominal $6 Billion annual budget with approximately 35% of the budget targeted to safe management, treatment and disposition of tank waste as well as closure of the tanks

The four sites include the Savannah River Site, Hanford, the Idaho National Laboratory (INL) and the West Valley Demonstration Project (WVDP). Progress in waste retrieval, treatment and disposition vary among the sites, with the treatment mission essentially complete at WVDP and not yet substantially started at INL and Hanford. To accomplish this tank waste cleanup mission, EM and its field sites utilize approximately 5,000 contractors employed by various corporations, usually in the form of limited liability partnerships. In addition, EM and its contractors rely on specific technical expertise from national laboratories, outside technical experts, and universities to provide assistance. This is particularly true in the technology development area, in which partnerships with other agencies, e.g., NASA for robotics, and university partnerships are being utilized. This talk will address these aspects that are enablers to achieving the EM tank waste mission.

468758 Solar Photocatalytic Degradation of 2,4-D at Semi-Pilot Scale in a CPC Reactor Using TiO_2

Abstract:

The 2,4-dichlorophenoxyacetic acid (2,4-D) is a broad-spectrum herbicide of chlorophenoxy type, and is used worldwide on a large scale for general weed control on grain crops, pastures, and non-crop areas. It is a pollutant of great environmental concern because of its relatively high water solubility and mobility in aquatic systems. In order to avoid dangerous accumulation, the removal of this kind of herbicides from aqueous solutions has been investigated using powerful oxidation methods to achieve the degradation and mineralization of these compounds.

In recent years, advanced oxidation processes (AOPs), which are all characterized by the production of â€¢OH radicals, have received considerable attention for herbicides degradation because they are able to oxidize and mineralize almost any organic molecule, yielding CO_2 and inorganic ions.

Among the AOPs, heterogeneous photocatalysis is currently recognized as an effective technique to treat recalcitrant pollutants, such as 2,4-D in aqueous environments. Heterogeneous photocatalysis is based on direct or indirect absorption of radiant energy (visible or UV) by a solid (photocatalyst), which is typically a wide band gap semiconductor. Anatase-TiO_2 is the most studied catalyst and it is capable of producing electronic transitions by absorption of light in the near ultraviolet region (UV).

The use of solar irradiation to power this process is environmentally attractive, and has a potential to reduce the cost of this technology. In addition to small scale studies, several pilot plant experiments have been performed in order to test the applicability of solar photocatalysis for wastewater treatment. One of the best options of reactors for photocatalytic applications using sunlight is the compound parabolic collector (CPC) reactors. The advantages of CPC reactors are:

the possibility of using solar UV radiation coming from all directions in the sky (global UV radiation), simplicity of construction and operation, turbulent flow regime (which improves mass transfer), and high reduction of the vaporization of volatile pollutants.

In this research, the solar photocatalytic activity of Degussa P25 TiO_2 was investigated in the degradation of 2,4-D herbicide as a commercial formulation Diamond® using a solar CPC reactor. The photocatalytic tests were performed in 30 L of water containing 50 mg/L of herbicide concentration. In order to obtain optimized conditions for the photocatalytic degradation, a fractional factorial experimental design was conducted, the investigated variables were catalyst dosage (0.3, 0.6, and 0.9 g/L), and flow in the reactor (2, 3, and 5 gal/L). The decay in herbicide concentration was followed by reversed-phase chromatography and the mineralization was followed by measurement of the total organic carbon (TOC). The best conditions were 0.6 g/L of catalyst loading and flow of 2 gal/L achieving almost complete degradation (98.68%) and 80% of mineralization of commercial 2,4-D with an accumulated radiation of 120 kJ/m^2. These results demonstrate the effectiveness of treatments environmentally friendly in order to degrade recalcitrant compounds in wastewaters.

481310 Simultaneous Removal and Oxidation of NO and SO_2 from Flue Gas by Combined Temperature and Fe^{2+} Activated Aqueous Persulfate Solutions

Abstract:

This work evaluates the simultaneous removal efficiency of NO (1000 ppm) and SO_2 (1550 ppm) from flue gas induced by the combined temperature and Fe^{2+} (0.01M) activation of aqueous persulfate (0.1M) and elucidates the chemistry and reaction pathways. The results indicated that both SO_2 in the flue gas and Fe^{2+} in solution improved NO removal. The conversion of NO (i.e., percent removal of inlet concentration) was 46.2% at 30°C and only slightly to 48.2% at 50°C; but sharply increased to 81.2% at 70°C, while SO_2 was completely removed at all temperatures. Also, a kinetic-mass transfer mathematical model utilizing the pseudo-steady state approximation technique for proposed comprehensive reaction pathways was developed. The model results, which appeared to fit the experimental results of the NO_x-S(IV)-$S_2O_8{}^{2-}$-Fe^{2+} system remarkably well, was used to predict species and product concentrations, NO and SO_2 mass transfer coefficients, kinetic data and other process parameters.

References:

1) Khan, N.E.; **Adewuyi, Y.G.** Absorption and Oxidation of Nitric Oxide (NO) by Aqueous Solutions of Sodium Persulfate in a Bubble Column Reactor. Industrial Engineering Chemistry Research. **2010**, 49 8749-8760.

2) **Adewuyi, Y.G.**; Sakyi, N.Y. Removal of Nitric Oxide by Aqueous Sodium Persulfate Simultaneously Activated by Temperature and Fe^{2+} in a Lab-Scale Bubble Column Reactor. Industrial Engineering Chemistry Research. **2013**, 52,

14687-14697.

3) **Adewuyi, Y.G.**; Sakyi, N.Y. Simultaneous Absorption and Oxidation of Nitric Oxide and Sulfur Dioxide by Aqueous Sodium Persulfate Activated by Temperature. Industrial Engineering Chemistry Research. **2013**, 52, 11702-11711.

4) **Adewuyi, Y.G.**; Khan, M.A., Sakyi, N.Y. Kinetics and Modeling of the Removal of Nitric Oxide by Aqueous Sodium Persulfate Simultaneously Activated by Temperature and Fe^{2+}. Industrial Engineering Chemistry Research. **2014**, 53, 828 â€“ 839.

5) **Adewuyi, Y.G.**; Khan, M.A. Nitric Oxide Removal by Combined Aqueous Persulfate and Ferrous-EDTA Reaction Systems. Chemical Engineering Journal **2015**, 281, 575-587.

6) **Adewuyi, Y.G.**; Khan, M.A.; Removal of Nitric Oxide (NO) by Combined Aqueous Persulfate and Ferrous-EDTA Systems: Effects of Persulfate and EDTA Concentrations, temperature, pH and SO_2. Chemical Engineering Journal **2016**, 304, 793-807.

469340 Generation of Oxidative Chemistries By Cold-Atmospheric Plasma

Abstract:

Devices to produce non-thermal ("cold") atmospheric pressure plasma, or CAP, are gaining increasing attention for their ability to inactivate bacteria and viruses[1–3], and to oxidize numerous organic pollutants[4] and specifically industrial dyes[5,6]. CAP discharges in direct contact with an aqueous solution are known to produce a cocktail of reactive oxygen and nitrogen species in solution, including $OH^{\bullet}$ and $O_2^{\bullet -}$. The type and geometry of the discharge employed, the properties of the treated solution, all affect the quantity and ratio of species produced. These factors thus play an important role in optimizing such devices for both industrial and more specialized applications.

In this work, we first examine the factors affecting the generation and transport of highly reactive species produced by cold atmospheric plasma across the plasma-liquid interface. We present the results from studies on positive and negative polarity corona discharges in which the dye indigo carmine is used as a measure of oxidation. We comment on the origin of the reactive species produced (gas phase, boundary layer, or bulk solution) and the relative importance of kinetics versus convective and diffusive forces in optimizing the modification of a target.

References:

1. Traylor, M. J. *et al. J. Phys. D. Appl. Phys.* **44,**(2011).
2. Pavlovich, M. J., Chang, H.-W., Sakiyama, Y., Clark, D. S. & Graves, D. B. *J. Phys. D. Appl. Phys.* **46,**145202 (2013).
3. Aboubakr, H. a. *et al. Appl. Environ. Microbiol.* **81,**3612–3622 (2015).
4. Jiang, B. *et al. Chem. Eng. J.* **236,**348–368 (2014).
5. Brisset, J.-L., Benstaali, B., Moussa, D., Fanmoe, J. & Njoyim-Tamungang, E. *Plasma Sources Sci.*

Technol. **20,**034021 (2011).
6. Vautier, M. *J. Catal.* **201,** 46–59 (2001).

460300 Plasma-Based Water Treatment: Targeted Application and Guidelines for Process Scale-up

Abstract:

Electrical discharge plasma is an effective and versatile advanced oxidation process due to the formation of reactive species such as hydroxyl radicals. The technology requires no chemical additions, can degrade a broad range of contaminants, and produces no residual waste. Despite promising results of bench-scale studies and evident potential advantages of plasma water treatment, the technology has not been used in practice. A major obstacle hindering the development of plasma reactors is associated with the absence of general principles to guide the design of new reactors with higher efficiency. In an attempt to improve the feasibility of plasma-based water treatment technology and develop basic guidelines for reactor design and optimization, we have determined that the size of the contact area between the plasma and the treated solution is the single most important factor affecting the plasma reactor performance. Building on this knowledge, we have modified the traditional gas-discharge plasma reactor to maximize this contact area, which resulted in the development of the foaming plasma reactor. The foaming electrical discharge plasma reactor rapidly and efficiently degrades many chemicals including pharmaceuticals, personal care products, disinfection byproducts, and several perfluoroalkyl acids (PFAAs). Of all the compounds evaluated, perfluorooctanoic acid and perfluorooctanesulfonic acid were degraded the fastest. Experiments conducted with groundwater containing PFAAs demonstrated that co-contaminants do not appear to influence the efficiency of the reactor.

We have determined the efficacy of the plasma process for treatment of a wide range of different compounds of environmental importance and used the results of this investigation to construct a model to predict the approximate treatability of any compound based on just a few of the compoundâ€™s physical properties. We have determined design guidelines for plasma reactor scale up as well as the efficiency and relative competitiveness of the scaled up process for selected compounds.

469820 Selenate Reduction from Mine Water By Redox Reaction and Adsorption Process

Abstract:

A treatment process was developed to reduce total selenium in mine water to ≤5 μg/l. The treatment system utilizes a filtration and adsorption process. A proprietary iron-based media was used as an adsorbent. Using a sample from an abandoned coal mine site containing about 23μg/l of selenate, column studies were conducted for a period of six months. The objectives were to validate the applicability of this media in removing selenate from mine water. Several process parameters were investigated, including pH, ORP, adsorbent dosage, contact time, and the impact of co-occurring contaminants. Using an ICP-MS, selenium in water was analyzed. A high resolution X-ray Photon Spectroscopy (XPS), under nitrogen blanketing, was used to characterize the spent media. Results reveal that the adsorbent is capable of removing selenium to ≤5 μg/l. The breakthrough curve showed that about 90,000 bed volumes of water were treated prior to reaching the target effluent selenium of ≤ 5μg/l. The adsorption capacity of the media was not affected by the presence of co-occurring contaminants. Scaling or fouling was not observed within the column. No significant pressure drop (ΔP) was observed. The treated effluent contained about 2 mg/l of iron, and after aeration followed by direct filtration iron was reduced to < 0.1 mg/l.

Several kinetics models including pseudo first-order, pseudo second-order, power function, Elovich equation, and parabolic diffusion were used to analyze the data. The high r^2 (square of the correlation coefficient) value for the Elovich model suggests that the data are generally best described by this equation for selenate adsorption at the experimental condition. The results suggest that the adsorption kinetics could potentially be a chemisorptions process. Literature data indicate that the Elevoch equation has general application to the chemisorption kinetics. Thermodynamics study results reveal that increasing temperature increased the kinetics and adsorption capacity. The calculated heat of adsorption value [ΔH] ranged between 130 and 150 KJ/mol. The X-Ray Photon Spectroscopy (XPS) results indicate that after adsorption, selenium onto the spent media existed in the form of elemental selenium [Se^0]. It is apparent that the elemental selenium is the product of the redox reaction. It is anticipated that the selenate removal by the media could be a combination of redox reaction and chemisorptions process.

Selenium concentration in the Toxic Characteristics Leaching Procedure (TCLP) extract of the spent media was less than the regulatory limit of 1 mg/l, the spent media can be disposed as a non-hazardous material.

475423 Sustainability Concepts in the Food-Energy-Water Nexus: Chemical Engineering Perspective

Abstract:

The idea of sustainability and/or sustainable development emerged from the World Commission on Environment and Development (WCED), generally known as the Brundtland Commission. It defined sustainable development in "Our Common Future" as that which meets the needs of the present without compromising the ability of future generations to meet their own needs. Environmental sustainability and/or sustainable development "occurs when we maintain or improve the material and social conditions for human health and the environment over time without exceeding the ecological capabilities that support them". As such, sustainability has three underlying bases: (1) economic viability - the cost and business aspects of a project; (2) social concern - human health and social welfare; and (3) natural or ecological issues - depletion of natural capital and environment. Sustainable development has become a major driving initiative in engineering businesses throughout the world today. Our focus is on all aspects of sustainable development including environmental, industrial, business, economical, ethical, and social sustainability as they impact the availability of food, energy, and water. This is important because it ***has been generally acknowledged that the increasing use of non-renewable resources to support an increasing population has created an unsustainable or nearly unsustainable situation where the ability to provide adequate food, energy, and water is in question. Should this rate of use be maintained, two dire consequences will follow: first, the future generations will be unable to maintain a good standard of living, and second, developing countries will have less of an opportunity to bring their living standards to a level comparable to that of the affluent countries. This is important because the availability of food, energy, and water is an indispensable necessity for maintaining a good standard of living. The practice of chemical engineering, perhaps more than any other technical discipline, involves the use of both non-renewable and renewable materials, and energy and exergy (available energy) resources for the production of chemicals, value-added goods, services of commerce, and waste as***

pollution. This paper primarily deals with the ideas of environmental sustainability and/or sustainable development, methodologies to redesign sustainable chemical and allied processes and products development using the LCA approach. These principles are illustrated with a case study for the production of ethanol, an energy source, from molasses, a human food, in a combined agricultural and chemical process that requires substantial amounts of water.

461670 Optimizing Spatio-Temporal Sensor Placement for Nutrient Monitoring: Algorithmic Framework

Abstract:

INFEWS is a new initiative in NSF which deals with Innovations at the Nexus of the Food, Energy and Waters Systems. This project addresses the INFEWS: N/P/H2O solicitation. The food production system generates waste streams that are characterized by high concentrations of organic matter, nitrogen- and phosphorus-containing species in water. Therefore, monitoring nitrogen and phosphorous species is important for water quality requirements for agricultural as well as energy production. Currently these species are monitored via stationary monitoring stations. However nitrogen and phosphorous species move via agricultural run-off to other water systems, and requires portable sensors which can change the positions in real time. This type of dynamic sensing requires novel algorithms which decide sensor locations in real time in the face of inherent uncertainties in the fate and transport of the species. To develop such an algorithmic framework to solve the problem of sensor placement in real time is the objective of this proposal.

This paper presents the first part of the work which is the algorithmic framework for optimal spatial-temporal locations of the sensors in real time. The algorithmic framework is based on a novel algorithm called Better Optimization of Nonlinear Uncertain Systems (BONUS). We present a case study of an agent based model from social sciences for testing these framework. This model is based on behavior of cows in a field where there are grass patches. The cows move in a herd towards the greener pasture and eat grass till it finishes. The grass growth and cow movements are probabilistic. We are using four sensors for object tracking (e.g. cows) and these sensor move spatially and temporally in order to track maximum number of cows. Results show that the optimal movement of sensors allows to track maximum number of cows.

In the future, the algorithmic framework will be coupled with nutrient and fate transport models. A case study of a small watershed will be selected from literature for nutrient monitoring sensor location problem. Optimal spatial-temporal locations will be the result of this new framework for this case study.

Bandwidth Study on Energy Use and Potential Energy Saving Opportunities in Manufacturing Food and Beverages

Presenter:
Caroline Dollinger
Engineer, Energy Analysis
Energetics Incorporated

American Institute of Chemical Engineers
Annual Meeting
San Francisco, CA
November 14, 2016

Collaboration

Caroline Dollinger (Speaker)
Energetics Incorporated
cdollinger@energetics.com

Sabine Brueske
Energetics Incorporated
sbrueske@energetics.com

Joseph Cresko
U.S. Department of Energy
Advanced Manufacturing Office
joe.cresko@ee.doe.gov

Alberta Carpenter
National Renewable Energy Laboratory
Alberta.carpenter@nrel.gov

Energetics Incorporated is an energy consulting company, providing subject matter expertise to private and public sector clients

2

Overview

- An *Energy Bandwidth Study* is currently being performed to analyze energy use and energy saving opportunities in energy intensive manufacturing sectors
- This presentation focuses on the work being done for a *Food and Beverage Manufacturing Energy Bandwidth Study*
- Energy data will broken down by sub-sector, product, and unit operation to identify processes and energy-saving technologies with the greatest opportunities for energy savings

Results presented in this talk are draft data; the work is still ongoing and a final report is expected to be published next year.

3

ENERGETICS
A Subsidiary of VSE Corporation

Energy bandwidth studies can be a useful tool for assessing energy savings opportunities in manufacturing

Energy bandwidth studies frame the range (or *bandwidth*) of potential energy savings in manufacturing, and technology opportunities to realize those savings.

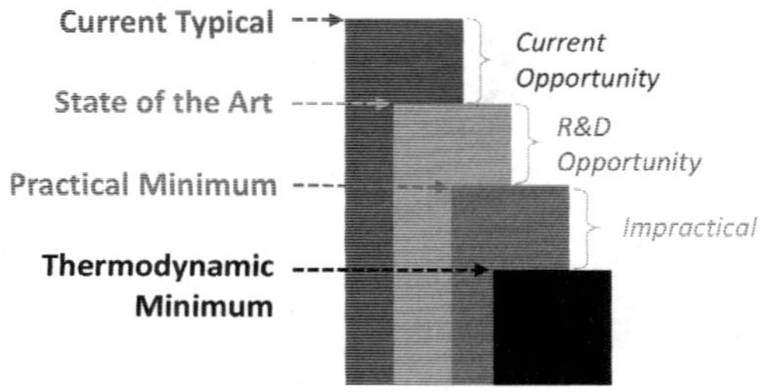

Measures of energy intensity studied:

Current Typical (CT)	State of the Art (SOA)	Practical Minimum (PM)	Thermodynamic Minimum (TM)
Determined from a literature review and stakeholder outreach, based on current typical manufacturing processes in the U.S.	Determined from a literature review and stakeholder outreach, based on the most energy-efficient technologies and practices available worldwide	Modeled based on plausible energy savings from identified R&D technologies under development worldwide	Calculated analytically using a Gibbs free energy approach assuming ideal conditions

4

ENERGETICS
A Subsidiary of VSE Corporation

Six new energy bandwidth studies on lightweight materials were recently completed in draft form

- These studies have previously provide a deep-dive analysis into manufacturing energy use and potential energy savings for:
  - Four energy-intensive manufacturing sectors
  - Manufacturing of six lightweight materials
- This talk will focus primarily on a new bandwidth on food and beverage manufacturing.
- Published bandwidths (including this one, once completed) are available on the DOE Advanced Manufacturing Office website

Summary thumbnail for bandwidth study on petroleum refining

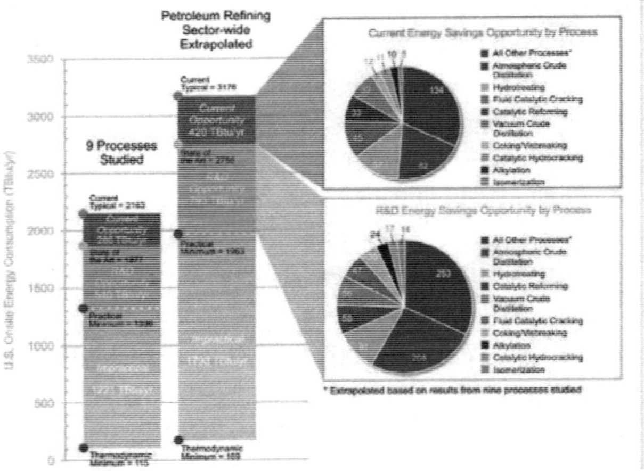

5

First Four Industry Bandwidths – US Results

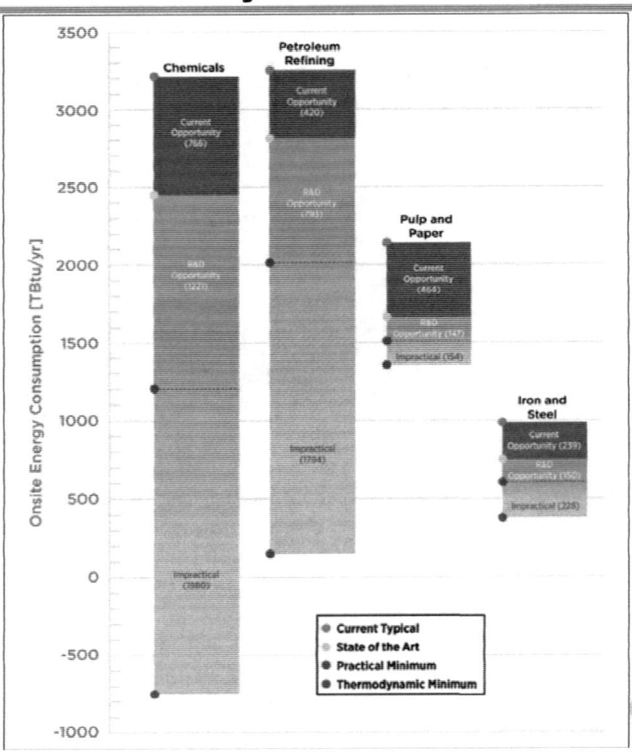

Challenges in Researching Energy Use and Savings in the Sector

- **Highly diversified sector that produces thousands of different products**
- **Processing facilities range from small to large units, normally producing more than one product**
  - **As a result energy intensities will vary from facility to facility and from product to product**
- **Less active in the R&D, with a focus on energy savings technologies, compared to other sectors**
- **Competitive marketplace, where energy efficiency may help to offer a competitive advantage**
- **Largely mechanized processes, difficult to ascertain and source energy data at the sub-process level**

ENERGETICS
A Subsidiary of VSE Corporation

Food & Beverage Manufacturing Bandwidth
Study Coverage

Food and Beverage Sub-sectors Studied	Food and Beverage Products Studied
Grain and Oilseed Milling	Wet corn milling- corn oil and by-products Soybean oil and meal Other vegetable oils and meals
Sugar Manufacturing	Beet sugar Cane sugar
Fruit and Vegetable Preserving and Specialty Food Manufacturing	Frozen fruits, vegetables, juice Canned fruits, vegetables Pickled vegetables Dried and dehydrated fruits and vegetables
Dairy Product Manufacturing	Milk Butter Cheese Dry, condensed, evaporated milk Ice cream/frozen desserts
Animal Slaughtering and Processing	Beef, pork, chicken, lamb Animal fat, lard, edible tallow
Beverage Manufacturing	Soft drinks Beer, wine, distilled liquors

ENERGETICS
A Subsidiary of VSE Corporation

Food & Beverage Manufacturing Bandwidth
Study Coverage

Food and Beverage Sub-sectors Studied	Energy Intensive Sub-processes
Grain and Oilseed Milling	Crushing Extraction Evaporation Dewatering and drying
Sugar Manufacturing	Dewatering and drying Evaporation Vacuum pans (crystallization)
Fruit and Vegetable Preserving and Specialty Food Manufacturing	Heat sterilization/retorting Cooking/blanching Freezing Concentration Evaporation Drying Canning
Dairy Product Manufacturing	Pasteurization Freezing Concentration
Animal Slaughtering and Processing	Rendering
Beverage Manufacturing	Beer processes (brewhouse, packaging) Wine processes (refrigeration)

Food and Beverage Manufacturing Bandwidth
Scope Description

What's Out- Agriculture Sector NAICS 11

Establishments primarily engaged in growing crops, raising animals, from a farm, ranch, or their natural habitats

111-Crop production

112-Animal production

What's In- Manufacturing Sector NAICS 31

Industries in the Food Manufacturing subsector transform livestock and agricultural products into products for intermediate or final consumption (basically, any food that is processed)

311 – Food manufacturing

What's Out- NAICS 42, 44, 72

424 – Food Wholesalers

445 – Food and Beverage Stores

722 – Food Services and Drinking Places

10

ENERGETICS
A Subsidiary of VSE Corporation

Food and Beverage Sector
U.S. Energy Consumption Overview

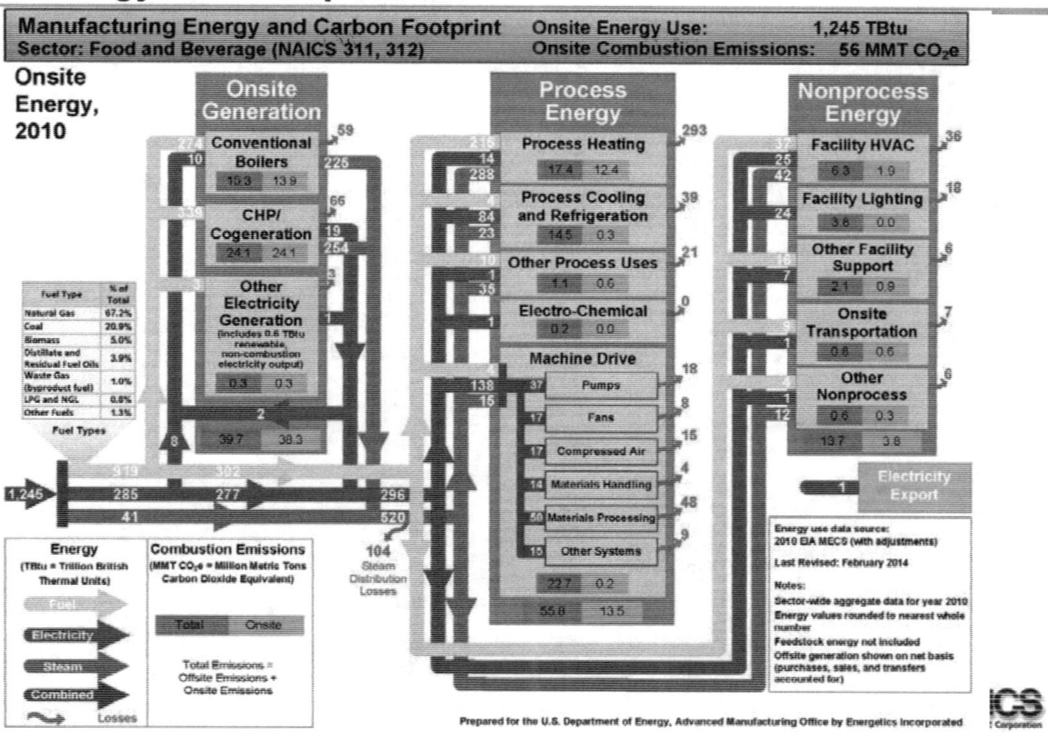

Food and Beverage Bandwidth
U.S. Baseline Energy Consumption, 2010

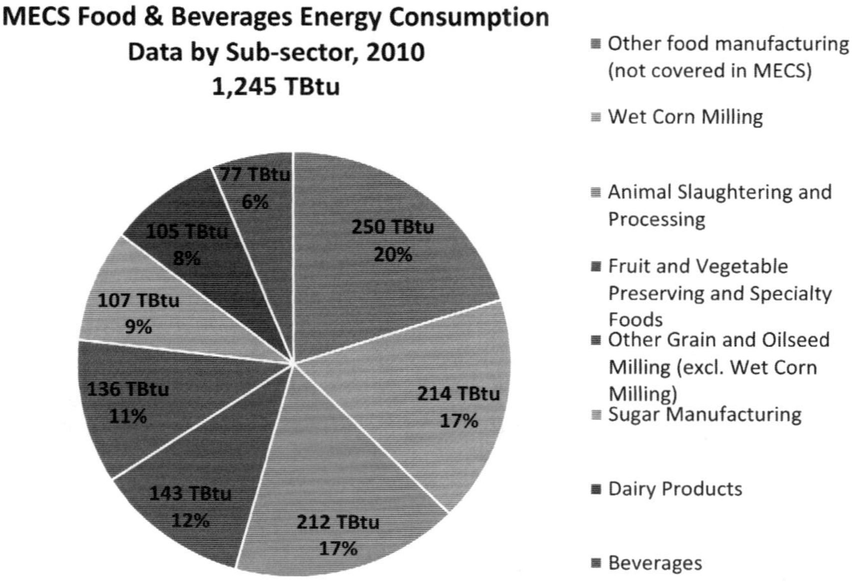

Source: EIA 2014, Manufacturing Energy Consumption Survey (MECS)

12

Food and Beverage Bandwidth
U.S. Production for Sub-sectors Studied, 2010

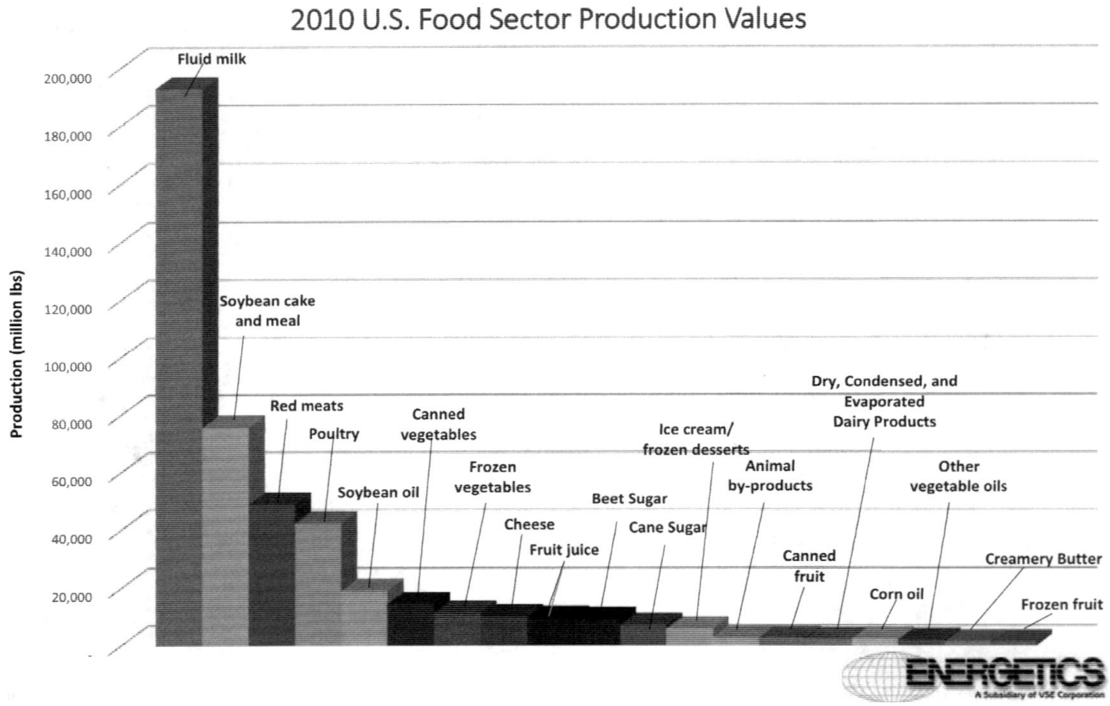

Dairy Products Manufacturing
Energy intensive sub-processes being analyzed

- Pasteurization
- Cooling/refrigeration
- Concentration
- Freezing
- Motors/pumps
- Spray Drying
- Canning

Manufacturing Process	CT (TBtu/year)	% of sub-sector
Pasteurization	20	18.6%
Cooling	16	15.6%
Motors, pumps	9	8.4%

14

Grain and Oilseed Milling

Energy intensive sub-processes being analyzed

Wet Corn Milling

- Starch dewatering and drying
- Gluten thickening and drying
- Gluten feed drying
- Steeping
- Steepwater evaporation
- Fiber and gluten recovery
- Germ dewatering and drying
- Grinding

Other vegetable oils/meals

- Crushing and extraction
- Neutralization
- Oil refining
- Deodorization

15

Fruit and Vegetable Preserving

Energy intensive sub-processes being analyzed

Canned Fruits & Vegetables

- Washing
- Retorting
- Blanching/cooking
- Canning

Frozen Fruits & Vegetables

- Washing
- Peeling
- Blanching/cooking
- Freezing

Other fruit/ vegetable products

- Pasteurization (fruit juice)
- Dehydrating
- Retorting (fruit juice)

16

Animal Slaughtering and Processing

Energy intensive sub-processes being analyzed

Red Meat Products

- Dressing and cutting
- Refrigeration
- Processing (curing, smoking, cooking)
- Packaging
- Splitting
- Scalding
- Singeing
- Blood processing
- Edible rendering

Poultry Products

- Scalding, slaughtering, defeathering
- Evisceration and cooling
- Refrigeration
- Liquid effluent treatment
- Packaging

17

Current Typical (CT) energy intensities and consumption for red meat products, U.S. (2010)

Manufacturing Process	CT (Btu/lb)	TBtu/year (% of sub-sector)
Chilling	468	23 (10.8%)
Processing (curing, smoking, cooking)	320	16 (7.4%)
Edible rendering	4,573	14 (6.5%)
Dressing and cutting	181	9 (4.2%)
Packaging	93	5 (2.1%)
Scalding*	149	3 (1.6%)
Splitting*	117	3 (1.2%)
Singeing*	82	2 (0.9%)

* Pork products

18

ENERGETICS A Subsidiary of VSE Corporation

Animal Slaughtering and Processing
Example of challenges in data and analysis

- **Lack of recent analysis done on energy usage in U.S. manufacturing facilities, must rely on internationally available data**
- **Most estimates of energy intensity are presented in terms of total energy/product or input (e.g., kWh/tonne), not broken down by sub-process**
- **Studies done on individual plants may not be applicable to be generalized sector-wide**
- **Energy data presented in different forms (consumption per head, carcass weight, dressed weight, product weight)**
  - **Involves accurate conversions to calculate total energy consumption using production values from USDA, other assumed factors**
- **If data is provided for sub-processes, only provided for a few, not allowing for calculation of balance**

ENERGETICS
A Subsidiary of VSE Corporation

Current Opportunity and *R&D Opportunity*

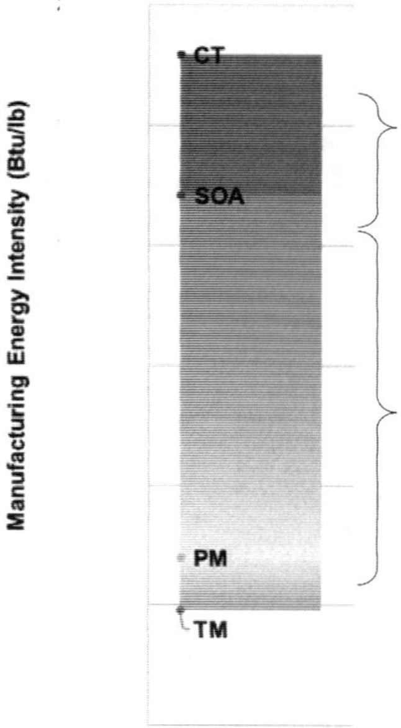

Current Opportunity

Energy savings if the best technologies and practices available were used to upgrade production

R&D Opportunity

Additional energy savings if applied R&D technologies under development worldwide were successfully deployed

<u>**SOA Technology Examples:**</u>

- **Steam system improvements**
- **Adjustable-speed drives**
- **Air compression improvements (reduce leaks, efficient vacuum pumps)**
- **Alternative technologies**
- ...

<u>**R&D Technology Examples:**</u>

- **Membrane filtration**
- **Microwave heating**
- **Advanced Ohmic Heating**
- **High pressure processing**
- **Cost efficient thermal management in motors using nanomaterials**
- ...

ENERGETICS
A Subsidiary of VSE Corporation

What about the Water part of the Nexus?

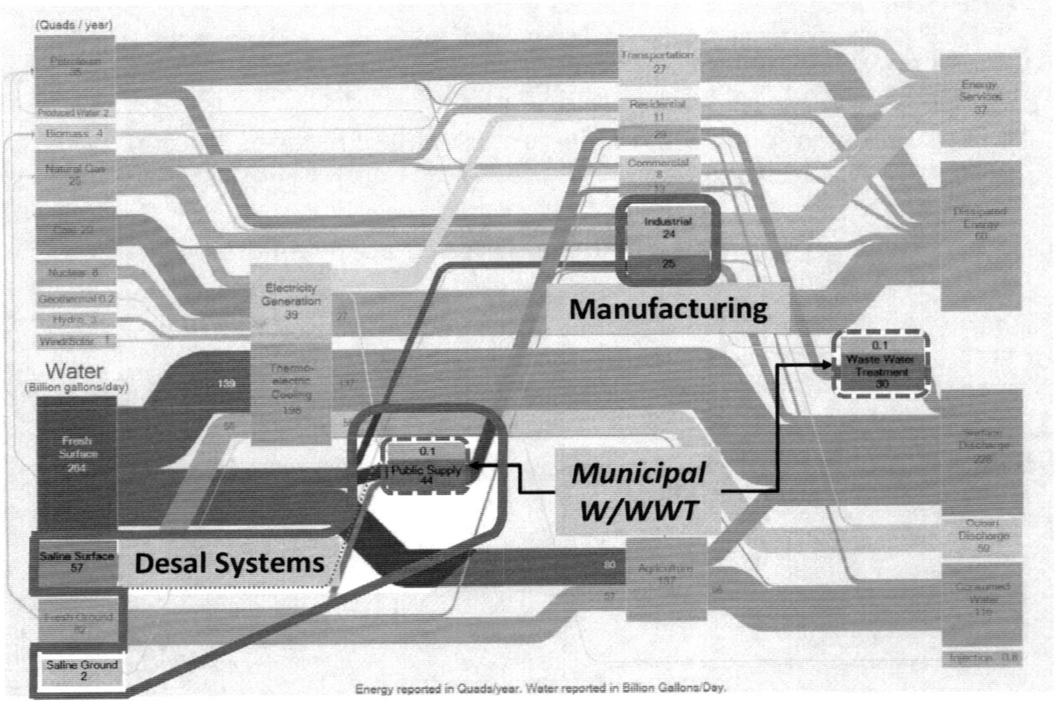

Energy reported in Quads/year. Water reported in Billion Gallons/Day.

Conclusions

- An *Energy Bandwidth Study* is being performed to analyze manufacturing energy use and energy saving opportunities for the food and beverage manufacturing industry
- Manufacturing energy data are broken down by sub-sector, product, and sub-process to identify processes and energy-saving technologies with the greatest opportunities for energy savings
- This type of study has not been done in the past, so data must been gathered carefully; also there is a lack of data in certain areas
- While more research is to be done, research to date implies that there may be more current rather than future energy savings opportunity, due to the fact that there are few areas with active R&D in food and beverage, while more traditional areas for energy efficiency savings (steam systems, etc.)

Results presented in this talk are draft data; studies are currently being peer reviewed. If you are interested in participating, please contact me!

22

Thank you!

Energetics Incorporated
Caroline Dollinger
(CDollinger@energetics.com)

ENERGETICS
A Subsidiary of VSE Corporation

468656 Insight-Based Design of Local Integrated Systems for Food, Energy and Water

Abstract:

The importance of Food-Energy-Water nexus (also referred to as Water-Energy-Food nexus) has been widely recognised by national governmental bodies and international organisations such as the United Nations, for its significance to sustainable development and national security. Though these three sectors are inextricably linked, as actions in one sector have impacts in one or both of the others, these areas have too often been considered individually. Thus, there is a need to look at them from a holistic perspective and develop tools that take into consideration their interdependencies. This opens up opportunities for process system engineering research. As such, the aim of this study is to develop a systematic insight-based approach for the design of an integrated local production system particularly to meet local demands for food, energy and water. One of the benefits of the insight-based approach is that it allows decision variables to remain tractable throughout the design process. Unlike mathematical optimisation approaches, intermediate results from insight-based approaches can be analysed, interpreted and used to inform the implications of a given design for the operation of the whole system, thus offering a more practical tool for decision-makers and local planners to use. Insight-based approaches which include techniques such as pinch analysis for heat integration and mass integration have proved particularly useful for the integration of energy and material streams in industrial production processes within eco-industrial parks and industrial symbiosis systems. They have been successfully implemented for enhancing resource efficiency and sustainability of processes involving heat, water and hydrogen, to name a few applications. Although the proposed study also pursues higher resource efficiencies through integration of resources, the design of the local integrated production system has a distinctive emphasis on local resources and demands and on the holistic consideration of all types of agricultural, industrial and municipal processes to take place at the locality of concern. This work will combine existing methods and adapt them to enable the identification of the efficient use of various locally available resources. A net gain indicator will be defined to quantify the avoided cumulative effort that would have been spent for importing resources to satisfy the demand.

Using the net gain indicator, parallel processing and utilisation options within and between different subsystems can be compared and the optimal actions identified. The proposed insight-based approach for the design of local integrated system of food, energy and water will be illustrated through a case study based on an eco-town in the UK.

468765 Recovering Runoff Particulate-Bound Phosphorus Via Fungal Bioextraction

Abstract:

Phosphorus (P) loss is a key issue affecting food-energy-water nexus sustainability. P in fertilizers is critical for ensuring food security for the continuously growing world population. Currently, P is obtained via energy-intensive mining and processing of phosphate rock, which is becoming more scarce and concentrated in a few geographical areas. Excessive fertilizer application due to slow diffusion and heavy fixation of P in soils have led to massive P losses amounting to almost half of all mined P due to agricultural soil erosion and runoff. P-rich runoff streams discharging into surface water bodies can cause eutrophication, toxic algal blooms, and overall quality deterioration of freshwater resources. In this work, we propose capturing and processing runoff sediments using a low-energy and inherently benign fungal bioprocessing method to unlock a previously untappped source of recoverable phosphorus for nutrient recycling and closing the loop of P use. This can potentially reduce the demand for mined sources while at the same time help protect watersheds from the devastating effects of eutrophication.

This paper presents the preliminary findings of our work involving the use of phosphorus-solubilizing fungi (PSF) for extracting and recovering particulate-bound phosphorus in erosion and runoff streams. PSF species can generate organic acids and phosphatase enzymes that can induce mobilization of matrix-associated or surface-adsorbed phosphate ions on inorganic minerals in runoff particulates, and degrade phosphate ester bonds in sedimentary organic phosphorus compounds to facilitate the release of phosphate ions, respectively. Samples of runoff sediments discharging into a eutrophic lake in Kalamazoo, MI were collected and characterized for total P content and P speciation. P extraction experiments on runoff sediment samples were performed using model organic acids and spent culture broth *Aspergillus niger* ATCC 15475. Phosphate solubilization and changes in P speciation of the runoff sediments during treatment were measured using a sequential chemical extraction method and colorimetric analysis. Changes in runoff sediment mineralogy was analyzed using X-ray Diffraction (XRD) spectroscopy. The results

indicate that oxalic and citric acids can extract and solubilize a significant fraction of phosphates bound to the iron, aluminum, and calcium mineral components of the runoff sediments. A kinetic model describing phosphate mobilization via ligand-exchange and induced desorption mechanisms was developed and fitted into the experimental data. We also observed a significant reduction in the organic P fraction of sediments treated with spent fungal culture broth, indicating the possibility of phosphatase enzyme production and activity. Thus, changes in organic P speciation in the sediments were monitored using solution ^{31}P nuclear magnetic resonance (NMR) spectroscopy. The findings demonstrate the feasibility of the proposed technology for P recovery from nonpoint nutrient pollution sources such as runoff sediments. It will help develop an effective, efficient, economical, and environmentally benign system that will benefit the community, agriculture and food production systems, and the environment in terms of resource security and environmental sustainability.

469658 Preliminary Study on Detroit's Urban Food-Energy-Water (FEW) Nexus

Abstract:

With 66% of the worldâ€™s population expected to live in urban areas by 2050, cities around the world will face numerous challenges in sustainable provision of food, energy, and water. These challenges are particularly complex due to the interconnectedness of the food system, energy system, and water system. For example, water is required for food and energy production; energy is needed for food production as well as water and wastewater treatment and distribution; and industrial agriculture and food production lead to oversupply of nitrogen and phosphorus in the urban water system through fertilizer use. Therefore, policy and technology solutions addressing urban food-energy-water (FEW) challenges need to be evaluated through the lens of FEW nexus to identify co-benefits and avoid unintended consequences. This calls for innovative approaches to model the urban FEW nexus as an integrated whole, instead of examining them individually.

Existing studies primarily investigated urban FEW systems individually (e.g., nutrient flows, energy flows, and water flows) or focuses on two of the three FEW systems (e.g., water-energy nexus and food-water nexus). Existing studies that consider all FEW systems primarily aim to understand the impact of one system on the other two, such as how urban water-sector innovations change food and energy flows. There lacks an integrated approach that characterizes the comprehensive interdependence of urban FEW systems.

In this study we developed a systems modeling framework for the urban FEW nexus. We also applied this framework to the Detroit Metropolitan Area (DMA) as a preliminary study. The urban FEW systems modeling framework we proposed is based on material and energy flow analysis (MEFA) which quantifies flows and stocks of a group of substances within a system during a given period of time. This framework includes three steps:

Step 1: Individual FEW systems are modeled as resource flows across major sectors (e.g., water treatment, wastewater treatment, food production) of the urban system. Given the variety of food crops and products, we used nitrogen and phosphorus flows as the currency to represent the food system. Quantities of resource flows (water, energy, nitrogen, and phosphorus) are estimated based on a variety of data such as population, water withdrawal, and wastewater generation that were collected for the DMA case study.

Step 2: We identified and measured interdependence of individual FEW systems. This was done through literature review for common FEW interactions and quantitative estimation of resource flows between individual FEW systems (e.g., the amount of water used for food production and energy production).

Step 3: Each individual FEW system is characterized as a network of resource flows. Based on the interconnections between individual resource flow networks obtained from Step 2, we integrated the individual FEW resource flow networks as a network of four networks (water, energy, nitrogen, and phosphorus). This provides a mathematic representation of the urban FEW nexus, which will be the base of further analysis and modeling.

475777 A Fail Safe, Passive Operation Small Modular Nuclear Reactor for Electricity Generation and Process Heat Production

Abstract:

A Fail Safe, Passive Operation Small Modular Nuclear Reactor for Electricity Generation and Process Heat Production Mohamed S. El-Genk

Regents' Professor, Nuclear, Mechanical and Chemical Engineering, and Director, Institute for Space and Nuclear Power Studies, School of Engineering, University of New Mexico, Albuquerque, New Mexico, USA

SMRs are those which provide for a total electrical power generation capacity from as little as 10 to 300 MWe. Depending on the installed electrical power and the type of the SMR plant, all or most reactor components could be built and assembled in the factory and shipped to the construction site by rail, heavy trucks or on barges. The modularity and the short construction time of the plant of 12 – 18 months reduce capital cost and enhances the quality assurance for the reactor and plant components. This flexibility avoids large financial obligations associated with building a large plant, which may require developing and managing a consortium of multiple utilities in more than one state. This is, in addition to paying interest on a large loan ($5B- $8B) for many years and the lack of revenues during 5-6 years construction time.Several countries are pursuing different SMR designs, building prototypes, doing actual construction and planning a near term deployment. These include the Russian Federation, Japan, Argentina, China, India, the Republic of Korea and USA. Nevertheless, there are many challenges associated with developing and deploying SMRs. Examples include: (a) developing a functional central control room for

coordinating multiple operating units on a single site; (b) developing operators training for managing multiple reactor units; (c) licensing multitude of design innovations; (d) reaching an economic competitiveness with large plants and renewable energy sources; (e) establishing future market demand; and (f) developing procedures for the handling, storage and processing of used fuel or reactor unitsThis paper introduces the University of New Mexico's Scalable LIquid Metal cooled small Modular (SLIMM) reactor (Figs. 1 and 2), capable of generating 10 to 100 MW of thermal power, for 55–6 full power year, respectively, without refueling. Natural circulation of in-vessel liquid sodium cools the SLIMM reactor core during nominal operation and after shutdown, with the aid of in-vessel tall chimney and a sodium-sodium (Na-Na), concentric helically coiled tubes heat exchanger (HEX) placed at the top of the down-comer in the reactor vessel.

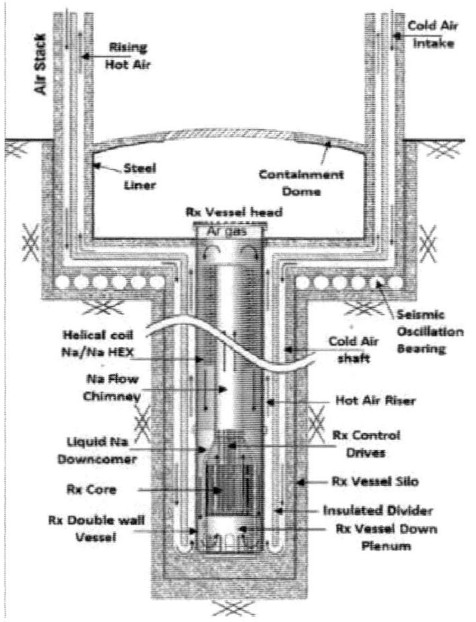

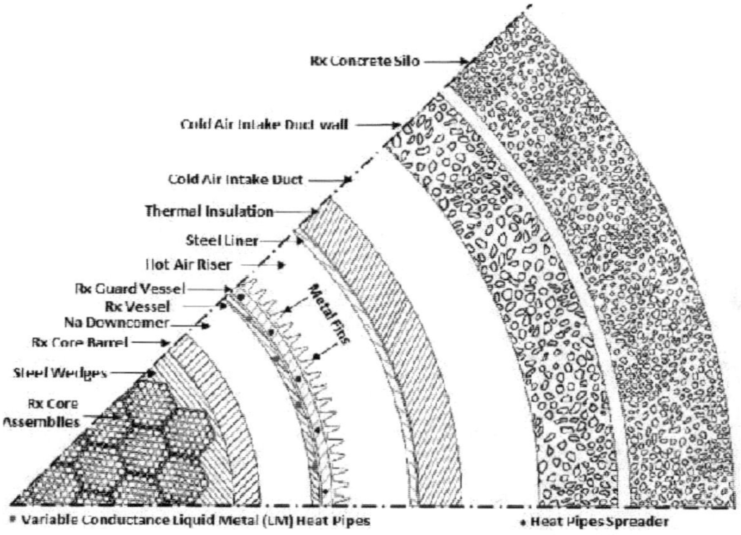

Fig. 2. A Pie Section of the SLIMM reactor Layout.

Fig.1 Axial Cross-sectional View of SLIMM, SMR Installed Below Grade

In case of a malfunction of the HEX, a backup cooling system (BCS) passively removes the decay heat generated in reactor core after shutdown, from the outer surface of the reactor vessel by natural circulation of ambient air (Figs. 1 and 2). This system is capable of removing 1-1.26 MW of thermal power, which together with the large mass of the liquid sodium in the reactor vessel (~ 38 MT), limit its temperature rise after reactor shutdown to < 900 K. Owing to the lower vapor pressure of liquid sodium (boiling temperature is ~ 1156 K at atmospheric pressure), the SLIMM reactor nominally operates slightly below atmospheric pressure. In addition, the high negative temperature reactivity feedback of the SLIMM reactor core would effectively shut it down, in case of a malfunction of the redundant reactor control and emergency shutdown systems of enriched B4C poison rods.The SLIMM reactor employs UN fuel, and HT-9 steel cladding, core structure and reactor vessel for operating at 600 K inlet and 800 K exit temperatures of the circulating liquid sodium through the reactor during nominal operation. At these temperatures, liquid sodium is compatible with the HT-9 steel with negligible corrosion concerns during the reactor long operation lives. The relatively high sodium exit temperature from the SLIMM reactor core makes it possible to generate high-pressure superheated steam

for achieving a plant thermal efficiency in excess of 40% and provide high temperature heat for a host of industrial uses and the thermochemical production of hydrogen fuel.The SLIMM reactor would be fully fabricated, assembled and sealed in the factory and shipped to the construction site by rail, a truck or on a barge. It would be installed below ground (Fig. 1), to avoid direct impact by missiles or aircraft, and mounted on seismic insulators to resist earthquakes. The variable conductance liquid-metal (LM) heat pipes, along the primary vessel wall (Fig. 2), transport thermal energy to thermoelectric modules to generate 10s of kWe of auxiliary power for operating the plant's vital functions during nominal operation and in case of a total loss of onsite and offsite powers, avoiding a Fukushima Daiichi type disaster.

474730 Session Keynote - Hybrid Renewable Energy Systems: State of the Art of the Impact of Design and Control Strategy on Performance and Costs

Abstract:

The pushing development toward a greater and greater penetration of Renewable Energy Sources (RES), which are characterized by intermittent availability, is giving new challenges through the delocalization of power production, aiming at consuming energy wherever it is originally converted. The use of HRES (Hybrid Renewable Energy Systems), i.e. characterized by the presence of power subsystems generating from renewables (i.e. PV and wind turbine systems primarily), coupled to batteries for short term storage, and Electrolyzers/Fuel Cells to mange storage over the long term, may be interesting to improve the flexibility of the solutions.

Such systems however present some challenges, as the efficiency depends strongly on the control strategy implementation, which in turn may be based on simple rules (RBS=Rule Based Strategy), or on more sophisticated elaborations (e.g stochastic, or model predictive based [1]) taking into account the probable evolution of the system status due to future consumption pattern management (Demand Response) and weather forecasts. Also, the typical small size (in relative terms to usual power plant size) of the systems give rise to a major drive to downsize at the utmost the components and simplify the control strategies (for example minimizing the amount of hardware required to optimize the system behavior, such as inverters, measurements probe, etc) [2].

The solutions are usually evaluated in terms of efficiency and Total Cost of Ownership: it is clear that these parameters must be defined on an energy basis, over a timeframe of at least one year to make sure to take into account the seasonal variation of the weather. It is also very important to highlight that the relation between power production and consumption is very much dependant on the specific application: for example, performances of off-grid systems [3] may be very different from grid-connected [4]; the consumption of energy in electric vehicles [5] or the use of FC as CHP systems may give very different results in terms of cost effectiveness and global efficiency.

The cited aspects will be properly highlighted and commented in this presentation, which will provide the audience with a description of the state of the art of system design, as well as with details on control strategy and main trade-offs.

References

[1] Bruni G., Cordiner S., Mulone V., Rocco V., Spagnolo F,. En Conv Magag., 2015.

[2] Erdinc O., Uzunoglu M., Renew and Sust. Energy

[3] Bruni G., Cordiner S., Mulone V. et al., Int. J. Hydr. Energy, 2014.

[4] Dagdougui H., Minciardi R., Ouammi A., Robba M., Saciel R., En Conv Manag., 2012.

[5] Fazelpour F., Vafaeipour M., Rahbari O., Rosen M.A., En. Conv. Manag., 2014.

475287 Modeling of reverse electrodialysis for waste heat

recovery

Abstract:

Reverse electrodialysis (RED) harnesses salinity gradient energy by controlling the ion transport through ion exchange membranes. Natural resources such as sea and river water can be used in the RED process, thus sodium chloride has been commonly used for a variety of the RED researches. However, those resources require the repetitive shipping and pretreatment cost and can act as a locational limitation for installation of the RED plant. A closed-loop RED, which uses thermolytic solution, can solve these issues as well as generate electricity. In recent years, ammonium bicarbonate (NH4HCO3), which can be separated by relatively low temperature around 60 „*f*, has been investigated as a promising solute that enables the system closed-loop. Therefore, the closed-loop system keeps the salinity gradient by solute separation and recapturing process.
In this study, we developed a numerical NH4HCO3-RED model for the investigation of the RED performance. Because of the unclarified electrochemical information of NH4HCO3 in solution, such as activity and equilibrium constant, the Planck-Henderson equation was used to approximate the membrane potential. And, solution resistance was experimentally calculated by the definition of the molar conductivity. Moreover, permselectivity that has been assumed either as a unity or some constant value in previous researches was also experimentally obtained to estimate the performance more precis ely.
We found the developed NH4HCO3-RED model was in great accordance with the experimental results under various operating conditions. Using the developed model, we also characterized the net power density, which considers the pumping loss. In our system, the maximum net power density of 0.84 W/m2 was obtained at an optimum operating conditions that the intermembrane distance of 0.1 mm, the concentration ratio of 200 (2.0 M/ 0.01 M), and the flow rate of 3 mL/min. We believe that this study will improve our understanding of the NH4HCO3-RED system for& nbsp;a novel waste heat recovery technology.

481270 Techno-economic Analysis (TEA) of Next-Generation Proton-conducting Solid Oxide Fuel Cells (SOFC)

Abstract:

This work develops preliminary techno-economic models for a highly-innovative, next generation proton-conducting, anode-supported solid oxide fuel cell (SOFC) that integrates a methane coupling catalyst into the anode. This cell is under development by Argonne National Laboratory (ANL) and the Illinois Institute of Technology. The cell is currently at technology readiness level (TRL) 1, i.e. proof-of-concept at a laboratory bench scale. This work summarizes (1) a specific physical embodiment of the ANL cell design modelled, (2) the bill of materials (BOM) for this modelled cell, and (3) preliminary cell, stack, and system cost estimates based on this cell. The techno-economic analysis (TEA) focuses on electrolyte-electrode assembly (EEA), or the 'heart' of the electrochemical cell, where the anode and cathode half-reactions occur.

The project methodology is as follows:

1. Develop a physical embodiment of the design for the EEA.
2. Develop a Bill of Materials (BOM) that specifies masses and/or quantities of components and materials in the EEA.
3. Obtain price quotes for materials in the EEA.
4. Estimate materials costs for the EEA to develop a 'price floor' for minimum costs.
5. Identify cost drivers within the EEA.
6. Estimate fuel cell stack costs for a 100 kWe net electric stack mass-produced at a rate of 50,000 systems per year.
7. Estimate fuel cell system costs for a 100 kWe net electric stack mass-produced at a rate of 50,000 systems per year.

Preliminary results indicate that EEA is estimated to cost 5.56 cents/cm^2 or $278/kW of gross electric power from the cell or stack. Within the EEA, the primary cost driver for the EEA appears to be the anode substrate, which is composed of yttrium-doped barium zirconate (BZY or $BaZr_{(1-x)}Y_xO_{3-d}$ or sometimes $BaZr_{0.8}Y_{0.2}O_x$ or $BaZr_{0.9}Y_{0.1}O_{3-d}$), with barium zirconate represented as ($BaZrO_3$), and nickel oxide (NiO). The BZY and NiO anode substrate material is estimated to cost either 2.88 cents/cm^2 [based on price quotes for BZY from TransTech Inc.] or 16.40 cents/cm^2 [based on price quotes from Praxair Specialty Ceramics]. The secondary cost driver for the EEA appears to be the anode's methane coupling catalyst. The methane coupling catalyst is modelled as a dual metal catalyst composed of Platinum (Pt), gallium(III) oxide (Ga_2O_3), and silicon dioxide (SiO_2). The catalyst material is estimated to cost 2.45 cents/cm^2. About 94% of its cost is due to the material cost of platinum as a catalyst.

Preliminary results also indicate that, for a 100 kW net electric stack mass-produced at a rate of 50,000 systems per year, the fuel cell stack subsystem cost is estimated to be ~$478/kW of net electric power from the system. This cost is approximately 74% higher than a 'plain vanilla' SOFC subsystem, analyzed by the author for DOE in a prior analysis. For a 100 kW net electric stack mass-produced at a rate of 50,000 systems per year, the overall fuel cell system cost including all BOP components is estimated to be ~$597/kW of net electric power from the system. This cost is approximately 48% higher than the entire 'plain vanilla' SOFC system, analyzed by the author for DOE in a prior analysis.

452586 An Optimization Approach to Thermodynamic Limitation on Convective Mass Transfer Enhancement

Abstract:

An optimization method to find the thermodynamic limitation on convective mass transfer for fixed viscous dissipation is developed. The method applies the principle of extremum entropy generation in the mass transfer problem. The mass transfer-induced entropy generation in the fluid flow is defined as the criterion. By maximization or minimization of the criterion under a constraint of fixed viscous dissipation, a velocity field and associate body force field can be obtained, which can enhance the mass transfer process to the thermodynamic limitation. By applying the optimization method, the dependence of the maximum extent of convective mass transfer enhancement on the viscous dissipation can be found. Such a limitation for mass transfer is shown to be useful in setting a target for mass transfer enhancement in practical applications. The analysis is illustrated by three examples.

474658 Session Keynote - Hybrid Fuel Cell System for Producing Chemicals and Electricity from Natural Gas

Abstract:

We are developing a hybrid intermediate temperature (500-700°C) proton-conducting fuel cell (PCFC) system that operates directly on natural gas or natural gas liquids to generate electrical power and to produce high value chemicals such as ethylene or propylene as part of a U.S. Department of Energy Advanced Research Projects Agency – Energy (ARPA-E) funded project under its REBELS Program. For natural gas, non-oxidative coupling of methane (NOCM) catalyst technology is integrated into the anode compartment of the fuel cell to produce C_2 and C_3 alkanes, olefins and H_2. For natural gas liquids, propane dehydrogenation (PDH) catalyst technology is integrated into the anode compartment to produce propylene and H_2. The benefit of integrating severely thermodynamically-limited reactions such as NOCM or PDH into a PCFC is that the H_2 produced by these reactions is oxidized within the anode, generating protons which migrate through the electrolyte to the cathode where they react with oxygen to form water. By removing H_2 from the product mixture using the fuel cell, higher conversion levels for these reactions can be achieved at lower temperatures than in a conventional thermal catalytic process (Le Chatelier's Principle). Additional benefits of our hybrid fuel cell system are that the lower operating temperature allows for a wider choice of less-costly materials compared to conventional reactor processes, the use of a proton-conducting electrolyte eliminates the loss of carbon due to the formation of CO and CO_2 that occurs in oxidative coupling of methane (OCM) when oxygen is co-introduced using an oxide ion-conducting electrolyte, and the lower operating temperature minimizes coke formation during propane dehydrogenation due to thermal cracking of propylene.

To develop this hybrid system, we have had to develop (1) a novel proton-conducting fuel cell that is capable of operating efficiently at 500-700°C, (2) NOCM and PDH catalyst technologies that operate with high activity and selectivity at these temperatures, and (3) a methodology for integrating the NOCM or PDH catalysts into the PCFC. This talk will summarize our efforts to develop the proton-conducting fuel cell technology based on a barium-cerium-yttrium electrolyte; the NOCM and PDH catalyst technologies based on single-site catalysts that exhibit high yield and selectivity while minimizing coke formation, methodologies for integrating the catalysts into the anode compartment of the fuel cell, and preliminary test results. Preliminary results from a techno-economic analysis (TEA) comparing of our technology to competing technologies will be presented.

This work is supported by the U.S. Department of Energy, Advanced Research Projects Agency – Energy (ARPA-E) under Contract DE-AC02-06CH11357

474790 A Comparative Techno-Economic Analysis of Renewable Hydrogen Production Using Solar Energy

Abstract:

A technoeconomic analysis of photoelectrochemical (PEC) and photovoltaic-electrolytic (PV-E) solar-hydrogen production of 10,000 kg H2 day-1 (3.65 kilotons per year) was performed to assess the economics of each technology. The systems analyzed include: (i) two PEC systems that are defined by a single unit operation with inputs of solar power and water and outputs of hydrogen and oxygen, and that are differentiated primarily by the extent of solar concentration (unconcentrated and 10x concentrated), (ii) two PV-E systems that are defined by two unit operations, solar panels that convert solar power to electricity and proton exchange membrane (PEM) electrolyzers that convert electricity to hydrogen, and that are differentiated by the degree of grid connectivity (unconnected and grid supplemented), and (iii) PEM electrolyzers supplied by grid electricity to serve as a commercially practiced benchmark.

For each system, a base-case design and bill of materials was developed and used to determine the untaxed, plant-gate levelized cost of hydrogen (LCH). Based solely on laboratory scale demonstrations and thus likely optimistic projections for system costs, the unconcentrated and 10x concentrated PEC systems base-case capital expense and LCH values were $205 MM ($293 per m2 of solar collection area (mS -2), $14.7 W-1 H2) and $11.5 kg-1, and $160 MM ($428 mS -2, $11.5 W-1 H2) and $9.2 kg-1, respectively. Consisting of significantly more mature technologies, the non-grid connected and grid connected PV-E systems base-case capital expense and LCH values were $260 MM ($371 mS -2, $18.8 W-1 H2) and $12.1 kg-1, and $66 MM ($441 mS -2, $4.9 W- 1 H2) and $6.1 kg-1. The benchmark PEM grid electrolysis systems capital expense and LCH were $34 MM ($2.4 W-1 H2) and $5.5 kg-1, respectively. A sensitivity analysis indicated that, relative to the base-case, increases in system efficiency could effect the greatest cost reductions for all systems,

due to the areal dependencies of many of the components. The balance-of-systems (BoS) costs were the largest factor in differentiating the PEC and PV-E systems.

Successful research and development, measured solely by achieving a system efficiency of greater than 20% within the current embodiments of solar H2 generators, is not sufficient to produce systems with LCH values comparable to that of fossil-fuel derived hydrogen. Panel mounting materials, labor, and other balance of systems costs, irrespective of the active materials, amount to LCH values in excess of current hydrogen and energy prices. For basecase PEC systems, a carbon tax of greater than $800 (ton CO2)-1 would be required in an unconstrained CO2 energy market for fossil-fuel derived hydrogen to reach price parity with these solar technologies.

Expected electricity prices from CO2-neutral nuclear and low-CO2 fossil fuel with CCS technologies are currently an order of magnitude lower than expected electricity prices from solar or wind systems with battery or fuel storage solutions that provide reliability similar to that of current base-load or dispatchable generation technologies. Given these low electricity prices, electrolytic hydrogen prices are also projected to be significantly lower than the solar hydrogen routes analyzed, requiring disruptive approaches to solar hydrogen generation relative to the present embodiments of the technology. Efforts to directly reduce CO2 photoelectrochemically or electrochemically could potentially produce products with higher value than hydrogen, but many as yet unmet challenges include catalytic efficiency and selectivity, and CO2 mass transport rates and feedstock cost.

As compared to other, commercial, energy production and storage technologies, the solardriven technologies analyzed herein are limited by an intrinsically diffuse solar power density of <1 kW m-2, as well as by a low capacity factor, of under 25%, for fixed-tilt panels in the optimal locations in the U.S., and by a conversion efficiency of under 20%. The capacity factor of presently known solar-based energy systems is their fundamental limitation; any capital item used only 25% or less of the day will be at a disadvantage to capital used more effectively.
Efforts to increase the low capacity factor of terrestrial solar power, as well as re-designed solar installation schemes that significantly reduce the balance of system costs and increase the efficiency to near the Shockley-Queisser limit without increasing the capital cost, will have the largest impacts on the economic competitiveness of the resulting technology implementations.
Radically new materials and system designs that achieve fully installed costs similar to simple material installations such as artificial grass are required to achieve the equally dramatic cost reductions needed for solar power to compete with current generation technologies.

467494 Cyclic Polyethylene Furanoate As a Monomer from Renewable Resources for Ring Opening Polymerization

Abstract:

Introduction

The replacement of oil based chemicals by renewable resource based chemicals is one of the major issues chemical industry is facing right now. The alternative polymer poly (ethylene furanoate), PEF, is a possible substitute for poly (ethylene terephthalate), PET. Among different process alternatives, PEF can be produced at high polymerization rates and molecular weights via ring opening polymerization of cyclic oligo (ethylene furanoate), cyOEF. On the other hand, the effective production of these cyclic oligomers at high purity is still the major obstacle towards the process industrialization. In this work, different strategies aimed to produce these cyclic molecules from the two monomers Furan dicarboxylic acid (FDCA) and ethylene glycol (EG) are explored. And a model is presented to describe the equilibrium behavior of said cyclic oligomers.

Three pathways towards cyOEF

The three synthetic strategies shown in figure 1 have been investigated. In all cases, high dilutions in organic solvents are required to achieve high yields; the final product is a distribution of cyclic oligomers with different number of repeating units.

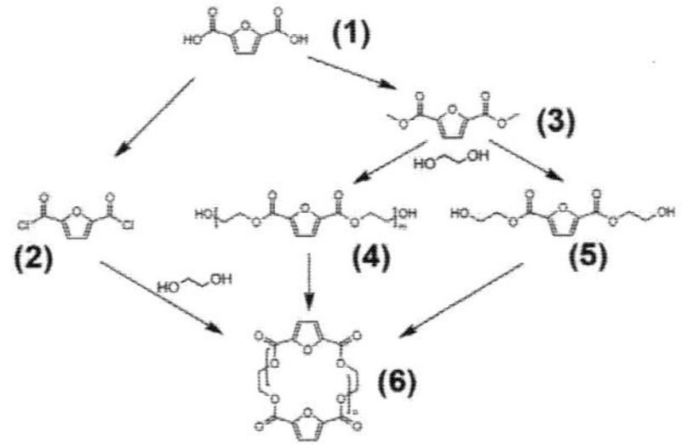

Figure 1 Reaction pathways for the production of cyOEF.

The first (scheme 1-2-6) is a fast cycle formation using the rapid esterification of acid chlorides with alcohols. Moderate selectivity towards cycles of around 50% are in contrast to a complex reaction setup and toxic chemicals involved for the chlorination. The product size distribution is kinetically controlled and consists mainly of C3, with C2 and C4 being the second most abundant species.

The depolymerization (scheme 1-3-4-6) and reactive distillation (scheme 1-3-5-6) are the two more promising routes towards cycles. For depolymerization FDCA and EG are converted to oligomeric chains of around 12 repeating units. This prepolymer is subsequently diluted in a high boiling organic solvent and converted to a mixture of cyclics and linears at 180-200 °C. In case of reactive distillation the FDCA and EG (in excess) are converted to very short chains (1-2 repeating units). The chains are then diluted in a high boiling organic solvent and finally converted to cyOEF by distilling a portion of the solvent together with the excess EG.

Both routes yield a similar distribution of cyclic species. Possible purification routes include precipitation and adsorption processes, which have both been applied to yield cycles with high purity starting from linear species (>99% by HPLC area). The yield of both reactions lies in the range of 60 to 80 % depending on the process conditions. One striking feature of this type of reaction is that it is possible to recycle the solvent and the linear side products in order to increase the yield of the process and furthermore make it more sustainable.

Modeling of the ring chain equilibrium

The equilibrium composition of cyclics and linears in 2-Methylnaphthalene and Dichlorobenzene (DCB) could be modeled using the Jacobson-Stockmayer model (Figure 2). Under high dilutions this equilibrium shifts towards the formation of cyclic oligomers. It is based on computing the probability of ring closure using the assumption of randomly coiled hydrocarbon chains for linear polymers. Two fitting parameters are involved: one is the equilibrium constant for the step growth polymerization, which is in equilibrium with the cyclisation reactions, and the second one is the effective length of the bonds in the polymer chains.

As shown in the figure below, our own results from depolymerization and reactive distillation experiments were fitted using the model. These results suggest that it is indeed possible to use this approach to describe the ring chain equilibrium. The model is currently under refinement in order to describe system-specific effects such as the complete absence of cycles with six repeating units.

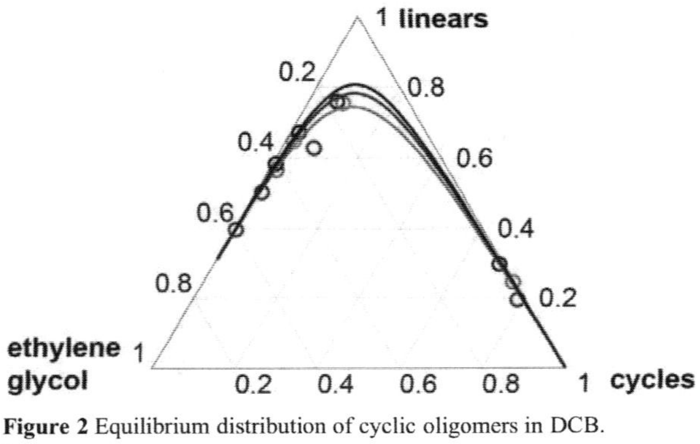

Figure 2 Equilibrium distribution of cyclic oligomers in DCB.

References

[1] Gomes M, Gandini A, *J. Polym. Sci., Part A: Polym. Chem.*, 49, 3759-3768 (2011).

[2] Burch, RR, Lustig, SR, Spinu, M *Macromolecules*, 33, 5053-5064 (2000).

[3] Stockmayer, WH, Jacobson H, *J. Chem. Phys.*, 18, 1600-1606 (1950).

453695 Polymeric Nano-Metal Composite Membranes for Water Remediation

Abstract:

Responsive polymers react to external stimuli like temperature, pH, radiation, ionic strength or chemical composition. These polymers in combination with nanomaterials provide novel hybrid properties such as bactericidal, electrical, hydrophilic, catalytic, selectivity, etc., and they can be self-supported or in a convenient matrix such as a membrane.

These membranes, functionalized with polymers having pH responsiveness (hydrogels), is a promising research field for numerous applications. The functional groups of the hydrogels gain or lose protons based on the pH, collapsing or extending their network, making them swell or un-swell solutions. These nanostructured materials can be exploited to synthesize metal nanoparticles and as ion exchangers. They also prevent the loss of nanomaterials to the environment and improve reactivity due to larger surface areas, expanding its range of applications to selective catalysis, remediation and mining.

The present work describes "green" techniques used to create nanocomposites based on polyvinilidene fluoride (PVDF) hollow fiber (HF) and flat sheet (FS) membranes, both sponge-like and thin. Hydrophobic HF membranes were synthesized with 57.2% porosity and 0.72 μm mean pore size. These membranes were then hydrophilized by a water-based process of cross-linking polyvinylpyrrolidone (PVP) by persulfate (PS). Though, hydrophilization of PVDF membranes using PVP is known in industry, as a post-process has only been implemented at lab-scale.

The PVP cross-linking is confirmed but also a ring-opening is present due to etherification. Hydrophilization was measured by contact angle changes (CA) (76.3° to 60.3°) and surface free energy calculations (40.3 to 49.8 $mJ \cdot m^{-2}$). HF and FS were subsequently functionalized with a poly(acrylic acid) (PAA) hydrogel through free radical polymerizations (thermal or redox) using PS as initiator. This work advanced FS membrane functionalization from lab-scale to full-scale by modifications of the polymerization procedures.

Swelling tests were performed to hydrogels/functionalized membranes finding responsiveness up to 1676% swelling at pH 9.0. Also, mesh size values (ξ) were calculated showing an inverse relationship with membrane permeability, tuning the flux through the membrane. An increase of ξ from 5.89 to 8.33 nm (pH 3.0 to 9.0) represented a 93% permeability decrease (53.9 to 3.4 $L \cdot m^{-2} \cdot h^{-1}$) at pressures between 1.0 and 4.0 bar.

Throughout a double ion exchange of sodium/iron and a subsequent reduction, bimetallic Fe/Pd nanoparticles (60 to 300 nm) were synthesized. The Fe exchanged was almost 500 mg Fe/g PAA. Similarly, it was possible to use the exhausted accelerants of Fe ions of the redox polymerization to synthesize nanoparticles, but having only 9.2 mg Fe/g PAA of loading.

Finally, depending on the Pd concentration in the membrane, the catalytic activity in the reduction of trichloroethylene (TCE) was studied, having enhanced reactivities (up to ten times) in sponge-like membranes compared to previous studies. Using hydrogels alone, similar results to nanoparticles in solution phase were found without nanoparticle loss. 80% of reduction was achieved within 2 hours with chloride production near to stoichiometric values (3/1), demonstrating absence of intermediates.

This research is supported by the National Science Foundation EPSCoR, and by the National Institute of Environmental Health Sciences (P42ES007380). HF membranes were synthesized at the Singapore Membrane Technology Center, NTU (Singapore). Nanostone Water (Oceanside, CA) provided support and collaboration in the development of full-scale flat sheet membranes.

454787 Thermodynamic and Economic Assessment of the Production of Ethylene and Propylene from Bioethanol

Abstract:

Introduction

The increasing worldwide pollution and the environmental problems associated with the petroleum extraction and use, have led to the development of new energy resources and sustainable raw materials [1]. The production of bioethanol from biomass offers a sustainable biotechnology solution at competitive costs with high availability. In Colombia (South America), bioethanol production has increased because the local policies encourage its production, being the third producer in America [2]. However, the focus on bioethanol production has been related to its use as biofuel. Alternatively, there are some high-commercial-value compounds —mostly produced from petroleum— that can also be obtained from renewable sources [3]. Recently, bioethanol has been proposed as a source for producing commodities such as light olefins in the framework of the biorefinery concept [4]. Among these biomaterials, ethylene stands out as one of the most produced chemicals worldwide, having a projected production of 400 million ton in 2020 [1]. In addition, the demand for propylene is growing faster, having a projected production of 100 million ton in 2021[5]. Therefore, this study aims to propose a conceptual design for producing ethylene and propylene from bioethanol samples procured in Colombia, integrating the reaction and separation stages and assessing its technical and economic feasibility. Aspen Plus and Aspen Hysys were used as process simulators.

Methodology

Aspen Plus V7.3 and Aspen Hysys V7.3 (Bedford, USA) were used as process simulators throughout the study. In both, the Soave-Redlich-Kwong (SRK) equation of state was used as thermodynamic model. First, the bioethanol dehydration reaction was analyzed in the RGibbs reactor module in Aspen Plus. Water/ethanol (S/E) feed ratio and reaction temperature were varied in the module in order to assess their effect on products selectivity. Ethylene, propylene and *n*-butene were selected as possible products, and a selectivity response surface was obtained in Matlab R 2014 (Salt Lake City, USA). Aspen Hysys was used to perform the sizing of the process units, heat integration, and process yields. Products purification was modeled in the Distillation Column module and the energetic analysis was performed using Heat Exchanger and Heaters/Coolers modules in Aspen Hysys. Finally, a class-V-order economic evaluation was done to the final flowsheets by determining the capital expenses for all units in the processes [6].

Results and analysis

Figure 1a shows the surface response for ethylene selectivity during bioethanol dehydration simulation, when varying S/E ratio and temperature. The higher the temperature and S/E ratio, the higher the ethylene yield, with a selectivity exceeding 60% beyond S/E = 5:1 and 600 °C. On the other hand, Figure 1b shows a maximum peak of 45% selectivity for propylene at 400 °C, and a slightly positive effect of the S/E ratio. *n*-butene formation also presented a similar trend (Figure 1c), showing a maximum selectivity of 14% at 300 °C. Heavier olefins are formed by ethylene condensation, which is thermodynamically favored at lower temperatures [7]. However, the absence of ethylene at these lower temperatures reduces the amount of this compounds.

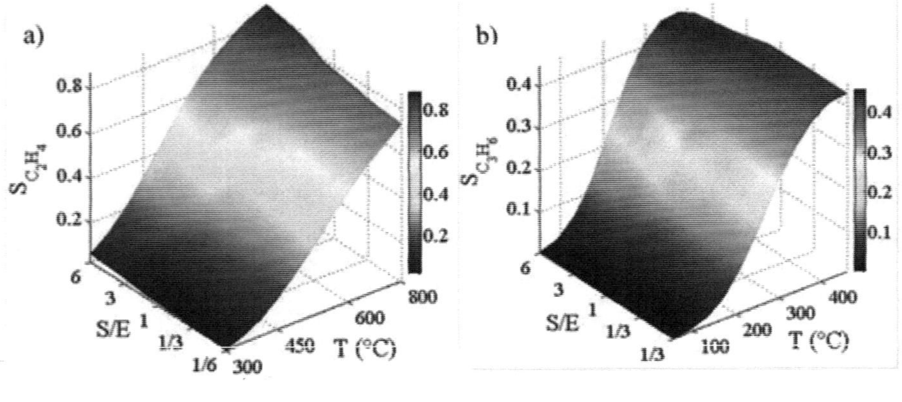

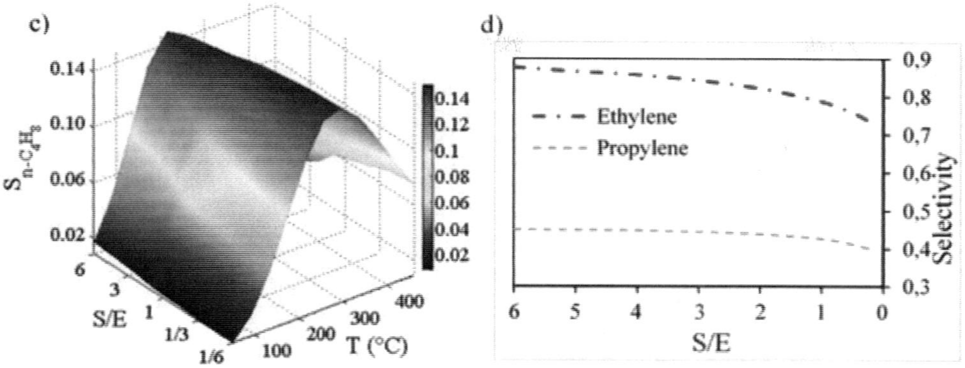

Figure 1. Matlab response surface from Aspen Plus simulated data for the production of (a) ethylene, (b) propylene, and (c) *n*-butene using different water:ethanol ratios (S/E) in the temperature range of 200–800 °C; and (d) the effect of the S/E ratio on the maximum ethylene and propylene yields.

Figure 1d shows the effect of the S/E on the maximum both ethylene and propylene selectivity. The excess of water is slightly positive for the ethylene formation because its acts as a quasi-inert diluent from a thermodynamic point of view [7]. In this way, a bioethanol dehydration reactor feed with a stream of S/E = 5 and operating at 615 °C was selected as the proper design to produce ethylene. In addition, a dehydration reactor operating at 400 °C was selected to produce propylene. These reactors were included in two different processes to analyze the conceptual design of two plants, one producing ethylene and another producing propylene from bioethanol.

Figure 2 shows the designed processes for producing (a) ethylene and (b) propylene from bioethanol. The processes begin with the feed stream of 300,000 ton/year of raw, fermented bioethanol (5 mol% ethanol, S1), which corresponds to the 10% of the national bioethanol production. This stream is heated to 40 °C before its further introduction to a flash separator V-100, which operates at 95 °C and 1 atm. These conditions were established according to the ethanol-water equilibrium to obtain an S/E = 5:1 (16.4 mol% ethanol). After V-100, the product stream rich in ethanol (S3) is heated to either 615 °C or 400 °C (reaction temperatures for ethylene or propylene production, respectively) by a recycle of the flash bottoms stream (S4). S4 is rich in water and is used to the process heat integration. The product streams (S7 in Figure 2a and S6 in Figure 2b) are pressurized to 19 atm in the compressor K-100. The pressurized streams are cooled to 18 °C by fresh water streams. These streams would deliver the energy duty for heating streams (S20 in Figure 2a and S19 in Figure 2b). Subsequently, the cooled streams are used as an inlet in a three-phases flash separator V-101 with the purpose of removing the residual water after the reaction. Finally, the product streams are introduced into distillation columns to separate the olefins.

The ethylene process delivered a stream of 99.7 mol% ethylene with an almost 100% recovery (S16 in Figure 2a), producing 0.60 mol ethylene/mol ethanol feed. In addition, propylene process delivers a stream with 92.0 mol% propylene in S15 in Figure 2b. This corresponds to the 70% of the propylene entering to the distillation tower, producing 0.43 mol propylene/mol ethanol feed.

A class-V economic assessment delivered investments of USD 2 million for producing ethylene and USD 2.23 million for propylene. These costs represent 159 USD/ton-year of ethylene and 327 USD/ton-year for propylene. True [8] reported investment costs of 1300 USD/ton-year for ethylene and 1450 USD/ton-year for propyylene produced from petroleum for a plant constructed in 2016. However, these costs include lumpsum, engineering, procurement, construction and commissioning turnkey contract; which not were included in the current analysis. Nevertheless, this class-V economic assessment shows that olefins production from bioethanol could be a profitable and sustainable alternative to petrochemical processes.

b)

a)

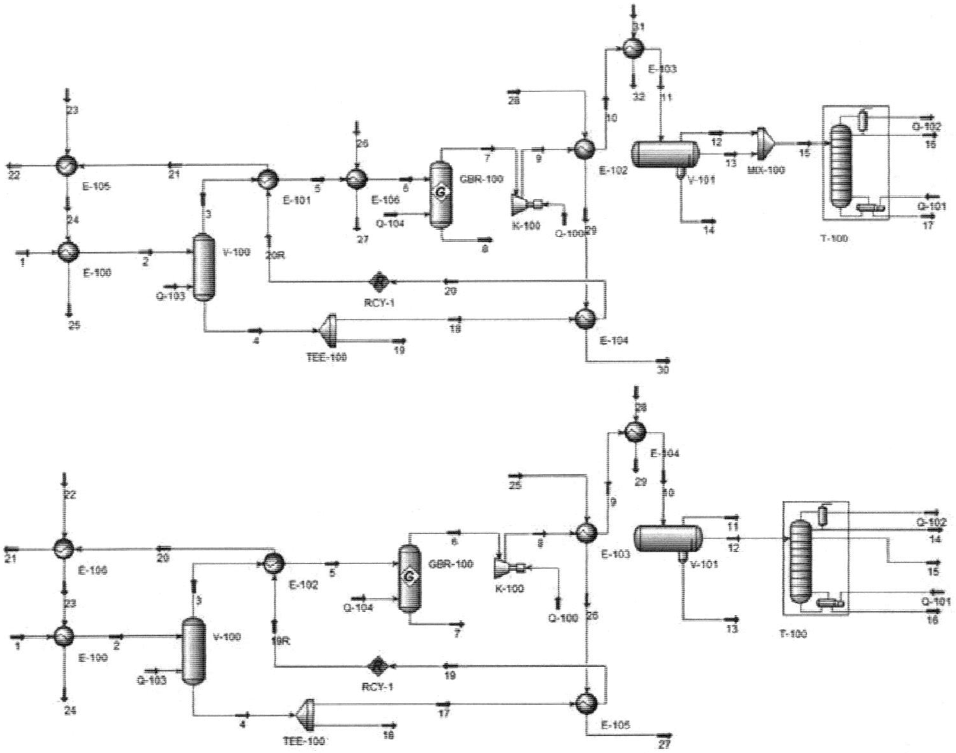

Figure 2. Process flowsheets designed in Aspen Hysys for producing (a) ethylene and (b) propylene from bioethanol. Blue lines represent mass streams and red lines represent energy streams. Equipment notation: E: Heat exchanger, V: Flash separator, K: Compressor, MIX: Mixer, T: Distillation tower, TEE: Splitter, and GBR: Gibbs reactor.

Conclusion

This study evaluated a conceptual design for producing light olefins from bioethanol base on thermodynamic predictions and using commercial software. To the best of our knowledge, it is the first time that reaction and separation stages are integrated and economically evaluated in two different processes, one to produce ethylene and another to produce propylene from bioethanol. Due to water showed a positive effect on the ethanol dehydration reaction, both processes start with a simple flash separator operating at 95 °C and 1 atm to reach the ethanol initial concentration needed in the reactors (S/E=5, 17.6 mol% ethanol). Two different reactors were designed, one operating at 615 °C for producing ethylene and another at 450 °C for producing propylene. The designed ethylene process delivered a stream of 99.7 mol% ethylene with a 60% yield and an investment cost of USD 2 million, equivalent to 159 USD/ton-year of ethylene. On the other hand, propylene production from bioethanol delivers a stream with 92 mol% propylene and 43% yield. Propylene process has an investment cost of USD 2.23 million (327 USD/ton-year propylene). These conceptual analysis show the technical, environmental and economic feasibility of the biorefinery concept development.

References

[1] H. Xin, X. Li, Y. Fang, X. Yi, W. Hu, Y. Chu, et al., J. Catal. 312 (2014) 204–215.

[2] G.P. Ortegón, F.M. Arboleda, L. Candela, K. Tamoh, J. Valdes-Abellan, Sci. Total Environ. 539 (2016) 410–9.

[3] O.J. Sánchez, C.A. Cardona, Bioresour. Technol. 99 (2008) 5270–95.

[4] R. Le Van Mao, T.M. Nguyen, G.P. McLaughlin, Appl. Catal. 48 (1989) 265–277.

[5] Ceresana, Market study: Propylene, (2014). http://www.ceresana.com/en (accessed May 1, 2016).

[6] American Association of Cost Engineers, (2015). http://www.aacei.org/ (accessed November 15, 2015).

[7] T. Lehmann, A. Seidel-Morgenstern, Chem. Eng. J. 242 (2014) 422–432.

[8] W. True, Oil Gas J. 110 (2012).

458322 Design and Economical Evaluation of Polygen Process to Co-Produce Synthetic Natural Gas (SNG), Methanol and Ethylene Glycol

Abstract:

Design and Economical Evaluation of POLYGEN process to co-produce Synthetic Natural Gas (SNG), Methanol and Ethylene Glycol

Bor-Yih Yu and I-Lung Chien*

Department of Chemical Engineering, National Taiwan University, Taipei 106, Taiwan

The steady state design and economical evaluation of poly-generation (POLYGEN) process to co-produce synthetic natural gas (SNG), methanol and ethylene glycol (EG) is studied in this work. POLYGEN has been a widely studied topic recently, in which several different chemicals can be produced parallel at the same time. In POLYGEN process, the relative production rates for different chemicals can be adjusted based on different market demands, daily usages, and also changing political strategies. This character enhances the flexibility of plant-wide operation and also the economic performance. Among the products, SNG holds very similar composition and heat value to typical natural gas, and can be used as a replacement in industrial and home usages. Methanol is one of the most important industrial feedstocks, which can be the basis of various kinds of chemicals. Ethylene glycol is also an important organic chemical, which can be used to synthesize polyesters, antifreezes, solvents, and so on.

The items covered by this work are illustrated in the following. In coal-based POLYGEN processes, syngas is first generated from coal gasification in a CO2-rich atmosphere, and the raw syngas is then passes the treating stage. Syngas treating stages contains two section, namely, sour water-gas-shift-reaction (SWGSR) and acid-gas-removal (AGR). In SWGSR, excess amount of steam is added to adjust the H2/CO ratio. After that it is cooled to low temperature to enter AGR. In AGR, SELEXOL method is used to capture 99.7% of H2S and 92% CO2. Part of the captured CO2 is sent back to gasification section in order to reduce the carbon emission, and the other part is compressed and sequestrated. With proper arrangement in configuration, three streams with qualified composition to be the feedstocks for downstream chemical plants are produced.

For downstream chemical plants, the products produced will meet the industrial standard. The SNG has heat value higher than 965 (BTU/scf), the methanol purity is 99.5 (mol%), and EG purity is 99.8 (mol%). EG is traditionally synthesized through petroleum route, a novel route from coal-derived syngas has been widely considered a potential alternative as the oil shortages continuing. Thus in this work, the production of EG from syngas is adopted. Firstly, CO and H2 in syngas are separated, and are used in different reaction stages. In the first stage, CO is reacted to form dimethyl oxalate (DMO), and in second stage, DMO is further reacted with H2 to form ethylene glycol.

After the design, optimization works are studied based on minimized total annual cost (TAC), and the economical evaluation is performed to investigate the overall economic performance. Besides, some waste heat could be recovered by generating steam, and the steam is used to drive turbine to produce electricity for usage inside the plant-wide process. Thus through this work, a better understanding of POLYGEN process, and the design and operation of coal to ethylene glycol route is reached.

458992 Preparation and Properties of Soybean Oil-Based Composites Containing Natural Fillers

Abstract:

Biobased composites have been prepared, which consist of a vegetable-oil based matrix, with ground nutshell or eggshell as natural, rigid, filler materials.

Crosslinked polymers of acrylated epoxidized soybean oil (AESO) and styrene were first successfully produced across the entire composition range via bulk free radical polymerization. Mechanically, these copolymers varied from hard and rigid at high styrene contents, to soft and rubbery at high AESO contents.

The intermediate 50:50 copolymer was chosen as the matrix for composites containing varying amounts of pecan nutshell flour or powdered eggshells. Composites were produced by pre-blending the fillers into the comonomer mixture before reaction and solidification. Through the AESO and filler contributions, these composites contain substantial fractions of biobased starting materials.

Copolymers and their composites exhibited good structural integrity, and were characterized by flexural testing to fracture, dynamic mechanical analysis, SEM imaging, as well as hardness and impact testing.

Addition of the natural fillers to the matrix resulted in substantial increases in hardness and stiffness, with a nearly six-fold increase in modulus observed at 35 volume% eggshell filler. These increases were accompanied by modest decreases in breaking strength and impact energy, and no observed effect on the glass transition temperature of the matrix. Use of coupling agents [1] to increase adhesion and potentially further modify mechanical properties will be discussed.

General comparison of the filled polymer properties to a baseline of neat polystyrene shows improved or similar individual mechanical characteristics, indicating viability of such biobased composites as a replacement material.

[1] Nachtigall, S.M.B., Cerveira G.S., Rosa S.M.L., Polymer Testing, 26(5), 2007.

460812 Solvent Free Sucrose Esters Production in Reactive Systems Containing Emulsifiers

Abstract:

Surfactants are chemicals used in a wide variety of applications, mainly in the manufacture of consumer goods, industrial cleaners, food-feed, and cosmetic products. In the past, efforts have been made to develop biobased surfactants from renewable raw materials in order to reduce the environmental impact of traditional petroleum-based surfactants. Currently, most biobased surfactants available in the market are lipids-based, but still they contain polar moieties generated from fossil resources (e. g. sulfonic acid, ammonium, or poly-ethoxylated groups) (1). In contrast, in completely biobased surfactants (e. g. acylglycerols, alkyl-poly-glucosides, glycosides of fatty acids) both, the polar and non-polar structures are of renewable origin.

Among biobased surfactants, fatty acid sucrose esters stand out as important additives for the food industry (food ingredient number: E473). These are generally produced by transesterification of sucrose (from sugar cane) with fatty acid methyl esters (from vegetable oils). In addition to their renewable character, sucroesters are high value added, biodegradable and biocompatible surfactants, which can also be used in specialized applications such as cosmetics, clean and care products and pharmaceuticals. Recent estimations (2) indicate that the current market of sucroesters is around than 6000 t/year, with prices around 4 to 20 USD/kg depending on the nature of the product (i.e. lipid chain, degree of substitution, etc).

Sucrose is a polyalcohol with eight hydroxyl groups (See Figure 1), so theoretically, sucrose esters could be monoesters, diesters, triesters, and, so on (up to octaesters). As a result, tuning of reaction selectivity to a specific product is a major challenge. Besides, sucroesters properties and application are heavily determined by the degree of esterification. For example, sucrose monoesters are good water-in-oil emulsifiers, while sucrose octaesters are good oil-in-water emulsifiers. In general, the degree of substitution depends of different process conditions such as temperature, reactants molar ratio, catalyst nature and concentration, and the use of other additives (e. g. solvents, emulsifiers).

In addition to controlling the degree of substitution, a major challenge of the process is the incompatibility between reactants; only 0.1 g of sucrose can be dissolved in 100g of FAME at 140°C (3). This reduced solubility limits the mass transfer and reduces the reaction rate and conversion. A processing alternative to overcome this problem is the use of emulsifiers, which enhance contact and solubility of sucrose in FAMEs (4–7). Some reports indicate that fatty acid soaps (mono- and divalent), sucrose esters (the product itself), and mono- and diacylglycerols can be used as emulsifiers for the reaction. However, it is necessary to study the influence of such emulsifiers (nature and concentration) on the transesterification process (i.e. rate of reaction and esterification degree).

In this work, conversion and selectivity of transesterification reaction to produce sucroesters was experimentally evaluated using three different emulsifiers (potassium palmitate, sucrose palmitate and glyceryl monoester) at different concentrations (up to 20%w). Reactions were carried out in a 150 ml glass-jacketed reactor coupled to a methanol total condenser. The system was homogenized with an Ultraturrax® at 15000 rpm prior to reactions, and maintained under agitation at 600 rpm at constant temperature (110-140 °C) until no further evolution of methanol was observed. FAMEs consumption and sucroesters production were tracked by HPLC analysis of samples. Observed conversion were around 40% and mono substitution increased with the concentration of the emulsifying agents. Higher conversions were obtained when potassium palmitate was used as emulsifier, and reaction times were shorter when emulsifier's concentration was increased. Degree of substitution of the final product was higher when a mixture mono- and diacylglycerols was used as emulsifier and lower when sucrose palmitate was used. Results of this study can be used for preliminary process design in a solvent-free production of sucrose esters with different emulsifying properties.

Figure 1. Emulsifiers used to provide contact in sucrose and fatty methyl estermixtures (Molecular structures taken from ChemSpider)

References

1. HAYES, Douglas G. ***Biobased Surfactants : A Useful Biorefinery Product That Can Be Prepared Using Green Manufacturing***. Knoxville, TN, USA, 2012.

2. OTOMO, Naoya. Basic properties of sucrose fatty acid esters and their applications. In : HAYES, Douglas G., KITAMOTO, Dai, SOLAIMAN, Daniel K. Y. and ASHBY, Richard D. (eds.), ***Biobased Surfactants and Detergents***. Urbana, Illinois : AOCS Press, 2009.

3. ZHAO, Rongrong, CHANG, Zhidong, JIN, Qiong, LI, Wenjun, DONG, Bin and MIAO, Xiaowen. Heterogeneous base catalytic transesterification synthesis of sucrose ester and parallel reaction control. ***International Journal of Food Science & Technology***. 2014. Vol. 49, no. 3, p. 854–860. DOI 10.1111/ijfs.12376.

4. PARKER, Kenneth J., JAMES, K. and HURFORD, J. Sucrose Ester Surfactants — A Solventless Process and the Products Thereof. In : HICKSON, John L. (ed.), ***Sucrochemistry***. Washington, DC : American Chemical Society, 1977. p. 97–114.

5. FITREMANN, Juliette, QUENEAU, Yves, MAITRE, Jean-paul and BOUCHU, Alain. Co-melting of solid sucrose and multivalent cation soaps for solvent-free synthesis of sucrose esters. ***Tetrahedron Letters***. 2007. Vol. 48, p. 4111–4114. DOI 10.1016/j.tetlet.2007.04.015.

6. GALLEYMORE, Harry R, JAMES, Kenneth, JONES, Haydn F and BHARDWAJ, Chaman L. Process for the production of a surfactant containing sucrose esters. 4298730. 1981. United States.

7. RIZZI, George P and TAYLOR, Harry M. Synthesis of higher polyol fatty acid polyesters. 3963699. 1976. United States.

461686 Synthesis and Characterization of Biochar-Based Carbon Supported Metal Nanoparticles

Abstract:

Synthesis and characterization of biochar-based carbon supported metal nanoparticles

Sai Teja Neeli and Hema Ramsurn

Russell School of Chemical Engineering, University of Tulsa, Tulsa, OK 74104

Biochar is a carbon-rich solid formed after biomass has been carbonized. It is also a residue from a number of thermochemical conversion processes whereby biomass is converted to bio-oil and/or syngas. Therefore, this material is environmental-friendly and carbon-neutral. Traditionally, biochar has been used as soil amendments for the healthy growth of plants due to its retention capability of nutrients. This investigation aims to use biochar in a different manner: as a support for metal catalysts to promote green chemistry and engineering. The biochar used in this study is obtained from hydrothermal carbonization of model compounds of biomass namely lignin, cellulose and hemicellulose. This means that high pressure and temperature water is used as a benign solvent to depolymerize the model compounds to biochar. The biochar so produced is chemically activated with 0.1 M HNO_3 solution resulting in activated carbon with defined micropore size distribution, high micropore volumes and large specific surface areas. Metal precursors (like iron nitrate) are then added to the biochar and dried in a convection oven at 110 ^{o}C for one week to obtain metal (iron) impregnated biochar. The dried mixture is thermally-treated in a quartz tubular reactor at 1000 ^{o}C for one hour in a nitrogen flow of 500 ml/min. These synthesized biochar based metal catalysts will then be characterized using a number of techniques including scanning electron microscope (SEM), x-ray powder diffraction (XRD), inductively coupled plasma atomic emission spectroscopy (ICP-AES) and Brunauer-Emmett-Teller (BET) surface area analyzer. Other metal catalysts such as platinum, gold and nickel can also be impregnated on the biochar in the same manner. It is expected to observe distinct XRD patterns corresponding to graphite for the thermally treated metal-impregnated biochar, due to the conversion of the carbonaceous materials to graphite. SEM images will help not only in sizing the metal nanospheres (about 20 -60 nm) but also in determining how the metal is supported: half embedded or completely enclosed in the porous char matrix. Based on the characterization results, these metal-impregnated biochar catalyst could be tested for a number of reactions. These supports can withstand high temperatures and can even be used for gasification reactions and upgrading bio-oil during hydrogenation reactions. There is also the possibility of using bifunctional metal/acid catalysts on the biochar nanoparticles for dehydration and hydrogenation reactions and for aqueous phase processing to produce selectively targeted alkanes from glucose. The ultimate goal is to use these green catalysts to upgrade the oxygenated biofuels by removing the oxygen through various mechanisms and hence increasing the quality and heating value of the desired product.

447633 Characterization and Management of Some Oil Fields Produced Water of Upper Assam Basin

Abstract:

Several toxic elements are present in oil field produced water (PW). This paper tries to characterize the physical and chemical parameters of PW with an aim to manage the huge quantity of oil field PW. The oil fields of Upper Assam Basin have been producing with high water cut sometimes even exceeding 90%. The PW discharge large volumes of liquid waste containing several toxic elements adversely affect both biotic and aquatic life forms. These wastes need to be treated because of the presence of several harmful constituents. These PW are complex mixture of organic and inorganic compounds. We directly can't dispose it to the environment without proper treatment of these liquid wastes. PW discharges into the environment have made PW management a significant part of the industries. Several developing countries like India are facing the problem of fresh water scarcity. Proper management, treatment and reuse of the PW for the beneficial purpose of the mankind and environment can mitigate the scarcity of fresh water. This paper attempts to describe about characteristics of PW, their pollutant profile, and management option to mitigate the adverse environmental impacts. Also tries to describe the isotopes, major and minor elements presents in PW.

452032 3-Dimensional Disordered Silica As a Dual Pore Amine Functionalized Sorbent for Carbon Dioxide Adsorption

Abstract:

Recent work in the capture of carbon dioxide on porous solids has focused on how to increase capture capacity and capture speed, in order to make these materials more applicable to industrial conditions. One popular method is to load amine groups onto these porous supports, for instance onto zeolite or metal organic framework sorbents. While these materials can show high capacities with enough amine group loaded onto their surface, there is a tradeoff that exists between capture speed and amine loading, which occurs because the pores where amine is impregnated into the support is often the only channel available for gas diffusion. In order to overcome this challenge we have begun investigating sorbents that contain multiple pore channels which can be preferentially loaded with amines via size selection, allowing for a material which contains high contents of amine groups but still retains fast diffusion speed. One material, known as 3-Dimensional Disordered Silica, is composed of agglomerated spheres of zeolite beta that when loaded with polyethylenimine show fast capture kinetics and high capacities. We have shown that this material can be preferentially loaded in either the micropores present within spheres, or mesopores between spheres, allowing for an increased control in the pore-amine interaction and resulting capture performance.

452617 Response of Activated Sludge to the Presence of Phenol in Batch and Continuous Flow Systems

Abstract:

The activated sludge process probably the most important wastewater treatment technology used for both municipal and industrial wastewaters. The process utilizes microorganisms to decompose organic matter in wastewater. The activated sludge process is frequently subjected to various inhibitory pollutants, such as toxic organics, which have inhibitory effect on the biochemical reactions and decrease the efficiency of the wastewater treatment plants. This study presents the experimental results on the inhibition effects phenol on the activity of urban recycle activated sludge. The inhibition the biological processes was examined in batch activity tests by measuring the specific oxygen uptake rate (sOUR) and in continuous flow aerobic reactors equipped with porous membrane in order to retain the suspended solids, while allowing treated water to pass through. The continuous flow system was monitored by analysing the reactor mixture for total suspended solids (TSS), dissolved oxygen (DO) and pH and the treated effluent for total carbon (TC), total nitrogen (TN), phenol, NH_4^+, NO_3^-, NO_2^- and several other ions (Na^+, Ca^+, Mg^+, PO_4^{3-}, SO_4^{2-}, Cl^-).

454794 Model Reduction in Environmental Chemical Mechanisms

Abstract:

Numerical modeling of complex chemical reaction networks is an essential component of many environmental applications. In most cases the detailed network must be condensed (reduced), e.g., due to computational expense or practical considerations. We will discuss a state of the art approach for dynamic model updating and reduction, applied to the gas phase isoprene oxidation mechanism as an example.

457192 Capturing CO2 with Phase-Changing Polyamine/Ethanol Solutions

Abstract:

Sequestration of carbon dioxide has been intensively studied in recent years as the global warming becomes a serious concern among the public. Absorption processes based on aqueous alkanolamines are often used as a benchmark for CO_2 capture technologies. The major drawback of such processes is the large energy consumption required for the regeneration of the absorbent. One of the alternative ways to tackle this challenge is to use a new class of non-aqueous amine solutions called phase-changing solvents, which can form two phases after absorbing CO_2. One phase is rich of CO_2, while the other is lean of CO_2. Significant energy saving can be achieved as only the CO_2 rich phase needs to be treated to obtain regenerated absorbents. In this work, we observed that polyamines dissolved in ethanol can produce solid precipitate after CO_2 absorption which can then be easily separated. The four polyamines investigated here are ethanediamine (EDA), piperazine (PZ), diethylenetriamine (DETA), and triethylenetetramine(TETA). The reaction kinetics of CO_2 and polyamines in ethanol solutions are also examined. Phase changing transition of solutions only occurs when CO_2 is absorbed into polyamine/ethanol solutions, but not monoamine/ethanol solutions. Results show that polyamine/ethanol solutions offer both higher absorption rate and larger absorption capacity for CO_2 than aqueous amine solutions. More than 80% of the absorbed CO_2 was captured in the solid phase. Moreover, solid precipitates can be regenerated at 90°C. The polyamine/ethanol solutions also show a stable performance after multiple absorption/regeneration cycles.

452121 Synthesis of Immobilized Amine Sorbent Pellets from Poly (chloroprene) and Fly Ash Binders for Post-Combustion CO_2 Capture

Abstract:

Synthesis of Immobilized Amine Sorbent Pellets from Poly (chloroprene) and Fly Ash Binders for Post-Combustion CO_2 Capture

Walter Christopher Wilfong[1], McMahan L. Gray[2], Brian W. Kail[3], Thiago F. De Aquino[4], Bret H. Howard[2], Sabrina T. Estevam[4]

[1]National Energy Technology Laboratory (NETL), USA, Pittsburgh, PA; Oak Ridge National Laboratory, Oak Ridge, TN

[2]National Energy Technology Laboratory (NETL), USA, Pittsburgh, PA

[3]National Energy Technology Laboratory (NETL), USA, Pittsburgh, PA; AECOM, Pittsburgh, PA

[4]Beneficent Association of the Santa Catarina Coal Association (SATC), Brazil, Criciúma, Santa Catarina.

Post-combustion CO_2 emissions primarily from coal-fired power plants constitute 31% of total CO_2 greenhouse gas emissions, where CO_2 represents 82.7% of all emitted greenhouse gases.[1] With the predicted use of coal as an energy source to continue for decades to come, it is paramount to remove these CO_2 emissions in parallel with the current development of cleaner energy sources. Basic Immobilized amine sorbent (BIAS) processes are a promising technology to solve this CO_2 problem. However, problems such as both the large pressure drops across tall sorbent beds and the failure of mechanical valves and conveyors due to particle agglomeration require pelletization of sorbents into larger millimeter-sized materials for practical application. Pelletization of these immobilized amine sorbents has been achieved by different methods [2, 3], including combining the sorbents with an inorganic strength additive, such as fly ash, and bonding the dry mixture with different polymer binders, namely poly (vinyl chloride) (PVC).[4] High crush strength of the FA/PVC/BIAS pellets in the study resulted from chemical interactions of a PVC polymer network with the BIAS and FA particles, where the smaller FA particles interlocked the larger BIAS particles.

Although PVC with -Cl functional groups produced the strongest pellet of all polymers tested, these pellets degraded during semi-long (intermediate) term testing from reactions between PVC and BIAS. Therefore, an alternative chlorinated polymer is needed to produce strong and stable CO_2 capture pellets. In this work immobilized amine sorbents were pelletized with a fly ash additive and poly (chloroprene) (PC, chlorinated rubber) polymer binder, which produced pellets with high attrition resistance and stable CO_2 capture capacity during adsorption-desorption cycling under humid conditions.

Pelletization of a dry mixture of FA/[ground 50 wt% ethylenimine E100 (Huntsman)/silica]-20/80 with a poly (chloroprene) polymer binder (PC) (Sigma-Aldrich, Mooney viscosity=40) was accomplished in four steps. Step 1 was preparing multiple batches of polymer binder solutions containing 10-13 wt% PC dissolved in 1,4-dioxane at 105 °C for 20-90 min. In step 2, 1.1 g of each PC solution was mixed with 1.0 g of the FA/sorbent dry mixture to form a putty-like paste, which was extruded into ropes in step 3. In step 4, the ropes were dried at room temperature for 1 hr then dried at 105°C for 1 hr. The final 1.5 to 2.5 mm diameter pellets contained 17-18 wt% FA, 10-13 wt% PC, and the balance of BIAS. CO_2 adsorption of the pre-treated pellets (105°C, N_2, 10 min) was performed in a thermogravimetric analyzer (TGA) with a 60 mL/min flow of 14 vol% CO_2/N_2 at 55°C. Figure 1 shows that slightly increasing the PC content of the pellet from 10.9 to 12.6 wt% diminished the relative CO_2 uptake kinetics (normalized TGA weight profiles) and overall CO_2 capture capacities from 3.20 (particle sorbent) to 1.46 mmol CO_2/g (12.6 wt% PC), respectively. This reduction is attributed to pore blockage by PC. Here, PC likely interacted with the hydroxyl and amine groups of both the FA and BIAS inside the pore on the external particle surface, which imparted CO_2 diffusion limitations within the pellet and blocked previously accessible amine sites of the initial particle sorbent.[4] The procedure yielding the optimum combination of strength, relatively minimal structural flaws, and good CO_2 adsorption behavior (1.56 mmol CO_2/g) for the 12.2 wt% PC pellet was further improved upon to give a strong pellet, FA/E100-S_(20/80)_12.2PC, with a higher CO_2 capture of 1.76 mmol CO_2/g capture capacity.

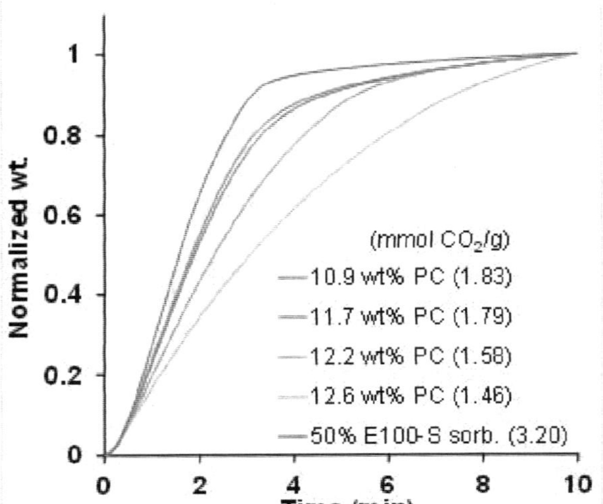

Figure 1: Normalized TGA weight profiles during CO_2 adsorption over PC/FA/BIAS pellets containing 17-18 wt% FA and different amounts of PC. The numbers in parenthesis indicate the total CO_2 capture capacity of the sorbents.

Figure 2 (a) shows the scanning electron microscope (SEM) cross-section of the optimized FA/E100-S_(20/80)_12.2 wt% PC pellet along with energy-dispersive X-ray spectroscopy (EDS) elemental maps. These images and maps reveal that the attrition resistance/strength of the pellet is attributed to the interactions among a well-dispersed poly (chloroprene) network (Cl), and the FA (Ca, Fe, Al) and sorbent (Si) particles. The flexibility of this robust PC (chlorinated rubber)/FA/sorbent network, in contrast to the rigidity of the commercial pellets, could allow our pellets to absorb and dissipate the impact force experienced during attrition testing.

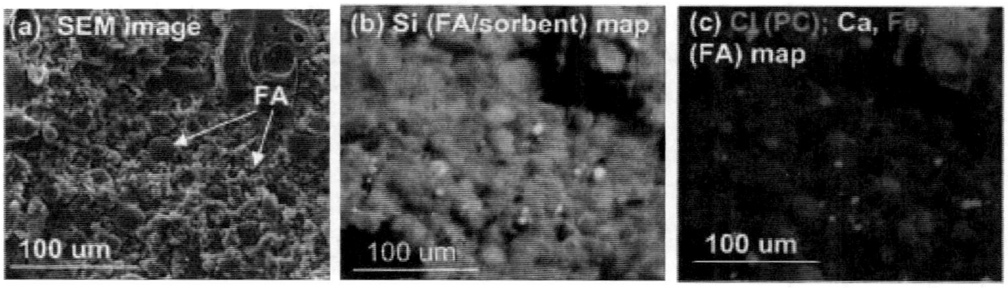

Figure 2: (a) SEM image of the FA/E100-S_(20/80)_12.2PC pellet cross section along with EDS elemental mapping of key sorbent, FA, and PC components.

Attrition resistance of FA/E100-S (20/80)_12.2PC pellets was ascertained from 24 hour attrition testing of 2.0 g of the pellets in an in-house constructed ball mill, utilizing a 39 rpm rotating jar (D=7 cm, Vol=230 mL) affixed with a 0.5" Al baffle and filled with 46, 3/16" steel ball bearings. A <0.5 wt% attrition for our pellets was significantly smaller than attrition of two commercial silica pellets (14.5-80.0%, 1 hour) and one commercial brand zeolite 13x pellet (36.6%, 24 hour). These results suggest that flexibility of our pellets, imparted by the PC network, facilitated their high attrition resistance. Both the good CO_2 capture stability of our pellets under practical conditions in the fixed-bed system and the high pellet attrition resistance make PC/FA/BIAS materials a promising candidate for larger-scale CO_2 capture processes.

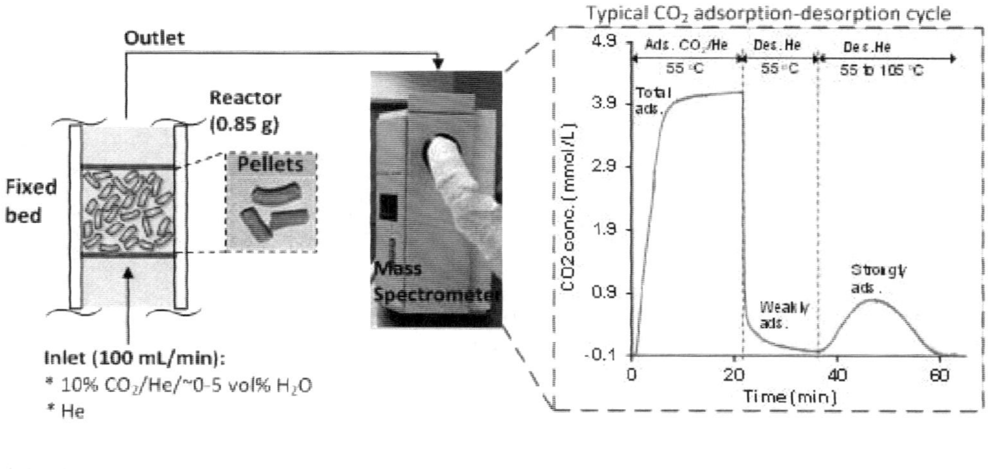

Figure 3: Fixed-bed reactor system loaded with the FA/E100-S_(20/80)_12.2PC pellets, along with a typical dry CO_2 adsorption-desorption cycle.

To assess the pellet performance under practical humid CO_2 adsorption-desorption cycling conditions, the pellets were placed into a fixed bed reactor system illustrated in Figure 3 and were first pre-treated at 105°C in flowing N_2 then cooled. In a typical cycle, CO_2 was adsorbed by the pellets at 55°C from a 10% CO_2/He flow for 20 min, where the adsorbed species were subsequently desorbed by flowing He at 55°C for 15 min (pressure swing desorption) and then by heating at 105°C in He for 20 min (temperature swing desorption). Two dry cycles (D1, D2) were followed by two wet cycles (~5 vol% H_2O), and finally two dry cycles (D3, D4); total testing time was 13 hrs (6 hrs humid). Results of the CO_2 capture tests, shown in Table 1, reveal the stability of the pellets during CO_2 adsorption-desorption cycling in the presence of steam.

Table 1: Results of the fixed-bed (FB) reactor system tests on FA/E100-S (20/80)_12.2PC pellet. PCR values were calculated from the CO_2 MS profiles (1 g system) and weight change (TGA) methods.

Cycle	CO_2 ads. (mmol CO_2/g) (cycle 1)	Percentage of CO_2 capture retained /PCR* (%,from two-cycle averages)
Dry 1 (fresh, FB)	1.74	
Dry 2 (fresh, FB)	1.74	
Dry 3 (steam-treated, FB)	1.69	96
Dry 4 (steam-treated, FB)	1.65	
Fresh/before Dry 1 (TGA)	1.78	
Steam-treated/after Dry 4 (TGA)	1.58	88

* Calculated as, PCR=CO_2 capture,steam-treated/CO_2 capture,fresh x 100%.

Excellent agreement between the fresh CO_2 capture of the pellets as determined from the CO_2 MS profiles (fixed-bed, 1.74 mmol CO_2/g) and from the TGA (1.78 mmol CO_2/g) methods highlight the reliability of calculating CO_2 capture capacity from the MS gas profiles. The high CO_2 capture capacity of the pellets after testing (~1.67 mmol CO_2/g) along with no change in the gray color of the pellet confirms the stability of the PC/FA/BIAS pellets under practical adsorption-desorption conditions. The low 8 percentage point difference in these FA/PC/BIAS pellets PCR values calculated by the CO_2 MS profiles (96%) and TGA (88%) further highlights reasonably good agreement between the two CO_2 capture capacity analysis methods. Overall the pellets exhibited excellent stability after 13 hrs of adsorption-desorption cycling, where exposure of the pellets to a humid environment was about 6 hrs.

References

1. US. Environmental Protection Agency, *U.S. Greenhouse Gas Inventory Report: 1990-2014*. 2014.

2. Knowles, G.P., Z. Liang, and A.L. Chaffee, *Shaped polyethyleneimine sorbents for CO_2 capture.* Microporous and Mesoporous Materials, 2016.

3. Rezaei, F., et al., *Shaping amine-based solid CO_2 adsorbents: Effects of pelletization pressure on the physical and chemical properties.* Microporous and Mesoporous Materials, 2015. **204**(0): p. 34-42.

4. Wilfong, W.C., et al., *Pelletization of Immobilized Amine Carbon Dioxide Sorbents with Fly Ash and Poly(vinyl chloride).* Energy Technology, 2016, DOI: 10.1002/ente.201500419

453790 Regeneration of Activated Carbon Using Dimethyl Ether for Water Treatment

Abstract:

Regeneration of Activated Carbon Using Dimethyl Ether for Water Treatment

Mitsuhiro Matsuzawa* and Tadashi Sano*

*Center for Technology Innovation – Mechanical Engineering, Research & Development Group, Hitachi, Ltd.

The water industry, which includes systems such as water supply and sewage systems, seawater desalination systems, and reclamation water systems, has been expanded with the increasing growth of the world's population and progress of industrialization. In the water treatment process, using activated carbon has been expected to remove the organic matters dissolved in the wastewater. The most common process for the regeneration of activated carbon is the thermal method. In this method, the used activated carbon is heated up to approximately 800 ºC to 900 ºC for several hours and the organic matters absorbed by the activated carbon are decomposed and detached from it. However, the thermal method needs the much time, energy, and costs to regenerate the activated carbon, so the used activated carbon generated in the water treatment process has frequently been disposed without being regenerated in reality. Therefore more efficient method to regenerate the activated carbon has been needed.

Dimethyl ether (DME) is an organic compound which is gas at normal temperature and pressures. DME is liquefied when it is pressurized up to approximately 0.5 MPa at normal temperature and liquefied DME has unique characteristic to dissolve both water and organic solvents [1][2]. Therefore, we investigated the method to regenerate activated carbon by using the characteristic of DME in this study. Especially, both batch and flow method to regenerate the used activated carbon using the liquefied DME were investigated and the regeneration rates were evaluated.

In batch method, the regeneration rate of the used activated carbon was approximately 70% and it was confirmed that the regeneration rate was kept at 70% when the activated carbon was regenerated at least 5 times.

On the other hand, the regeneration rate of the used activated carbon was approximately 80% in flow method when the contact time between DME and the activated carbon was 300 s. Moreover, the regeneration rates did not almost change when the regeneration process was repeated at least 5 times.

Finally, we estimated the cost needed in the regeneration process of the activated carbon using DME and compared it to that needed in the process using virgin activated carbon. The estimates indicated that the cost needed in the regeneration process using DME could be decreased by at least 79% compared to that needed in the process purchasing virgin activated carbon.

References

[1] H. Holldorff, and H. Knapp, Binary vapor–liquid–liquid equilibrium of dimethyl ether – water and mutual solubilities of methyl chloride and water: Experimental results and data reduction, *Fluid Phase Equilibria*, 44, 195–209 (1988)

[2] H. Kanda, H.Makino, and M. Miyahara, Energy–saving drying technology for porous media using liquefied DME gas, *Adsorption*, 14, 467–473 (2008.10)

458032 Development of a Pore-Scale Transport Assay for Protist-Facilitated Transport of Plant Growth-Promoting Bacteria

Abstract:

The rhizosphere, the area of soil that surrounds plant roots, is critical to the health of soil and the productivity of crops. Microbes active in the rhizosphere can promote plant growth by fixing atmospheric N_2 to bioavailable nitrogen forms, protecting plants from pathogens, and promoting the retention of soil moisture. Bacterial inoculants and other biological additives are a growing billion-dollar industry being used more often despite the well-known difficulty in transporting live bacteria through soil. Prior work in our lab has demonstrated that naturally-occurring soil protists can actively transport live bacteria and other micro- or nano-scale agricultural payloads. The objective of this project is to screen the particle-transport capabilities of a group of soil protists. Here, a high-throughput microfluidic screening assay was developed. The microfluidic device was comprised of an input well connected by a micro-structured region to a target well. The micro-structured region featured a pseudo-2D representation of the physical microstructure of a sandy loam soil. The target well contained a protist chemoattractant immobilized in a hydrogel plug. Devices were constructed out of polydimenthyl siloxane (PDMS) using photolithography and soft lithography, and were operated under quiescent (no-flow) conditions. Preliminary results show a wide range of soil protists are well-adapted to navigating porous media, and will take up a micro-scale payload. Ultimate performance of protist-facilitated transport as a sustainable agriculture biotechnology is a function of many factors including ease of protist culturing and storage, the rate and extent of swimming motility through porous media, the extent of chemotaxis for root-targeting functionality, payload uptake and egestion rates, and carrying capacity.

461013 Investigation of the Emission of Nitrogen-Sulfur Compounds As Persistent COD and Nitrogen Components from Flue Gas Desulfurization Wastewater

Abstract:

Some thermal power plants experience problems due to the increasing chemical oxygen demand (COD) and total nitrogen concentration (T-N) in the flue gas desulfurization (FGD) wastewater. Nitrogen-sulfur (NS) compounds present in FGD wastewater are speculated as the cause of increasing COD and T-N. However, the emissions of NS compounds are not yet fully understood. In this study, we present the method and results of analysis of NS compounds in real FGD wastewater from four units of coal-fired power plants. In addition, the emissions of NS compounds from FGD wastewater from two units of coal-fired power plants were investigated for a longer period of time.

Hydroxylamine-*N*, *N*-disulfonate (HADS), formed from the reaction between nitrite and bisulfite in FGD wastewater, is the precursor of NS compounds. HADS can be converted to various NS compounds by hydrolysis, oxidation and reduction, depending on the wastewater conditions. All 8 types of NS compounds contribute to T-N, while five contribute to COD. For the analysis, standard reagents of six different NS compounds, which are not commercially available, were synthesized. The conditions necessary for the measurement of NS compounds using ion-pair chromatography (IPC) and ion chromatography (IC) were examined. An analytical procedure was established for seven different NS compounds using a combination of two IPC and one or two IC measurement conditions, respectively.

NS compounds present in the real FGD wastewater from four units of coal-fired power plants were analyzed. When the COD or the T-N was high, hydroxylaminetrisulfonate (HATS), imidodisulfate (IDS), or amidosulfonate (AS) was generated in the FGD wastewater. On the other hand, when the COD or the T-N was low, these NS compounds were either not detected, or were detected in low concentrations.

The dependence of the generation of NS compounds on wastewater conditions was investigated by taking measurements over a long period of time, in two units of coal-fired power plant. The main type of NS compound was HATS in both the case. FGD wastewater of the first unit showed a strong negative correlation between the HATS concentration and the total organic carbon (TOC), while it showed a weak negative correlation between the HATS concentration and the oxidation-reduction potential (ORP). The FGD wastewater of the second unit showed a strong negative correlation between the HATS concentration and the ORP, while it showed a strong positive correlation between the HATS concentration and the nitric acid concentration. Therefore, maintaining the ORP of the absorbing solution for the FGD at a higher level can be effective in managing the COD and the T-N of FGD wastewater.

463639 Do Analog Fluids Resemble Hydrodynamics of Convective Mixing Involved in Dissolution of CO_2 in Deep Saline Aquifers?

Abstract:

Typical analog fluids such as mixtures of ethylene glycol and methanol (EG-MeOH) and propylene glycol (PPG) have recently been used as an analog system in recent experiments in the context of convective dissolution of CO_2 in deep saline aquifers. We have conducted a linear stability analysis of a gravitationally unstable diffusive boundary layer as well as direct numerical simulation of convective mixing involved in dissolution of these analog fluids in water. We provide new evidences that neither EG-MeOH nor PPG analog systems do resemble the dynamics of convective instabilities and subsequent mixing associated with dissolution of CO_2 in water. It is found that there are fundamental differences in the evolution of the buoyancy-driven instability and dynamics of convective mixing between CO_2/water and these analog systems. Our results show that for a constant Rayleigh number the onset of convective instabilities for these analog fluids can be different by an order of magnitude as compared with CO_2/water. We have shown that the base state density behavior of analog systems has a significant effect on the stability characteristics of the diffusive boundary layer and convective mixing beyond the onset of instability. According to the results of this study, the choice of an analog system is critically important in correct representation of the onset of convective instabilities of CO_2/water and design of experiments. This study improves our understanding of the instability behavior of analog systems, their proper selection, and motivates further experiments.

Rapid Fire Session: Environmental Division

Determination of Concentration Dependence of CO_2 and CH_4 Permeances through Zeolite Membrane Utilizing CFD technique

T.Kato ,Y.Kan-no,T.Sato,N.Itoh*
Dept.Applied Chemistry, Utsunomiya University

Date: Monday, November 14, 2016

Introduction

Securing fossil fuel becomes more and more important in progress of global resource depletion

Resources of CH_4
- Low-grade natural gas field
- Land-fill gas

Mixture of CH_4 & CO_2

CO_2 removal technology
By membrane separation is promising

Zeolite membrane
A high CO_2 separation capability

The separation performance is affected by some factors and such as
1) Concentration of feed gas
2) Feed rate
3) Pressure drop
4) Configuration of membrane separator

Must be all considered to determine

More accurate permeances for CO_2 and CH_4

CFD(Computational Fluid Dynamics) analysis is a useful tool to analyze the phenomena occurring inside the separator

2

Result

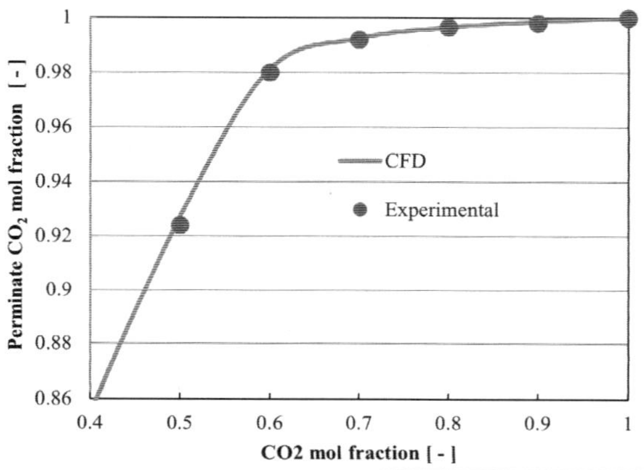

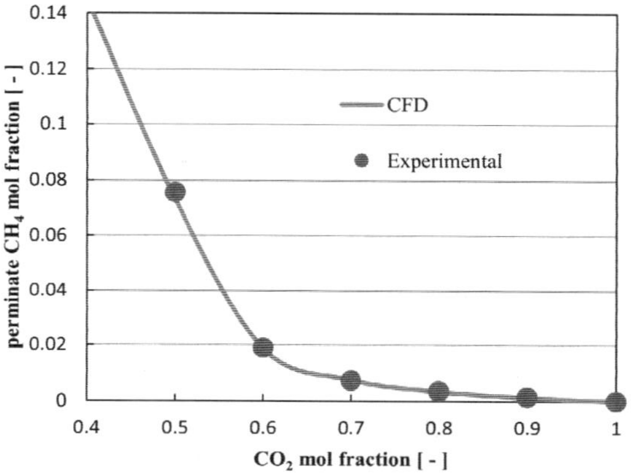

The both are in a good agreement

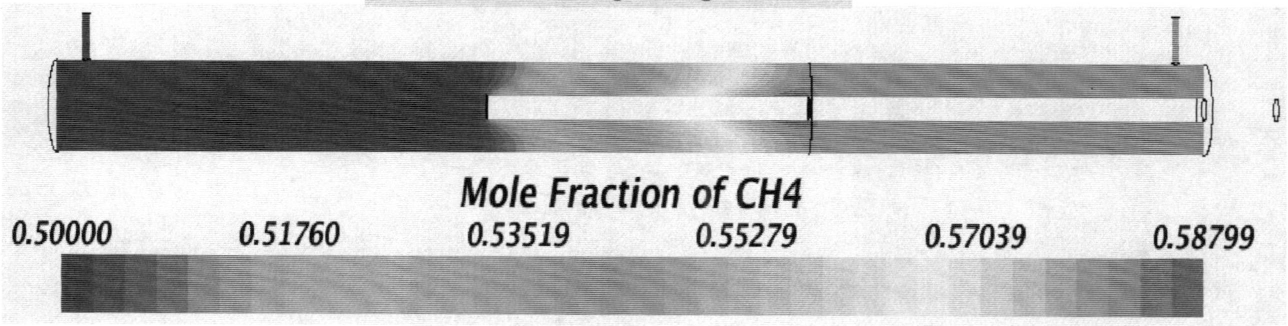

461565 Water and Emissions Trade-Offs in the Operation of Biogas-Fueled Combined Heat and Power Systems

Abstract:

Combined Heat and Power systems (CHP) can achieve much higher resource utilization efficiencies than conventional power plants and thus reduce fuel cost and emissions and losses associated to electricity transmission (Mago et al., 2009). CHP systems can also provide different energy products (steam, hot water, cold water, and eletricity) to commercial buildings, hospitals, housing complexes, and manufacturing facilities (El-Khattam and Salama, 2004). Modern systems can also run on multiple fuels such as natural gas, diesel, coal, biomass, and biogas (Dong et al., 2009).

The ability to achieve high energy integration in CHP systems, however, results in significantly more complex operations. In particular, it is necessary to capture strong and highly dynamic interactions between carriers and resources (Geidl and Andersson, 2007; Cho et al., 2009a), it is necessary to trade-off energy market conditions and local demands (Illerhaus and Verstege, 1999), one must consider dynamics and limited storage (Henning, 1998; Malysz et al., 2014), and one must consider inefficiencies associated to partial load operation (Bracco et al., 2014).

Greenhouse gas emissions and water usage are major factors driving the development of new CHP technologies and fuels. A key issue in the design and operation of advanced CHP systems is that environmental impact associated to water consumption and emissions are controlled by monetization (i.e., one associates a price/penalty to them). We argue that such an approach is highly inefficient when local environmental policies and market conditions are not well developed and this results in limited control of environmental impact and blocks adoption of new technologies. For instance, in some cases one might only be able to control environmental impact using extremely high and unrealistic water prices. We show that adaptive scalarization of objectives is necessary to ensure that water and emissions impact match the actual limits of a wide sprectrum of technologies and under a wide range of operating conditions. We demonstrate our developments using a CHP system for a real residential building complex that incorporates biogas produced from food waste and thermal storage. In particular, we demonstrate that existing carbon emission penalties and water prices do not incentivize the use of biogas and have limited impact on water utilization.

Geidl, M. and Andersson, G. (2007). Optimal power flow of multiple energy carriers. Power Systems, IEEE Transactions on, 22(1):145–155.

Mago, P., Chamra, L., and Hueffed, A. (2009). A review on energy, economical, and environmental benefits of the use of chp systems for small commercial buildings for the north american climate. International Journal of Energy Research, 33(14):1252–1265.

Owen, A. D. (2006). Renewable energy: Externality costs as market barriers. Energy policy, 34(5):632–642.

Illerhaus, S. and Verstege, J. (1999). Optimal operation of industrial chp-based power systems in liberalized energy markets. In Electric Power Engineering, 1999. PowerTech Budapest 99. International Conference on, page 210. IEEE.

Malysz, P., Sirouspour, S., and Emadi, A. (2014). An optimal energy storage control strategy for grid-connected microgrids. Smart Grid, IEEE Transactions on, 5(4):1785–1796.

Cho, H., Luck, R., Eksioglu, S. D., and Chamra, L. M. (2009a). Cost-optimized real-time operation of chp systems. Energy and Buildings, 41(4):445–451.

Dong, L., Liu, H., and Riffat, S. (2009). Development of small-scale and micro-scale biomass-fuelled chp systems–a literature review. Applied thermal engineering, 29(11):2119–2126.

463490 Electrochemical Behavior of $Ni_2bpy(TPTZ)_2$ Complex on Carbon Paste Electrode Modified with Nanocrystalline ZnO Powder for Usability in Dye-Sensitized Solar Cells

Abstract:

Recently, nanocrystalline ZnO powder has been used as promising alternative photoanode material in dye-sensitized solar cells DSSCs due to suitable band gap energy, electron affinity, high electron mobility and stability against photocorrosion[1,2]. The most efficient DSSCs demonstrated to date have all been based on Ru(II) bipyridyl (Bpy) complex dyes as sensitizers in which metal-ligand charge transfer takes place at a much faster rate than the back recombination reaction[3,4]. Alternative cost-effective transition metal bipyridyl complexes have been developed as sensitizer for photoanodes as well as redox mediator instead of almost exclusively employed iodide/ triiodide redox couple[5].

In this study, chemical structure of Ni(II) complex with the mixed N-donor ligands 2,2′-bipyridyl-1,3,5-tripyridyl-striazine (TPTZ) and Bpy was electrochemically determined using different metal/ligand ratios. The square-wave voltammetry (SWV) results on glassy carbon electrode (GCE) confirmed that $Ni_2Bpy(TPTZ)_2$complex was stable at pH 7.8 in phosphate buffer. The electrochemical behavior of the complex was investigated using a conventional three-electrode system, consisting of carbon paste (CPE) and ZnO modified carbon paste (ZnO/CPE) working electrodes to understand its possible use as a redox mediator and/or sensitizer in DSSCs.

The mechanisms of the electrode processes were elucidated by using SWV and cyclic voltammetry (CV) techniques. The number of electrons transferred in the electrode processes (n) and transfer coefficients (α) was determined from the effect of frequency on the SW voltammograms. Two reduction peaks were observed on the SWVs at both electrode surfaces whereas a quasi-reversible redox couple appeared

corresponding to anodic oxidation of central Ni(II) ion on ZnO powders. The scan rate dependence of the peak currents and potentials showed that electrode process was adsorption controlled. The rate constant for oxidation process was found to be much faster than the reduction one suggesting that electron injection from the excited complex to the ZnO nanoparticles may also be greater than the recombination with the oxidized complex in DSSCs.

It can be concluded that the complex can be employed as a photosensitizer for ZnO powder photoanode as well as the redox mediator in DSSCs used CPE cathode.

REFERENCES

1. Seema Rani, Poonam Suri, P.K. Shishodia, R.M. Mehra, Synthesis of nanocrystalline ZnO powder via sol–gel route for dye-sensitized solar cells. Solar Energy Materials & Solar Cells 92 (2008) 1639–1645.
2. Lanlan Lu, Renjie Li, Ke Fan, Tianyou Peng. Effects of annealing conditions on the photoelectrochemical properties of dye-sensitized solar cells made with ZnO nanoparticles. Solar Energy 84 (2010) 844–853.
3. Seigo Ito (2011). Investigation of Dyes for Dye-Sensitized Solar Cells: Ruthenium-Complex Dyes, Metal-Free Dyes, Metal-Complex Porphyrin Dyes and Natural Dyes, Solar Cells - Dye-Sensitized Devices, Prof. Leonid A. Kosyachenko (Ed.), ISBN: 978-953-307-735-2, InTech, Available from: http://www.intechopen.com/books/solar-cells-dye-sensitized-devices/investigation-of-dyes-for-dye-sensitizedsolar-cells-ruthenium-complex-dyes-metal-free-dyes-metal-co.
4. Jooyoung Chung, June young Lee, Sangwoo Lim. Annealing effects of ZnO nanorods on dye-sensitized solar cell efficiency, Physica B 405 (2010) 2593–2598.
5. Joel T. Kirner, C. Michael Elliott Are High-Potential Cobalt Tris(bipyridyl) Complexes Sufficiently Stable to Be Efficient Mediators in Dye-Sensitized Solar Cells? Synthesis, Characterization, and Stability Tests, J. Phys. Chem. C 2015, 119, 17502−17514.

MICROWAVE ASSISTED ADVANCED OXIDATION OF PETROCHEMICAL WASTEWATER

Alicia L. Garcia-Costa, Juan A. Zazo, Jose A. Casas, Juan J. Rodriguez, Universidad Autonoma de Madrid, Madrid, ES.

Abstract

The degradation of five different chemicals found in petrochemical wastewater (Phenol, Benzene, Toluene, Naphthalene and o-Xylene) was accomplished with Microwave assisted Catalytic Wet Peroxide Oxidation (MW-CWPO) using activated carbon (AC) as catalyst. Optimum operating conditions were found to be T:120ºC, C_{AC}:1g·L^{-1}, pH:3, $t_{reaction}$:15 min. Under this conditions, hot spot formation on the surface of AC allows complete pollutant elimination, practically complete H_2O_2 consumption and mineralization rates above 90% in all cases. Additionally, for phenol, toluene, naphthalene and O-xylene, no aromatic intermediates were detected, indicating there is a quick hydroxylation and breakdown of the aromatic ring yielding carboxylic acids. Therefore, MW-CWPO with AC has been proved to be a fast and effective process for petrochemical-like wastewater treatment.

Introduction

New environmental policies, with more restrictive discharge limits, demand new or enhanced technologies for wastewater treatment, especially when facing bio-refractory pollutants, such as those present in petrochemical wastewater.

Advanced Oxidation Processes (AOPs) have been widely used for this purpose. Among them, Catalytic Wet Peroxide Oxidation (CWPO) is one of the most successfully implemented. This AOP uses H_2O_2 as a source of HO_X• (HO• and HOO•) radicals for organic matter oxidation at mild working conditions (T: 25-120ºC; P: 1-5 bar) [1]. The main point of concern in CWPO is efficient H_2O_2 consumption. Metal catalysts are commonly used in CWPO, nevertheless, activated carbon by itself presents catalytic activity [2-7].

Recently, microwave (MW) heating has been used for CWPO intensification [8-12]. Previous work showed that MW absorbing properties of the catalytic support play a crucial role on MW-CWPO [13]. MW absorbers such as AC are able to increase its temperature, this effect results on hot spot formation on the surface of the AC. This hot spots may reach 1200ºC [10] and enhance the oxidation processes enhancing HO_X• formation, oxidation rate and inhibiting the polymerization or condensation of byproducts on the surface of the catalysts [13].

This work addresses the degradation of typical compounds found in petrochemical wastewater such as phenol, benzene, toluene, naphthalene and o-xylene using AC as catalyst in MW-CWPO. Prior to the degradation of the pollutants, operating conditions (temperature, C_{AC} and pH) were optimized.

Material and methods

Reactants

Aqueous pollutant and H_2O_2 solutions were prepared with phenol (99%, Sigma-Aldrich), benzene (99%, Sigma-Aldrich), toluene (99,99%), o-xylene (98%, Sigma-Aldrich) and naphthalene (99%, Sigma-Aldrich) and H_2O_2 (30% w/v; Panreac), respectively, at pH 3 using HCl (37% w/v;

Panreac). AC (Merck) was grinded and sieved below 100μm. H_2SO_4 (96 wt.%; Panreac), acetonitrile (Sigma-Aldrich) and H_3PO_4 (85 wt.%; Sigma-Aldrich) were used in the analytic procedure. All reagents are analytical grade and they were used as received without further purification. Milli-Q water was used throughout the work.

MW-CWPO experiments

MW-CWPO runs were performed in a high pressure 100 mL PTFE reaction vessel located in a microwave (flexiWAVE, Milestone). Reactor was initially loaded with aqueous pollutant solution, H_2O_2 and AC as catalyst. Initial heating rate was set at 80°C/min (at 1800 MW), after reaching 120°C, temperature plateau was maintained for 1 h for pH and catalyst concentration studies, afterwards reaction time was set in 15 minutes.

Analytical methods

Samples were periodically withdrawn from the reactors and immediately analyzed after filtration through fiber glass filters (Albet FV-C). Phenol, BTNX and aromatic intermediates were identified and quantified by means of an Ultra HPLC (Thermo Scientific Ultimate 3000) with a Diode Array detector (Dionex Ultimate 3000). An ion-exclusion column (ZORBAX Eclipse Plus C18, 100 mm, 1.8μm) was used as stationary phase. As mobile phases, a mix of 50% acetonitrile and 50% 4 mM H_2SO_4 aqueous solution at 1 mL/min was used for BTNX and only H_2SO_4 aqueous solution at 1 mL/min as mobile phase for phenol and aromatic intermediates. UV detector at 210 nm wavelength was used for BTNX, phenol, resorcinol, catechol and hydroquinone and at 246 nm for p-benzoquinone. Short-chain organic acids were analyzed in an ion chromatograph with chemical suppression (Metrohm 790 IC) using a conductivity detector. A Metrosep A supp 5-250 column (25 cm long, 4 mm diameter) was used as stationary phase and 0.7 mL/min of a 3.2 mM / 1 mM aqueous solution of Na_2CO_3 and $NaHCO_3$, respectively, as mobile phase. Total Organic Carbon was measured using a TOC analyzer (Shimadzu TOC-VSCH).

Results and discussion

Preliminary studies

Initially influence of temperature, AC concentration and pH were studied using phenol (100 $mg\cdot L^{-1}$) in order to determine the optimal operating conditions. Temperature was studied in the range between 50 and 120°C as displayed in Figs. 1 and 2. Analysis on temperature influence on the system is collected in Fig.1, where TOC and H_2O_2 evolution are depicted. For these runs 500 $mg\cdot L^{-1}$ of H_2O_2 were used, corresponding to the stoichiometric amount needed for complete degradation of phenol. In an ideal process, H_2O_2 consumption would match TOC conversion and experimental data would gather around the diagonal. Two regions have been differentiated, the white one, below the ideal performance, would correspond to a non-efficient consumption of H_2O_2, as there would be a high H_2O_2 decomposition which is not being employed to oxidize organic matter. On the contrary, the gray region would match pollutant adsorption on the AC surface, as there cannot be higher organic matter disappearance than HO_X• formation. As may be observed, temperatures below 90°C point towards adsorption, whereas for 120°C the system approaches the ideal behavior towards $t_{reaction}$:15 min. Adsorption process is deprived when increasing temperature, moreover, hot spots produced on the AC surface also inhibit phenol adsorption and play a key role on the oxidation of adsorbed contaminants [13]. This effect can be clearly observed working at 120°C.

As shown in Fig 2. TOC initial reaction rates are similar below 120°C, as the process is controlled by adsorption. Nevertheless, for H_2O_2 decomposition initial rates, an exponential rise is observed. So as to avoid adsorption limitations, following experiments were conducted at 120°C.

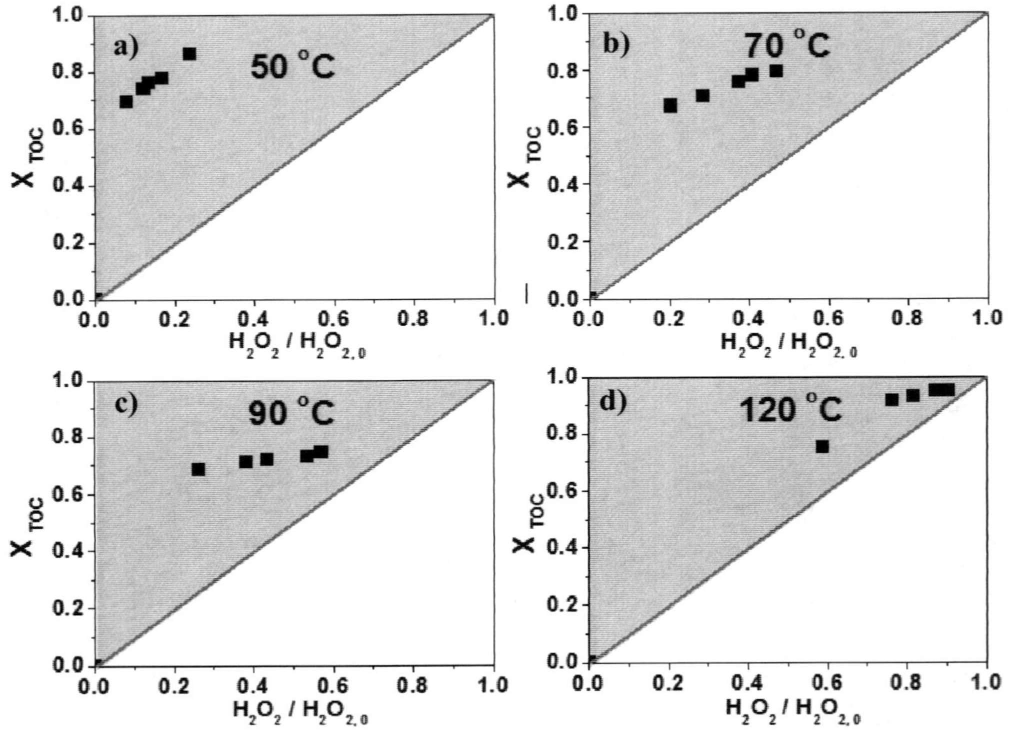

Figure 1. Temperature influence on H_2O_2 and TOC conversion at a) 50°C, b) 70°C, c) 90°C, d) 120°C. $C_{PHENOL,0}$:100mg·L^{-1},$C_{H2O2,0}$:500 mg·L^{-1}, C_{AC}:1 g·L^{-1}, pH:3, t:15 min.

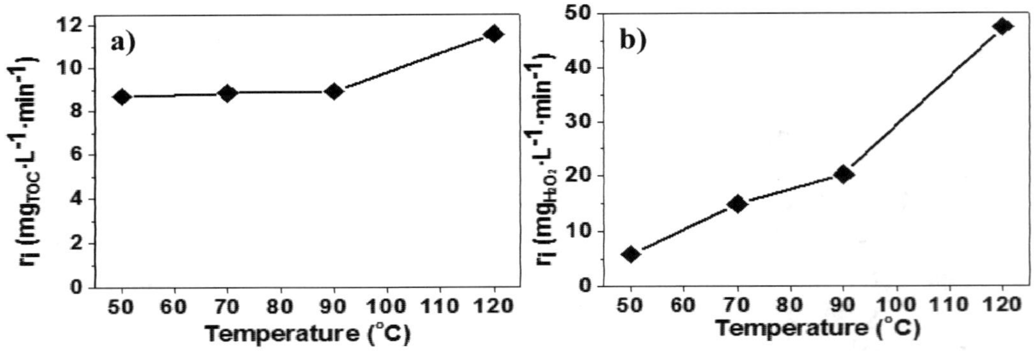

Figure 2. Temperature influence on a) TOC and b) H_2O_2 initial reaction rates. Operating conditions: $C_{PHENOL,0}$:100mg·L^{-1},$C_{H2O2,0}$:500 mg·L^{-1}, C_{AC}:1 g·L^{-1}, pH:3.

AC concentration was studied in the range from 0 to 1g·L^{-1}. Fig.3 shows the influence of catalyst concentration on TOC conversion and initial reaction rate. As can be depicted, even though MW radiation can produce $HO_x^{\bullet}$ radicals, when AC is introduced in the reaction system, hot spots boost H_2O_2 decomposition, thus mineralization rate is remarkably enhanced. Increase in AC concentration should imply a linear rise on the rate, nonetheless hot spot formation causes a synergic effect, causing a major growth on the initial reaction rate and leading to a 98% mineralization degree in only 15 min reaction for 1 $g \cdot L^{-1}$ of AC.

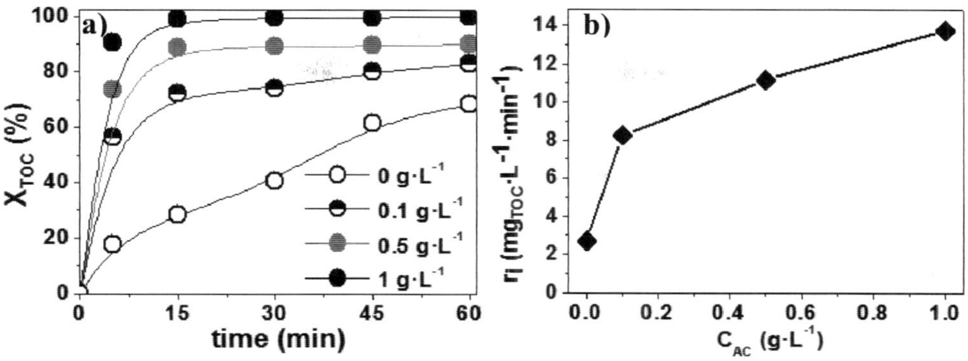

Figure 3. C_{AC} influence on a) TOC evolution and b) initial reaction rate. Operating conditions: $C_{PHENOL,0}$:100mg·L^{-1},$C_{H2O2,0}$:500 mg·L^{-1}, pH:3, T:120°C.

CWPO has been proved to have its better performance with pH around 3 [1]. Nevertheless, some studies affirm MW assisted AOP's may work on broader pH range than the traditional ones [14]. pH influence on the MW-CWPO system was studied between pH 2 and 7, as exhibited in Figure 4. According to initial reaction rate, pH has a less drastic effect in comparison to C_{AC} or temperature. Nonetheless, although operating at initial neutral pH is possible, pH 3, achieves a slightly higher mineralization rate and degree.

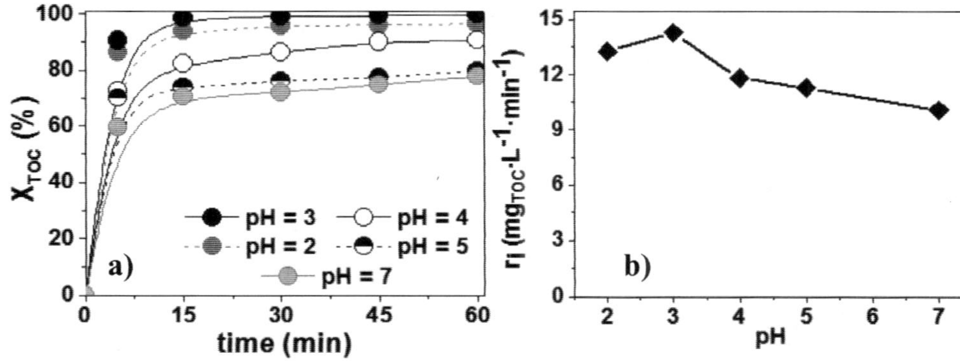

Figure 4. pH influence on a) TOC evolution and b) initial reaction rate. Operating conditions: $C_{PHENOL,0}$:100mg·L^{-1},$C_{H2O2,0}$:500 mg·L^{-1}, C_{AC}:1 g·L^{-1}, T:120°C.

In order to ensure hot spot formation and optimize H_2O_2 and energy consumption, the following experiments were conducted with: C_{AC}: 1 g·L^{-1}, pH:3, T:120°C and $t_{reaction}$:15 minutes.

BTNX and phenol degradation

After selecting the optimal operation conditions, aqueous samples of benzene (B), toluene (T), naphthalene (N), O-xylene (X) and phenol (P) where subjected to MW-CWPO. Pollutant and H_2O_2 conversion show different trends, as may be seen in Fig. 5. The MW-CWPO process achieves complete pollutant removal in all cases. However, it should be noted that toluene degradation presents a significantly slower rate in comparison to the rest of pollutants, implying that the nature of the pollutant plays an important role. B, P and X show a similar behavior. N is easily oxidized but T hydroxylation towards cresol is hindered.

In all cases, almost complete decomposition of H_2O_2 is reached in 15 mins, which reduces significantly the reaction duration in comparison to traditional CWPO [1, 15]. Considering the wide range of $C_{H2O2,0}$ employed (from 220 to 675 mg·L^{-1}, depending on each pollutant), it can be concluded that H_2O_2 decomposition follows the same mechanism, regardless of the contaminant.

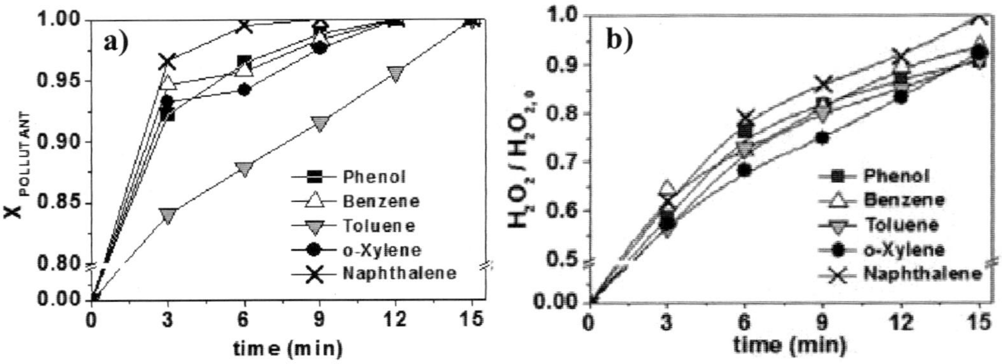

Figure 5. a) Pollutant and b) H_2O_2 evolution for $C_{B,0}$:100mg·L^{-1}, $C_{T,0}$:100mg·L^{-1}, $C_{N,0}$:100mg·L^{-1}, $C_{X,0}$:100mg·L^{-1}, $C_{P,0}$:100mg·L^{-1}, using $C_{H2O2,0}$: stoichiometric, C_{AC}:1 g·L^{-1}, pH:3

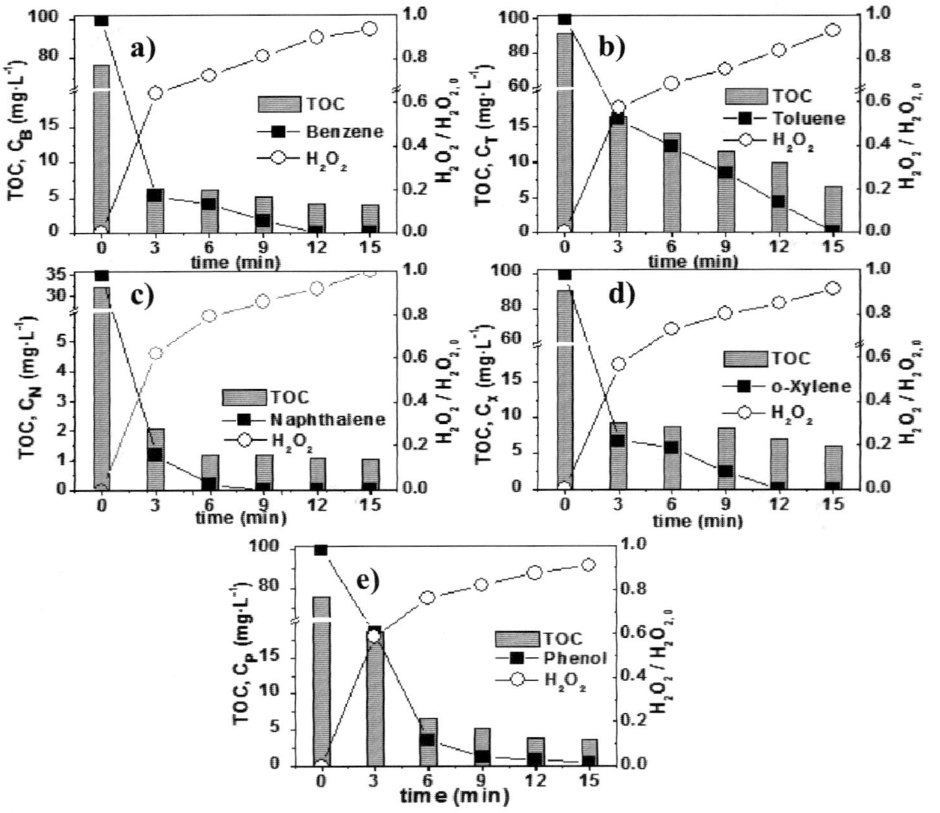

Figure 6. Pollutant, TOC and H_2O_2 evolution for a) $C_{B,0}$:100mg·L^{-1}, b) $C_{T,0}$:100mg·L^{-1}, c) $C_{N,0}$:100mg·L^{-1}, d) $C_{X,0}$:100mg·L^{-1}, e) $C_{P,0}$:100mg·L^{-1}, using $C_{H2O2,0}$: stoichiometric, C_{AC}:1g·L^{-1}, pH:3

Pollutant depletion, TOC removal and H_2O_2 consumption for each chemical are shown in Figure 6. TOC removal varies considerably depending on the initial pollutant. The methyl groups of toluene and o-xylene are hydroxylated by $HO_X^{\bullet}$ radicals to formic acid, which is refractory to this kind of treatment. Therefore, this compounds present a lower mineralization degree.

Intermediates analysis (data not shown) provide a new oxidation route in relation to CWPO. Intermediates evolution point towards a quick aromatic ring opening for T, N, X, and P, where no typical intermediates such as catechol, benzoquinone, hydroquinone or resorcinol were detected in reaction. On the other hand, B is slowly hydroxylated to P and further oxidation leads, as for T, N, X and P to short chain acids such as malonic, maleic and oxalic acid. These may be oxidized to acetic and formic acid, which are refractory to CWPO, or completely eliminated as CO_2 and H_2O as illustrated in Fig. 7.

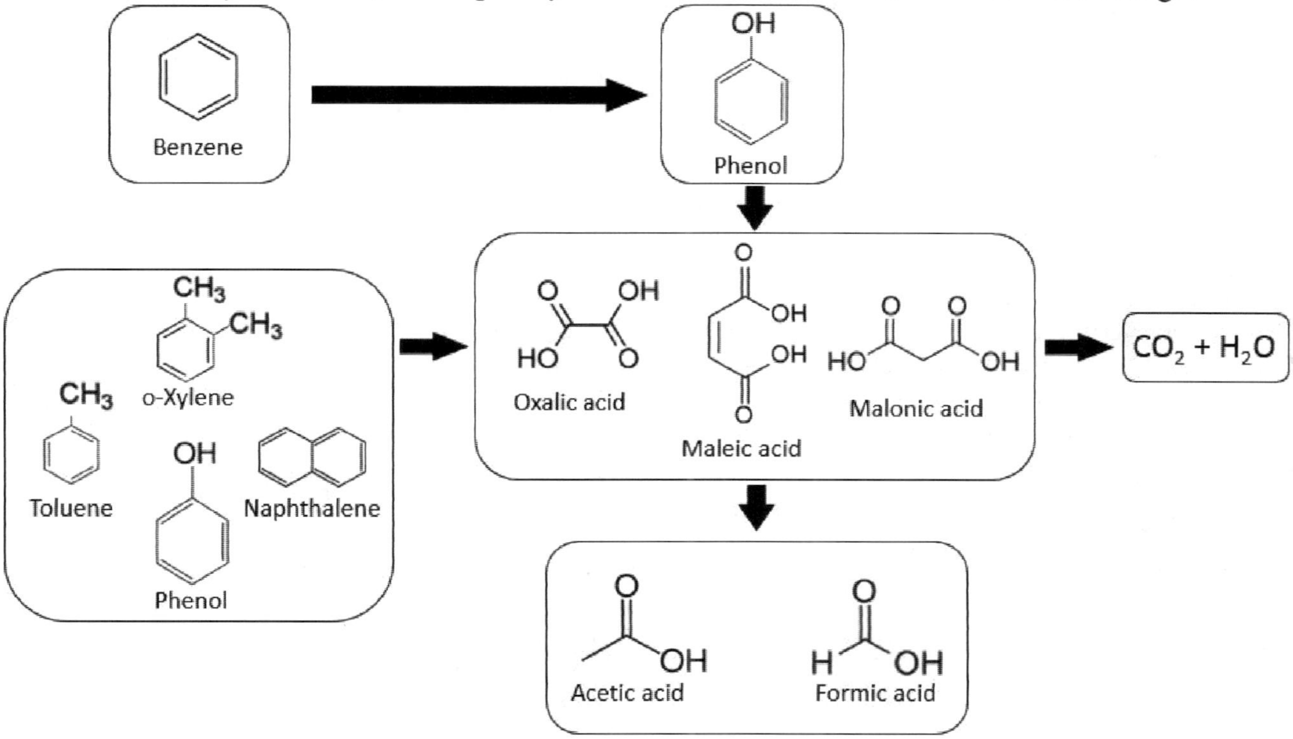

Figure 7. Oxidation route for P, B, T, N, X in MW-CWPO using $C_{H2O2,0}$: stoichiometric, C_{AC}:1g·L^{-1}, pH:3

Conclusions

MW-CWPO with AC has proven to be an effective process for petrochemical-like pollutants degradation in aqueous phase. Synergic effect of hot spot formation and catalytic activity of AC allows to operate in mild conditions (T: 120ºC, pH: 3) and short reaction time, 15 min, with outstanding results: complete contaminants depletion and high mineralization degree up to 97%. Intermediates analysis show a fast oxidation towards carboxylic acids.

Acknowledgements

Authors would like to thank the Spanish MINECO for financial support through project CTQ2013-41963-R. A.L. Garcia-Costa acknowledges the ESF and MINECO for PhD. grant BES-2014-067598.

References

1. Zazo, J.A., J.A. Casas, A.F. Mohedano, and J.J. Rodriguez (2006), "Catalytic wet peroxide oxidation of phenol with a Fe/active carbon catalyst". *Applied Catalysis B-Environmental*. 65, 3-4: p. 261-268.
2. Bautista, P., A.F. Mohedano, J.A. Casas, J.A. Zazo, and J.J. Rodriguez (2010), "Oxidation of cosmetic wastewaters with H2O2 using a Fe/gamma-Al2O3 catalyst". *Water Science and Technology*. 61, 6: p. 1631-1636.
3. Dominguez, C.M., P. Ocon, A. Quintanilla, J.A. Casas, and J.J. Rodriguez (2013), "Highly efficient application of activated carbon as catalyst for wet peroxide oxidation". *Applied Catalysis B-Environmental*. 140, 663-670.
4. Martin-Martinez, M., R.S. Ribeiro, B.F. Machado, P. Serp, S. Morales-Torres, A.M.T. Silva, J.L. Figueiredo, J.L. Faria, and H.T. Gomes (2016), "Role of Nitrogen Doping on the Performance of Carbon Nanotube Catalysts: A Catalytic Wet Peroxide Oxidation Application". *Chemcatchem*. 8, 12: p. 2068-2078.
5. Munoz, M., Z.M. de Pedro, N. Menendez, J.A. Casas, and J.J. Rodriguez (2013), "A ferromagnetic gamma-alumina-supported iron catalyst for CWPO. Application to chlorophenols". *Applied Catalysis B-Environmental*. 136, 218-224.
6. Rey, A., M. Faraldos, J.A. Casas, J.A. Zazo, A. Bahamonde, and J.J. Rodriguez (2009), "Catalytic wet peroxide oxidation of phenol over Fe/AC catalysts: Influence of iron precursor and activated carbon surface". *Applied Catalysis B-Environmental*. 86, 1-2: p. 69-77.
7. Ribeiro, R.S., A.M.T. Silva, J.L. Figueiredo, J.L. Faria, and H.T. Gomes (2016), "Catalytic wet peroxide oxidation: a route towards the application of hybrid magnetic carbon nanocomposites for the degradation of organic pollutants. A review". *Applied Catalysis B-Environmental*. 187, 428-460.
8. Atta, A.Y., B.Y. Jibril, T.K. Al-Waheibi, and Y.M. Al-Waheibi (2012), "Microwave-enhanced catalytic degradation of 2-nitrophenol on alumina-supported copper oxides". *Catalysis Communications*. 26, 112-116.
9. Jibril, B.Y., A.Y. Atta, Y.M. Al-Waheibi, and T.K. Al-Waheibi (2013), "Effect of copper loadings on product selectivities in microwave-enhanced degradation of phenol on alumina-supported copper oxides". *Journal of Industrial and Engineering Chemistry*. 19, 6: p. 1800-1804.
10. Remya, N. and J.-G. Lin (2011), "Current status of microwave application in wastewater treatment—A review". *Chemical Engineering Journal*. 166, 3: p. 797-813.
11. Tan, D., H. Zeng, J. Liu, X. Yu, Y. Liang, and L. Lu (2013), "Degradation kinetics and mechanism of trace nitrobenzene by granular activated carbon enhanced microwave/hydrogen peroxide system". *Journal of Environmental Sciences*. 25, 7: p. 1492-1499.
12. Zhao, G.H., B.Y. Lv, Y. Jin, and D.M. Li (2010), "P-Chlorophenol Wastewater Treatment by Microwave-Enhanced Catalytic Wet Peroxide Oxidation". *Water Environment Research*. 82, 2: p. 120-127.

13. Garcia-Costa, A.L., Zazo, J.A., Casas, J.A., Rodriguez, J.J.,(2016) "Microwave enhanced Catalytic Wet Peroxide Oxidation. Role of Catalytic Supports.", *6th EUCheMs*: Seville, Spain.
14. Wang, N. and P. Wang (2016), "Study and application status of microwave in organic wastewater treatment – A review". *Chemical Engineering Journal*. 283, 193-214.
15. Rey, A., A.B. Hungria, C.J. Duran-Valle, M. Faraldos, A. Bahamonde, J.A. Casas, and J.J. Rodriguez (2016), "On the optimization of activated carbon-supported iron catalysts in catalytic wet peroxide oxidation process". *Applied Catalysis B: Environmental*. 181, 249-259.

465070 Optimization of Sorption-Enhanced Glycerol Steam Reforming Reaction Process for High-Purity Hydrogen Production

Abstract:

As global demand for environmental-friendly energy has been increasing, many studies on biodiesel were conducted. With rapid growth of biodiesel market, a large amount of glycerol which is a byproduct in biodiesel production is inevitably flowed into the global market, and now became a serious surplus problem. Conversion of glycerol to hydrogen is one of attractive ways to use an excess amount of glycerol since hydrogen is a clean energy carrier that does not produce any pollutants during energy generation. Generally, conventional conversion of glycerol to hydrogen has been conducted through catalytic glycerol steam reforming (GSR) reaction. Previous studies, however, revealed that the conversion of glycerol to hydrogen is thermodynamically limited and the purity of produced hydrogen is low, so that additional purification steps are required for direct application. Note that the thermodynamic limitation can be overcome by a new concept, sorption-enhanced reaction process, which combines reaction and separation in a single reactor. In the sorption-enhanced glycerol steam reforming (SE-GSR) reaction process, byproduct CO_2 is removed by solid adsorbent, simultaneously with GSR reaction. In this manner, glycerol conversion can be dramatically increased by Le Chatelierâ€™s principle, leading to the direct production of high-purity hydrogen. In this study, numerical simulation was carried out to elucidate the effects of various operating parameters on reaction performance and to optimize the SE-GSR reaction process. The results showed that high-purity hydrogen (>99.99%) production with good hydrogen selectivity (>99.99%) is achievable through the SE-GSR reaction process at high temperature, low pressure and high steam content.

465545 Effect of MnO_2 Catalyst on Nonthermal Plasma Reactor Combined with Ceramic Filter for Trichloroethylene Decomposition

Abstract:

This paper attempts to improve the energy efficiency and to reduce by-products generation of the authors' nonthermal plasma reactor for VOC treatment by combining with MnO_2 catalyst. In the experiments, trichloroethylene is used as a model VOC and two location of setting the MnO_2 catalyst as "inside the reactor" and "downstream of the reactor" was tried. The results demonstrated that energy efficiency to decompose TCE was improved by combining with the MnO_2 catalyst for both location conditions of "inside" and "downstream" of the reactor. As to by-product reduction, although without catalyst case, small amount of phosgene, dichloroacetylchloride (DCAC), pentachloroethane (PCE), tetrachloroethane (TCE) generation was observed, tetrachloroethane generation was under the detection limit and other by-products generation was also reduced with the MnO_2 catalyst for "downstream" case. Moreover, when the catalyst was placed "inside" the reactor, phosgene and DCAC was also under the detection limit. These results indicated the significant effect of MnO_2 catalyst to our nonthermal reactor for VOC treatment.

464010 Photocatalysis in a Packed Bed: Â Wastewater Treatment with Immobilized Nanoparticles

Abstract:

Complete photocatalytic degradation of methyl orange, methylene blue and rhodamine chloride was accomplished using immobilized silver nanoparticles as catalysts and tethered calix[7]hydroquinone molecules as nano-reactors. The large interfacial area afforded by the packed bed geometry, presentation of the nanoparticles as a single, tethered layer, and the optically transparent packing material facilitated intimate interaction between the catalyst, dye solution and radiation. The degradation was accomplished using either UV or visible radiation and the nanoparticles displayed self-cleaning and anti-poisoning properties thereby demonstrating excellent durability and catalytic efficiency. The method addresses multiple issues that constrain scale-up of nanophotocatalytic systems.

466144 Identification of Antibiotic Resistance Genes from Microbes of Water Reservoirs

Abstract:

The development and spread of antibiotic resistance among bacteria is considered a universal threat to human, animal and environmental health. The effect of clinically relevant antibiotic resistance genes and antibiotic resistant bacteria that are released from anthropogenic sources, together with the excessive use of antibiotics in both human and veterinary settings, is currently considered to be a serious environmental problem. The environmental settings (e.g. water or soil) is important on the cycling of antibiotic resistance in nature. The antibiotic resistant genes carried by microbial communities can multiply in their hosts, pass on to other bacterial populations and is subject to further evolution. Therefore, environmental antibiotic resistant genes are an abundant reservoir of potentially transferable resistance for pathogens. In this study, 239 antibiotic resistances bacteria are isolated from water reservoir samples. Antibiotic resistance profiles were examined by employing heterotrophic plate counts (HPC) as a culturing method to obtain counts of bacteria resistant to 7 different classes of antibiotics. Many of the 239 isolates had resistance to ampicillin (Î²-lactam class), vancomycin (glycopeptide class), lincomycin (lincosamide class) and trimethoprim (dihydrofolate reductase inhibitor class). Furthermore, the antibiotic resistance genes are identified by screening from genomic library of multi antibiotic resistance bacteria. The most common types of antibiotic resistant genes were efflux pumps and genes conferring resistance to vancomycin, tetracycline, or Î²-lactam antibiotics used in veterinary and human healthcare. In conclusion, distinct antibiotic resistant bacteria phylotypes and a variety of antibiotic resistant genes are present in water reservoirs, suggesting that proper water treatment approaches should be applied for safe drinking water.

465809 Source Identification for Non-Volatile Particulate Matter By Laser Derivatization

Abstract:

According to the EPA, soot from diesel sources is responsible for more than 8,300 premature deaths, 360,000 asthma attacks, 23,000 cases of bronchitis, 10,000 hospital admissions and 1.5 million lost work days in the US per year [1]. Nearly 50% of all soot from aircraft is not released at ground level but in the upper troposphere. Though constituting only ~ 1% by mass of jet engine emissions, the black carbon emission presents the largest aerosol surface area [2]. New findings show soot may be contributing to changes happening near the North Pole, such as accelerating melting of sea ice and snow and changing atmospheric temperatures [3]. Although black carbon is hydrophobic, chemical and physical changes occurring at its surface can create functional groups that attract and retain water, thereby causing it to become hydrophilic [4]. Such changes determine whether soot acts as cloud condensation nuclei, its susceptibility to washout and participation in atmospheric heterogeneous reactions, thereby significantly impacting the atmospheric radiative balance [5]. Recent estimates place black carbons' radiative forcing as ~ 0.5 to 0.8 W/m^2, nearly 50% that of CO_2 [3].

The "bottom-up" or "source model" is one of the common soot identification methods. This approach begins by identifying pollution sources and their emission strengths, which are converted to emissions (via emission factors by category) and then by utilizing meteorological patterns predicting pollution advection (movement) and compositions over time and space [6]. With an inventory of known sources such models have great potential yet presently are limited because a) a complete inventory of sources is not usually known and b) meteorological dispersion is not accurate. The "top-down" or "receptor model" approach begins by sampling air in a given area and inferring the likely pollution sources by matching common chemical and physical characteristics between source and air pollution samples [6]. Top-down methods offer the promise of quantifying the relative contributions of the different sources to ambient air pollution, where rather little may be currently known. Additionally, top-down methods may require few atmospheric measurements and relatively simple analysis.

In a regulatory world, the top-down approach is more acceptable, primarily due to the involvement of direct pollution measurements at hot spots, analyzing the samples in a lab, and determining (statistically) the contribution of various sources to the pollution at that particular spot. On the downside, the measurement points are few, due to the costs involved in monitoring and chemical analysis, which is (and can be) compensated by a comprehensive bottom-up approach to cover as many hot spots, and a better mapping of the pollution sources in the city. Nevertheless the proposed work should be applicable to source apportionment studies by either a source-model or receptor-model based approach. Both approaches are used to understand air pollution at urban, regional, and global levels [7].

The main objective of the particulate matter (PM) pollution source apportionment is to identify the potential contribution of various sources to the ambient pollution levels (indicators), which can be utilized for effective informed decision making (planning). Though the methodology focuses primarily on the PM pollution, it includes the contribution of other pollutants in the secondary form. PM and in particular, soot can serve as a tracer for other pollutants.

Soot varies in physical structure and chemical composition. Over the past two decades it has become understood that different physical and chemical processes contribute to soot nucleation versus growth, that a variety of species contribute to soot growth and that nascent fuel molecular structure and composition, timescales and temperature all affect its chemical makeup and physical nanostructure [8,9]. Soot is a complex aerosol formed under non-equilibrium conditions, highly dependent upon the particulars of the combustion process. Though there is commonality of radical reactions driving soot formation, the variability of species identities, relative concentrations and associated rates conspire to create soot with system dependent nanostructure. That soot from practical scale power generation and engine combustion varies in composition and structure should not be surprising [10,11]. That nanostructure varies is well known and quantification tools for analyzing high-resolution transmission electron microscope (HRTEM) images are well developed [12,13]. These image analysis algorithms have been vetted by comparison to traditional single parameter analytical characterization methods such as X-ray diffraction for lamella spacing [14], and Raman spectroscopy for the in-plane lattice dimension L_a [12].

In this context it has been shown that the nanostructure of soot depends upon its formation and growth conditions for both applied systems [15] and by fundamental laboratory studies [16]. We define soot nanostructure as the degree of atomic order as manifested by graphitic layer plane segments and their physical relation to each other. Different physical measures can describe nanostructure: a) lamellae length, b) lamellae curvature, and c) lamellae spacing. Soots from different sources are distinguishable based upon structure and composition, as published elsewhere[10,11,17]. Yet often differences are small and variability from real sources and large-scale facilities is high, given both the complexity and variations in the combustion process in large-scale systems. It would be advantageous to accentuate recognizable structural differences and bring out subtle chemical differences.

The implications are significant: soot nanostructure is dependent upon the source and it can be quantified. This would permit definitive resolution of the soot source and its contribution to any particular receptor site. Our initial results using HRTEM validated this approach for the soots at-hand. Yet three parameters (length, separation and curvature) is not an exhaustive dataset, nor a unique set of parameters. Fuel-rich combustion conditions producing soot offer sufficiently similar chemistries that render distinctiveness in nanostructure subtle. This, coupled with variability in image processing details related to thresholding and contrast level, and extraction of the nanostructure parameters suggests that a more robust set of parameters is needed for distinctive recognition and identification of soot based upon physical structure.

How best to bring out small differences in nanostructure and other seemingly subtle differences in chemistry? The derivitization of analytes is very important in several branches of analytical chemistry, in fact often necessary to make a measurement possible [18]. It can bring out small differences in composition; render subtle structural differences distinguishable by providing recognition for and sensitizing differences to well known analytical techniques. Unconsidered to-date is the process of derivitization as applied to combustion produced soots, let alone by pulsed laser annealing. The goal of this work is to exploit the non-equilibrium formation history of combustion soot and bring out the subtle details of chemical and physical variations that while obscured by bulk analyses, can be highlighted by laser-based derivitization. With details as dependent upon combustion conditions, subtle compositional and structural differences may be brought to light thereby distinctively and uniquely identifying the source of the soot. Thus the proposed technique is referred to as, Soot Source Identification by Laser Derivitization, (SSILD).

References

[1] U.S. Environmental Protection Agency (EPA). (2002) Health assessment document for diesel engine exhaust. Prepared by the National Center for Environmental Assessment, Washington, DC, for the Office of Transportation and Air Quality. EPA/600/8-90/057F.

[2] Aviation Aerosols, Contrails and Cirrus Clouds (A2C3), (2000) Proceedings of a European Workshop, Sachem German, July 10-12, Air Pollution Report 74, Report EUR 19428, European Commission, Brussels.

[3] J. Hansen, L. Nazarenko, Proc. Natl. Acad. Sci. U. S. A. 101 (2003) 423-428.

[4] S. Picaud, P.N.M. Hoang, S. Hamad, J.A. Mejias, S. Lago, J. Phys. Chem. B 108 (2004) 5410-5415.

[5] V. Ramanathan, P.J. Crutzen, J.T. Kiehl, D. Rosenfeld, Science 294 (2001) 2119-2124.

[6] S.K. Guttikunda, Handbook on Particulate Pollution Source Apportionment, (2008). World Bank. (www.urbanemissions.info/pmsa)

[7] S.K. Guttikunda, R.V. Kopakka, Air Qual., Atmos. Health 7 (2014) 195-207.

[8] C.S. McEnally, L.D. Pfefferle, B. Atakan, K. Kohse-Hoinghaus, Prog. Energy Combust. Sci. 32 (2006) 247-294.

[9] M. Frenklach, Phys. Chem. Chem. Phys. 4 (2002) 2028-2037.

[10] M.D. Hays, R.L. Vander Wal, Energy Fuels 21 (2007) 801-811.

[11] R.L. Vander Wal, V.M. Bryg, M.D. Hays, Anal. Chem. 83 (2011) 1924-1930.

[12] R.L. Vander Wal, A.J. Tomasek, K.W. Street, D.R. Hull, W.K. Thompson, Appl. Spectrosc. 58 (2004) 230-237.

[13] K. Yehliu, R.L. Vander Wal, A.L. Boehman, Combust. Flame 158 (2011) 1837-1851.

[14] R.L. Vander Wal, A.J. Tomasek, M.I. Pamphlet, C.D. Taylor, W.K. Thompson, J. Nanopart. Res. 6 (2004) 555-568.

[15] R.L. Vander Wal, C.J. Mueller, Energy Fuels 20 (2006) 2364-2369.

[16] R.L. Vander Wal, A.J. Tomasek, Combust. Flame 136 (2004) 129-140.

[17] R.L. Vander Wal, V.M. Bryg, M.D. Hays, J. Aerosol Sci. 41 (2010) 108-117.

[18] J.M. Rosenfeld, TrAC, Trends Anal. Chem. 22 (2003) 785-798.

466349 Measurement of Emissions of Svocs and Other Low Molecular Weight Compounds from Polymeric Materials

Abstract:

Measurement of Emissions of SVOCs and Other Low Molecular Compounds from Polymeric Materials

The emission rates of semivolatile organic compounds and other low molecular weight compounds from polymeric materials were measured with the passive flux sampling (PFS) methods. Eights kinds of widely-used polymeric materials were selected as a base material, some of which contains additives such as plasticizers. The polymeric materials used were newly-made polyethylene (PE), polypropylene (PP), poly (vinyl chloride) (PVC), poly (methyl methacrylate) (PMMA), ethylene-vinyl acetate copolymer (EVA), poly (oxy-methylene) (POM), and polycarbonate (PC). The passive flux sampler was made from a glass petri-dish of which the inner diameter was 30 mm and the height was 5 mm. Tenax GR, an adsorbent for gaseous organic compounds was put on the bottom of the petri dish. The sampler was placed on the surface of a sheet of the sample polymer for a given sampling time under various conditions; the organic compounds emitted from the surface will diffuse in the sampler and captured by the adsorbent. Then the adsorbent was taken out from the sampler, and the captured components were thermally desorbed and measured with GC/MS. The emission experiments were performed under normal temperature (75°C) and elevated temperature at 50 and 75°C, and with or without irradiation of simulated solar light.

Emission of a number of low molecular weight compounds were observed at the elevated temperature conditions and the higher temperature conditions resulted in the higher number of the chemicals and the emission rates. The chemicals emitted which can be attributed to the compounds generated by the decomposition of the base polymeric materials and additives as well as the additives themselves. Different polymeric materials showed different set of chemicals emitted. In general, the numbers of chemicals emitted form PVC sheets were much higher than those from other polymeric materials. For the long time experimental studies showed emission rates deceased with time.

Emissions of chemicals under the irradiation of artificial sunlight are similar to those under elevated temperatures mainly due to the temperature increase by the irradiation. However, emissions of some chemicals characteristic to sunlight irradiation were also observed including oxygenated substances such as benzaldehyde or hexanal.

466594 Experimental Studies on CO_2、NO_X、so_X Adsorbing Capacity of Polyaniline-Based Materials

Abstract:

Experimental studies on CO_2、NO_X、SO_X Adsorbing Capacity of Polyaniline-based materials

Liyuan Shan[1], Li Hui[1], Huang jia[1], Gao lin[1],Binglu Meng[1], Delong Xu[1], Youhai Yu[2,b,*]and Yong Min[1,a,*]

[1.] Institute of Powder Engineering, Shaanxi Techno-Institute of Recycling Economy, P.R.China

[2.]Xi'an Institute of Optics and Precision Mechanics of Chinese Academy of Science

[a] email: yong686@126.com, youhai.yu@gmail.com. *Corresponding author

Keywords: CO_2、NO_X、SO_X adsorption, PANI-based materials, composites

Abstract

Eigenstate polyaniline (PANI) were prepared by in-situ polymerization method and under the same experimental conditions, composites doped with attapulgite、vermiculite and diatomite were prepared respectively. The exist of additive materials of attapulgite、vermiculite and diatomite acted as the skeleton of PANI. The samples were characterized by FTIR、XRD、SEM、UV spectrophotometer and TGA. The results revealed that the composites which have the structure of attapulgite and the nature of the

polyaniline. The composites doped with attapulgite、vermiculite and diatomite had better excellent adsorption Capacity in 100-160 for acid gas such as CO_2、SO_2 and NO_2 under TGA than that of PANI modified with nothing. This property of PANI-based materials has a great importance significance for dealing with the fuel gas in thermal power industry to protect environment.

Acknowledgement

This work was financially supported by the Project of National Natural Science Foundation of China (NSFC Grant No. 51274159) and Special Funds for the Major Science and Technology Innovation of Shaanxi Province (No. 2012zkc06-2)

466752 Water Acidification By Atmospheric Pressure Microplasmas Operated in Air

Abstract:

Water acidification by plasmas is an important phenomenon that the pH value of water decreases upon plasma treatment. This is of great interest in plasma for biomedical application. It allows for decrease in pH without the need of chemicals. This work presents the acidification of water by atmospheric pressure microplasmas, which are plasmas with geometric dimensions less than 1 mm operated at one atmosphere. The microplasma used in this work is a dielectric-barrier-discharge-type plasma operated in air. The microplasma generation device (MGD) is made of double-sided printed circuit board with patterns defined by the toner transfer method, which is a low-cost, customized, and easy-to-fabricate process. It is generally accepted that water acidification by air plasma is a result of dissolved of NO and/or NO_2 gases generated in plasma. The acidification rate is greatly influenced by the generation, transport, and dissolution of NO and/or NO_2. Plasma operating conditions influence the generation rate, and plasma-induced flow (for example, ionic wind) can be an important factor that influences the transport and hence the dissolution. Preliminary studies are performed using AC power with an operating voltage of 2-2.5 kVp-p and a frequency of 20 kHz to ignite plasma on the MGD. When 0.002 mL of DI water droplet is sit adjacent to the plasma generation electrode of the MGD with a distance of 2 mm, the pH value decreases from 7 to 3 in 80 sec. By changing the plasma-droplet distance (2 mm-10 mm), MGD electrode geometry, and the plasma voltage, the acidification rate can be tailored, and desired pH valued can be obtained. In this work, we will test the influence of electrode arrangements and operating conditions on the acidification rate.

466195 The Experimental and Simulated Investigation on Carbon Dioxide Absorption into Aqueous Alkanoamines Aqueous Solution in the Hollow Fiber Membrane Contactor

Abstract:

The experimental and simulated investigation on carbon dioxide absorption into aqueous alkanoamines aqueous solution in the hollow fiber membrane contactor

Fan Cao[a], Ge Gao[a], Wichitpan Wong, Zhiwu Liang[a,b*], Raphael Idem[a,b]and Paitoon Tontiwachwuthikul[a,b]
[a]Joint International Center for CO_2 Capture and Storage (iCCS), ProvincialHunan Key Laboratory for Cost-effective Utilization of Fossil Fuel Aimed at Reducing CO_2 Emissions, College of Chemistry and Chemical Engineering, Hunan University, Changsha 410082, PR China

[b]Clean Energy Technologies Research Institute (CETRI), Faculty of Engineering and Applied Science

University of Regina, Regina, Saskatchewan, S4S 0A2, Canada

*** Tel.: +86-13618481627; fax: +86-731-88573033; E-mail address: zwliang@hnu.edu.cn**

Key Words:CO_2 absorption; Amine solution; Hollow fiber membrane contactor; wetting phenomenon; Membrane separation

Either nowadays or in the next decades, fossil fuels still play very important roles in energy consumption worldwide.

Ahuge number of carbon dioxide (CO_2) was discharged into atmosphere with combustion of fossil fuels, which was considered to be one of dominating greenhouse gas responsible for global warming. Accordingly, given that when confronted with increasingly severe situation of CO_2 emission, it is extremely urgent for human to take effective measures to capture CO_2 released from industrial processes, especially from coal fired power plant.

Current major research efforts for mitigating CO_2 emission have been focusing on the development of novel CO_2 separation techniques. Generally, the most common method for CO_2 capture is applied by amine-based absorption in various gas-liquid contactors including packed, bubble, and spray columns. However, aforementioned conventional mass transfer equipments usually are exposed to their inherent drawbacks considerably weakening mass transfer performance in operation process such as entrainment, foaming, flooding and channeling. What¡¯s worse, an enormous capital and operation costs will also impose restrictions on a large-scale application. In order to overcome these problems, researchers have already proposed several alternatives. Amongst varieties of alternatives, the application of hollow fiber membrane contactors (HFMCs) employed for chemical absorption into aqueous alkanolamines has been proved to be a promising alternative on account of distinguished advantages over conventional gas-liquid contactor. HFMCs are a hybrid process combing chemical absorption with membrane contactors. Porous-structured property in HFMCs guarantees a higher gas-liquid contacting area so as to improve mass transfer performance remarkably compared with conventional contactors. In addition, it is flexible for HFMCs to scale-up and save space to meet actual requirements in terms of modularity and compactness respectively. Regardless of advantages above, It is generally known that the disadvantage of membrane contactors is membrane wetting phenomenon due to the intrusion of liquid absorbent into membrane pores, which causes a sharp decline in gas absorption process due to great contribution of resistance to mass transfer within membrane to resistance to total mass transfer. Hence, numerous researchers have been devoting themselves to wetting problem of hydrophobic membrane micropores in HFMCs[1-5].

This paper mainly investigates CO_2 absorption performance into aqueous solutions of monoethanolamine (MEA), methyldiethanolamine (MDEA) Diethylaminoethanol (DEEA) and their blends using a commercial micro-porous poly(tetrafluoroethylene) (PTFE) hollow fiber membrane contactor. CO_2 absorption performance with various operational parameters inclusive of liquid flow rate, gas flow rate, CO_2 concentration, composition of solution, gas-liquid flow orientation as well as CO_2 loading was performed under non-wetted and partially wetted mode. A 2D mathematical model on the basis of finite element method was established for the CO_2 capture in the flue gas by employing COMSOL Multiphysics. The model can present concentration profiles and flux vectors of transferred species existing in the tube side, membrane, shell side respectively. When it comes to real operating conditions, the wetting phenomenon (partially wetted mode) was taken into consideration for ensuring the accuracy of the model. The experimental results revealed that the order of decarburization performance is DEEA>MDEA, the impact of the gas phase parameters on CO_2 capture and mass transfer performance is more significant than that of liquid phase parameters. The simulated results indicated the membrane pores were partially wetted in the process of gas absorption. which makes the mass transfer performance of hollow fiber membrane contactor suffer from enormous losses. Even though a very small region of membrane pore is wetted, the CO_2 absorption flux will drop significantly compared with non-wetted mode,

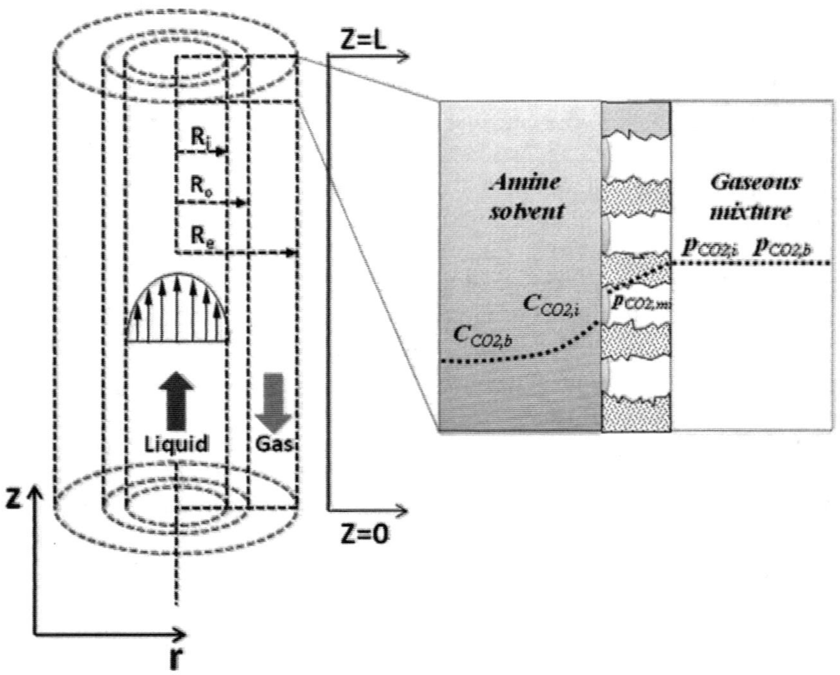

Fig.1 Scheme of Hollow Fiber

Acknowledgments

The financial supports from the National Natural Science Foundation of China (Nos. 21376067, U1362112 and 21476064), Innovative Research Team Development Plan-Ministry of Education of China(No. IRT1238), National Key Technology R&D Program (Nos. 2012BAC 26B01 and 2014BAC18B04), Specialized Research Fund for the Doctoral Program of Higher Education (No. 20130161110025), Key Project of International & Regional Scientific and Technological Cooperation of Hunan Provincial Science and Technology Plan (2014WK2037), and China¡¯s State ¡°Project 985¡± in Hunan University ¨C Novel Technology Research & Development for CO_2 capture are all gratefully acknowledged.

References

1. Dindore V Y, Brilman D W F, Geuzebroek F H, et al. Membrane¨Csolvent selection for CO_2,removal using membrane gas¨Cliquid contactors[J]. Separation & Purification Technolog 2004, 40(2):133-145.

2. Lv Y, Yu X, Tu S T, et al. Wetting of polypropylene hollow fiber membrane contactors[J]. Journal of Membrane Science, 2010, 362(s 1¨C2):444-452.

3. Mosadegh-Sedghi S, Rodrigue D, Brisson J, et al. Wetting phenomenon in membrane contactors -Causes and prevention[J]. Journal of Membrane Science, 2014, 452(452):332¨C353.

4. Rongwong W, Fan C, Liang Z, et al. Investigation of the effects of operating parameters on the local

mass transfer coefficient and membrane wetting in a membrane gas absorption process[J]. Journal of Membrane Science, 2015, 490(Supplement S1):236-246.

5. Goyal N, Suman S, Gupta S K. Mathematical modeling of CO2 separation from gaseous-mixture using a Hollow-Fiber Membrane Module: Physical mechanism and influence of partial-wetting[J]. Journal of Membrane Science, 2015, 474:64¨C82.

466654 Study upon Enhanced Removal of Cu(II), Ni(II), Pb(II), Zn(II), Mn(II), Cr(III) from Polluted Water By Self-Synthetic Struvite Treatment

Abstract:

Heavy metals existing in the water have brought much concern. The conventional hydroxide precipitation treatment is hard to achieve satisfactory removal efficiency, particularly under situation of low heavy metal concentration. The research has synthesized struvite crystals and utilized the self-synthesized struvite for the enhanced removal of six heavy metals (Cu, Ni, Pb, Zn, Mn, Cr(III)) from solution. The experiments have studied the removal efficiency under pH values from 6.0 to 10.0, and under a series of struvite addition. The research compared the performance of struvite treatment and mere hydroxide precipitation treatment as a control treatment. It was found that struvite treatment was far superior than the control treatment, with residual concentration of six heavy metals from copper to chromium at 0.01, 0.04, 0.00,0.00, 0.13, 0.004mg/L under 2g struvite addition and pH 10.0. For extremely low heavy metal concentration it achieved removal of 99.1%, 97.9%, 99.9%, 98.9%, 96.9%, and 98.3%, which was much higher than that of control treatment (10.5%, 8.7%, 13.2%, 14.1%, 6.7%, and 10.5%). Through XPS and SEM & EDS analysis, it was found the heavy metal hydroxides were precipitated on the struvite surface. Moreover, based on TEM observations, it was found the heavy metal hydroxides had a nucleation and development on the struviteâ€™s surface. The heterogeneous nucleation mechanism has been proposed for struvite treatment, which provides relationship of sorption, nucleation, and precipitation of heavy metal hydroxides on struvite surfaces and realize the quantification of the ability of struvite in enhancing heavy metal precipitate formation.

Â

466974 A First Way of Managing Humin Formation

Abstract:

In finding new ways of using a sustainable feedstock for high value chemicals, lignocellulosic biomass emerged as an excellent source. First plants for the industrial chemically catalyzed production of levulinic acid from biomass have been launched. However, a significant amount, between 40 and 60%, of widely branched polymers are produced. Those polymers, called humins, are dark colored, solid and insoluble in common solvents. So far the humins are solely used as a source of energy by being burnt directly on-site. Due to the insoluble and sticky character of the humins, the blockage of the reactor has to be avoided in a costly manner. Consequently, humins are not desired in in the process and lower both the E-factor and the profit of the process. Therefore it is of high interest to find a way to either minimize the humin formation in the process or to use the humins for the production of a new value added material. To obtain this, the kinetics, the thermodynamics and the mechanism of the humin formation as well as the structure of the humins are fundamental to know.

The focus of our research is on the extensive description of the humin formation with the use of suitable analytical methods, a deep analysis of the results and a simulation of the combined knowledge including results found in literature. The experiments are crucial for the analysis. They are planned with â€œDesign of Experimentsâ€ (DoE) and cover all known significant factors of the humin formation. A variety of analytical methods, including ATR mIR spectroscopy and elemental analysis, were used to gain insight into the structure of the humin samples. The results were combined with data from literature and give insights into the character of the mechanism, kinetic and thermodynamic of the humin formation and the structure of the product depending on the reaction conditions and sample treatment. The analysis of the results gained in scale by an extended use of van Krevelen diagrams. Therefore a new method was developed to obtain mechanistic information from the analytics. The method can find evidence for a certain mechanism or combination of more mechanisms that lead to one polymer. All information together was fed in a simulation that can be used as tool to control humin formation.

The work covers not only the control of the humin formation in amount but also a prediction of the structure and handling suggestions.

467032 Development of a Cellphone-Based Optical Emission Spectrometer for Analysis of Plasma Optical Emission

Abstract:

This work presents the development of a cellphone-based optical emission spectrometer for analysis of plasma optical emission. This spectrometer contains a slit, a grating, and utilizes the CCD camera of the cellphone to acquire spectra. The slit determines the amount of the light and defines the cut edge of the light that enters the spectrometer. The grating disperses the light into different wavelengths and greatly influences the spectral resolution of the spectrometer. The main structure of this device is constructed by a 3D-printer using polylactide (PLA) to obtain the customized shape and size. The slit is made by utility knife blade with a slit size of 0.05 to 0.2 mm. 500 and 1000 lines/mm gratings are tested. The spectral graph is acquired by taking images using the cellphone then is converted to intensity vs. wavelength plots by analyzing the image using either cellphone apps or computer-based programs such as Matlab. To calibrate the device and to quantify the full width at half maximum (FWHM) of the device, Hg-lamp is used as the standard light source. We test how device design influences the spectral resolution, the sensitivity, and signal-to-noise ratio of the device. The geometry of the current design is 26 mm * 23 mm * 27 mm with a FWHM of 17.65 nm. Furthermore, we use this device to analyze the optical emission emanating from the plasma. It is shown that with this device, major emission lines between 400 and 700 nm in Ar, He, and air plasmas are observed. Volatile organic compounds in the gas can be detected by the examination of the emission lines for carbon-containing species. We also show the device with extra accessories and designs that allow for more flexible and easier operation.

467698 Innovative Water Treatment Process Based on Electrochemical Oxidation for the Regeneration and Reuse of Water in Marine Recirculating Aquaculture Systems: Eloxiras

Abstract:

ABSTRACT

World aquaculture accounted for 90.4 million tonnes (Mt) of aquatic products in 2012, compared to 92.5 Mt reached by fishing [1, 2]. Considering the 24 Mt of fishing that will not be intended for human consumption, aquaculture is already the highest provider of food to worldwide population. Aquaculture keeps a significant growth rate that has increased by 8.9% in 2012. Asia is the largest aquaculture producer region, accounting over 88% of total production, followed by America (4.78%) and Europe (4.32%). According to FAO [3], world aquaculture production reached a market value of US$144.4 billion. Most of this aquaculture species are farmed in open-circuits, but entailing a lower process efficiency. The production trends are shifting to closed loop circuits to foster the productivity.

The economic feasibility of inland hatchery aquaculture activities requires high biomass culture densities (kg species/m^3 of water) obtained by means of the so called Recirculating Aquaculture Systems (RAS), in which water leaving the systems is recirculated after adequate treatment. Conventional treatments systems (mainly based on biofilters) are not able to provide satisfactory removal of highly concerned pollutants in RAS (ammonia, nitrite, organic matter, bacteria and viruses, etc.). This implies fast accumulation of toxic metabolized compounds in low water volumes, issue that requires large volumes of fresh water intake to avoid lethal concentrations, and the subsequent discharge of important wastewater volumes. In order to minimize the environmental impact and fulfil with law regulations, continuous efforts addressed to develop new water treatment technologies aimed to permit significant water reclamation and reuse with minimal cost.

Based on Electrochemical Oxidation Technology, and advanced oxidation process (AOP) this paper presents an innovative water treatment concept specially developed to improve the productivity and environmental impact of marine recirculating aquaculture systems (RAS) used by the exponentially growing aquaculture industry, ELOXIRAS [4]. The process allows the efficient destruction of all the highly concerned pollutants in RAS (ammonia, nitrite, organic matter, bacteria and viruses, etc.) without the addition of chemicals and with no change in water composition. It only requires electricity and the catalytic properties of the electrode materials to achieve water purification.

Removal efficacies up to 99% were validated at pilot plant scale that consisted in a 1 m^3 fish tank and a 170 cm^2 electrochemical reactor based on DSA electrodes. Post-electrochemical treatment were carried out in an activated carbon column. As an illustrated case study, the experimental conditions included 4.8 kg of gilthead bream fingerlings (3g each), a recirculating flow of 600 L h^{-1}, salinity of 33 g l^{-1}, inlet temperature of 24 ºC and initial pH of 7.4. Evolution of total ammonia nitrogen over time was kept in the range of 0-1 mg L^{-1} with variable current densities in the range of 150-450 A m^{-2} adjusted as a function of the hourly variable pollution rate from the fish metabolism. Similar behaviour was found for nitrites with values in the range of 0-0.2 mg L^{-1}. pH was kept in optimal values around 7 and electrolyte composition of the sea water was unaltered. 100% of fish survival was found after 72 h, proving nontoxic effects on the fish farmed.

ELOXIRAS introduces a new advanced marine RAS water quality control and purification system that overcomes the main limitations of the current conventional technologies (bio-filter technologies as main competitor in the market) such as: i) significant water uptake requirements; ii) limited efficacy in contaminant removal; iii) restricted culture density.

The novelty of ELOXIRAS system is the ability to increase the production of different marine species: 1) Allowing higher cultured densities within the purification controlled limits (30% more production); 2) Reducing up to 20% the new water intake consumption (and thus, proportional wastewater generation); 3) Eliminating almost completely the key pollutants (>90%), and 4) Increasing by a minimum of 30% the estimated potential benefit (â,¬/year) of the overall process.

The next innovative characteristics provide the key points for the market application of ELOXIRAS: a) High efficacy and removal of the targeted pollutants, including bacteria and virus; b) Compact and modular design; c) Low energy requirements; d) Easy to operate with adaptable capacity to required productivity. The following market applications with competitive edge have been developed based in this technology:

ELOXIRAS HYBRID focuses on new or existing large RAS facilities, offering an increment of common or actual productivity with lower water and energy use.

ELOXIRAS MINI focuses on small scale RAS facilities for final commercialization of adult aquaculture species, offering compactness and adaptable treatment capacity.

ELOXIRAS LOGISTIC focuses on typical truck or well boat transport operations from hatcheries to growing facilities, offering extended range of distances and bigger biomass capacity due to the compactness design of ELOXIRAS water treatment system that allows its installation in the trucks.

ELOXIRAS BIO focuses on quarantine and bio-security facilities of any RAS scale, offering control of pathogens (bacteria and viruses) due to the disinfection capabilities and contributing to high isolation levels.

Acknowledgements

This project has received funding from the European Unionâ€™s Horizon 2020 research and innovation programme under grant agreement No 698494. This contribution reflects only the author's view and that the Agency is not responsible for any use that may be made of the information it contains.

REFERENCES

[1] Food and Agriculture Organization of the United Nations, Statistics Division, 2016. http://faostat3.fao.org/, last accessed 6-5-2016.

[2] European Comission, Eurostat, 2016. ec.europa.eu/eurostat/, last accessed 6-5-2016.

[3] FAO, 2014.The State of World Fisheries and Aquaculture. Opportunities and Challenges. ISBN 978-92-5-108275-1; Rome.

[4] ELOXIRAS, 2016. http://www.eloxiras.com/, last accessed 6-5-2016.

468134 Environmental and Economic Efficiency Assessment of EU Manufacturing Sectors Via Input-Output Tables and Data Envelopment Analysis

Abstract:

The EU manufacturing sectors contribute significantly to its environmental footprint in a wide variety of categories, including global warming. Given the urgent need to reduce the impact of human activities in order to avoid exceeding the planetary boundaries, it is imperative to identify among these sectors the ones that contribute significantly towards the total impact but marginally in terms of economic wealth, as this information can assist in the design of more effective environmental regulations. Unfortunately, in today's globalized market, trade activities between sectors and countries makes it difficult to estimate their "true" footprint, as a wide variety of impacts are embodied in the goods and services exchanged between them.

In this article, the manufacturing EU sectors are assessed according to their impact in the global warming, acidification, and tropospheric ozone formation potential damage categories, and considering as well their contribution towards the EU's gross domestic product. To integrate all these metrics into a single meaningful score and facilitate their benchmark, we follow an approach that combines environmentally-extended multi-regional input-output tables (EEMRIO) and data envelopment analysis (DEA). EEMRIO tables are used to determine the production-based and consumption-based impact of each sector, while DEA is employed to identify efficient and inefficient sectors and establish improvement targets for those found to be inefficient.

A preliminary analysis is first carried out to identify the most polluting sectors and determine net flows of different impacts exported and imported via trade. The DEA linear programming models are next run in order to identify efficient and inefficient sectors and define targets for the latter through projection onto the efficient frontier that, if achieved, would make them efficient. Numerical results show that the most inefficient sectors are those related with minerals, metals, chemicals and petroleum products. Our approach provides valuable insight for public policy makers that seek to reduce anthropogenic impacts without sacrificing the competitiveness of the economic sectors.

469982 Effect of Regeneration Purge Gas Oxygen Impurity on Irreversible Adsorption of Volatile Organic Compounds

Abstract:

Irreversible adsorption or heel formation during cyclic adsorption/regeneration of high molecular weight volatile organic compounds (VOCs) onto activated carbon decreases its adsorption capacity and lifetime. The effect of regeneration purge gas oxygen impurity on activated carbon performance, specifically during successive adsorption/regeneration cycles was investigated. 5-cycle adsorption/regeneration tests were performed on microporous beaded activated carbon (BAC) using 1,2,4-trimethylbenzene (TMB) and 2-butoxyethanol (BE) as adsorbates. Nitrogen with different oxygen concentrations (â‰¤ 5, 625, 1250, 2500, 5000, and 10,000 ppm) was used as regeneration purge gas during thermal desorption of TMB. Cumulative heel formation increased from 0.5% to 13% as the oxygen concentration in the desorption purge gas increased from â‰¤ 5 ppm to 10,000 ppm, respectively. Thermogravimetric analysis of the regenerated samples showed extensive chemisorption of TMB when exposed to â‰¥ 625 ppm oxygen in the purge gas. Regeneration of BAC performed with the highest oxygen concentration in the purge gas (10,000 ppm) showed negligible heel formation for adsorbed BE (1.3%) compared to TMB (13%). The results suggest that the effect of regeneration purge gas oxygen impurities on the irreversible adsorption of VOCs is dependent on the nature of the adsorbate- likely its tendency to react with oxygen. The results from this study help explain the heel formation mechanism and how it relates to regeneration purge gas purity.

470026 Cellulose Acetate/TiO_2-Au Nanocomposite Reverse Osmosis Membrane for Water Purification and Desalination

Abstract:

Water is an integral part of our society and it is of the utmost importance that this water is clean and safe for portable use. Membrane technologies are regarded as some of the most prevailing devices to mitigate water challenges around the world. Here, an electrospun cellulose acetate (CA)/TiO_2-Au nanocomposite reverse osmosis membrane was developed for desalination and purification of water to overcome some of the main hurdles of membrane technology, such as waste generation, fouling, and requiring pressure. CA is used as the electrospun nanofiber scaffold of the membrane, which shows a great NaCl rejection ability. Nanoporous TiO_2/Au particles are prepared via a soft-templating technique, and then loaded on the CA nanofibers, which can provide heavy metals, hardness, sulfates, chlorides, and bacteria removal from water.

466971 Development of a Portable and Low Cost Atmospheric Pressure Microplasma Generation Device Driven By Mobile Power Pack

Abstract:

This work presents the development of a portable plasma system that integrates a microplasma generation device (MGD), a high voltage module, and a mobile power pack. The MGD is made of double-sided copper laminate with the 0.050-mm-thick FR-4 glass epoxy as the dielectric layer. The electrode patterns on this MGD are defined by the toner-transfer method, which enables the spatial resolution down to 0.150 mm and without the need of cleanroom facilitates. The MGD itself is low cost and disposable. MGD with different designs can be easily replaced and installed on this portable plasma system. The high voltage module is a printed circuit board that contains an oscillating circuit that converts DC to AC and a transformer that raises the voltage up to 168 times. The mobile power pack outputs 5 V DC and drives the high voltage module that gives a high voltage output: up to 2.5 kVp-p with a frequency of 20 kHz. The portable plasma system is low cost (<30 USD/system), compact (<70 mm * 40 mm * 40 mm), and portable, as opposed to conventional plasmas that are expensive and bulky. The system allows for generation of microplasmas for tens of minutes. We test several designs of the high voltage module and the MGD. Electrical and optical analyses are performed to obtain the system characteristics of each design with the goal of obtained a stable and controllable plasma with low power consumption. It will be demonstrated that by analyzing the optical emission spectrum emanating from this portable plasma system; ppm level volatile organic compounds in the gas phase can be detected.

Evaluation of Microorganism Resistant to Heavy Metal Isolated from Contaminated Water and Soil Used in the Agriculture Sector in Ecuador

Bertha E. Ibarra L. [1], *Marco E. Castillo*[1], *Daniel Reinoso*[1] *and Hernan Zurita*[1]
[1]*Faculty of Agriculture, Technical University of Ambato, Ambato, Ecuador*
elizabethibarra_che@yahoo.com

Abstract

Research was conducted to evaluate microorganism isolated from contaminated water and soil from an agricultural sector in Ecuador. The study was done in Cunchibamba were horticultural crop grows and the irrigation is taken from the water canal Latacunga-Salcedo-Ambato. This canal is polluted, most of its water is residual water which are effluents of hospitals, agricultural sector and factories; a small quantity of water comes from the snowy mountain, Cotopaxi. There is a need to have clean water and improve crop production. In this study, microorganisms collected from the water canal were isolated and purified. A total of forty bacteria were evaluated, gram staining and resistance assays were done to the strains. Solutions of mercury (II) oxide (HgO), chromium (III) oxide (Cr_2O_3), copper (II) sulfate ($CuSO_4$), zinc chloride ($ZnCl_2$), nickel (II) sulfate ($NiSO_4$), cadmium chloride ($CdCl_2$) were used to determine the bacteria that was resistant to metal. It was determined that the bacteria had similar growth and metal resistance for the two methods used, strains were resistant to mercury (Hg), Chromium(Cr) and Zinc (Zn) for the liquid and solid assays. Future studies include variation of metals concentration, pH resistance assays and identification of bacteria applying biomolecular techniques.

Keywords: bioremediation, salts, heavy metal, gram staining, strains

Introduction

Water is used in agriculture to increase crops production, contributing to rainfed and irrigated production. The majority of food is produced from soil moisture coming from precipitation (1). An adequate water supply is needed for irrigated crops where there is a requirement of water to be of usable quality. The quality of water supply can be defined by physical, chemical and biological characteristics (2).

Irrigation covers 20% of all cultivated land and accounts for about 40% of agricultural production (3). The pollution of water will affect water's demand and food production causing problems particularly in developing countries.

There are different forms in which water can be contaminated. For example, industrial effluents can carry different micronutrients (4). Trace elements occur in almost all water supplies at very low concentrations, usually less than a few milligrams per liter (mg/l) being the majority of less than 100 micrograms per liter (μg/l) (1). However, excessive quantities of trace elements will cause undesirable accumulations in plant tissue and reduction of the plant's growth. There have been few field experiments from which toxic limits could be established, especially for irrigation water. However, research dealing with disposal of wastewater has gained sufficient understanding to be useful in identifying the limitations caused by trace elements (1).

The use of microorganism to destroy or reduce the concentration of hazardous wastes on contaminated site is called bioremediation. Such biological treatment system has various applications, including, clean-up of contaminated sites such as water, soils and sludge (5). Microorganisms can be isolated from different environmental conditions. Microbes can adapt and grow at subzero temperatures, extreme heat, desert conditions, in water with an excess of oxygen, and in anaerobic condition, with the presence of hazardous compunds or any waste stream. The main requirements for microbes to be adapted are energy and carbon sources. Because of the adaptability of microbes and other biological systems, they can be used to degrade or remidiate environmetal hazards (6).

Microbes play key geoactive roles in the biosphere such as, biotransformations and biogeochemical cycling, metal and mineral transformation, decomposition, bioweathering and soil and sediment formation. Prokaryotes and eukaryotes can contribute actively to the geological phenomena. Microbes have a variety of properties that can affect changes in metal speciation, toxicity and mobility, these mechanisms are important components of natural biogeochemical cycles for metals as well as associated elements in biomass, soil, rocks and minerals (7).

This research focuses on the evaluation of microorganism taken from a water canal used for irrigation, and production of vegetables and fruits in Latacunga, Ecuador. This region supports a huge demand of crops. However, some of the water and soil used is contaminated by heavy metal. There is a need to find environmentally friendly techniques that can increase the production and quality of crops in order to meet growing water requirements from municipalities, industry and agriculture. The evaluation and identification of microorganisms will be the step to recover polluted zones and search for better remediation techniques. Ecuador is a country located in South America; it has an area of 283 520km^2 (8). The population of Ecuador is around 16 millions of habitants (8). The advantages of the geographical location of Ecuador are its agricultural and agro industrial production. This country is ideal for growing a wide of variety of fruits and vegetables. Several rivers, snowy mountains and paramo surround Latacunga. This province is recognized because of its industrial sector.

The project consisted on collecting water and soil samples contaminated by heavy metals from different locations in the canal Latacunga – Salcedo - Ambato. Selective media was prepared to isolate bacteria and fungi microorganism; Microorganisms were purified and bacteria were selected for resistance studies to heavy metals and microorganism characterization. Gram staining test, macroscopic and microscopic identification was prepared to the strains by Gram's Method (9). Resistance test to heavy metal was evaluated in liquid and solid phase (10). Results were analyzed and selected bacteria resist to a specific heavy metals. Further studies consist in evaluating the resistance to heavy metal in different range of concentration of the reactant.

Materials and Methods

Isolation and purification of Samples

Bacteria was isolated from samples of soil and water; each sample was diluted in distilled water. Microorganism were isolated in macconke agar (MCA), endo agar (ENDO), levine eosine methylene blue agar (LEM), letheen agar (LA), simmons citrate agar (SCA), ctr agar (CTR), malt extrac agar (ME), manitol salt agar (MSA), xld agar (XLD), tcbs agar (TCBS), mr-vp broth (MR), urea agar base (UA); Plates were incubated a temperature of 30°C for 48h.

Microorganism were selected and purified in potato dextrose agar (PDA), plates were incubated a temperature of 30°C for 72 h and stored. A total of 40 bacteria were purified and conform a bank of microorganisms of the faculty.

Gram Staining Test, Macroscopic and Microscopic Characterization

Gram staining tests were done by Gram's method. A drop of distilled water was suspended in a slide. Then a drop of the suspended culture to be examined was transferred to a slide with an inoculation loop. Crystal violet stain was added to the fixed culture and it was allowed to stand for a few seconds. Iodide was added for rapid discoloration and then ethanol acetone was added; the reactants were placed there for 60 seconds (9). The slides were examined under a microscope for macroscopic and microscopic characterization. Filamentous and pleomorphic forms were observed and strains were classified as Gram positive and Negative.

Resistance test to heavy metals

The resistance of bacteria was analyzed at a solid and liquid phase, at a concentration of 2mM of heavy metal like, mercury (Hg), chromium (Cr), cooper (Cu), zinc (Zn), nickel (Ni), and cadmium (Cd).

Reactants used to establish the concentration were: mercury (II) oxide (HgO), chromium (III) oxide (Cr_2O_3), copper (II) sulfate ($CuSO_4$), zinc chloride ($ZnCl_2$), nickel (II) sulfate ($NiSO_4$), cadmium chloride ($CdCl_2$), the solutions were stored in a glass bottle. The resistance to heavy metals was evaluated in the liquid and solid phase.

Liquid phase assay

Nutrient broth was prepared and bacteria was incubated for 24 hours at 30°C. To determine bacteria growth, spectrophotometry (ThermoSpectronic) was used and the absorbance was measured at a specific wavelength OD_{595nm}. The resistance of bacteria to heavy metal was calculated in the solutions of control and heavy with bacteria (10).

Solid phase assay

Difco™ Potato Dextrose Agar was prepared and heavy metal was added to the plate; the final concentration of the sample was 2mM. Microorganisms were inoculated for 24 hours, 48 hours at 30°C. At every 24 hours, growth of the strain was visualized to determine the resistance to heavy metal.

Results and Discussion

Gram staining, macroscopic, microscopic and resistance to heavy metal assays were done to the forty strains located in the bank of the Faculty of Agricultural Science in Ecuador in the Technical University of Ambato.

Most of the strain were cocci and 80% were Gram Negative, Table 1. shows the results of Gram Staining suggesting that gram negative bacteria are beneficial to remediation processes since it degrades heavy metals. Studies conducted by Shankar shows that cocci bacteria isolated from contaminated areas represent a good range of tolerance for bioremediation process contaminated with heavy metal for soil and water (11).

Table 1. Gram Staining Test

Gram Identification	Total of bacteria (%)
G-	80
G+	20

Both methods solid and liquid assays, show approximately the same results, most of the resistant metals were Hg, Cr and Zn. Figure 1 shows the percentage of bacteria resistant to heavy metal. Solid Assay shows that 82 % resisted mercury (Hg), 80% resisted chromium (Cr), 80% resisted zinc (Zn), 55% resisted cupper (Cu), 47% resisted cadmium (Cd), 20% resisted nickel (Ni). For the liquid assay 80% resisted mercury (Hg), 72% resisted zinc (Zn), 40% resisted chromium (Cr), 30% resisted cadmium (Cd), 22% resisted cupper (Cu) and 22% resisted Nickel (Ni). Leung W. et al (2000), describes the removal of five heavy metals such as copper, nickel, lead and chromium which can be removed using gram negative and cocci bacteria. Pseudomonas and micrococcus are capable of removing significant amounts of copper and lead, Langmuir adsorption isotherms shows the removal of copper and lead (12).

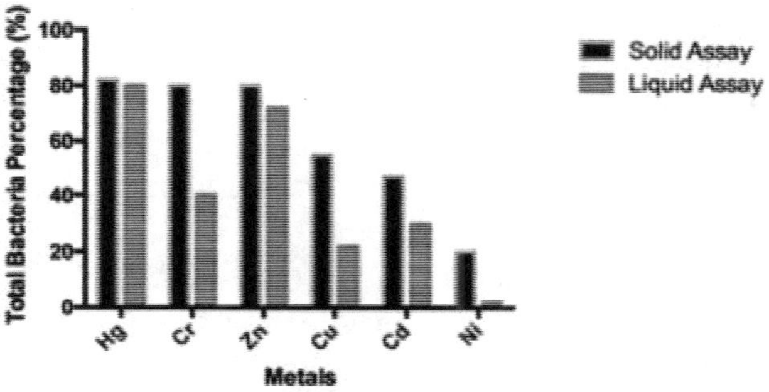

Figure 1. Resistance to heavy metal by liquid and solid assays

Conclusions

Clean water and its quality is important in crop production. Bioremediation is a technique that can be applied to remediate heavy metals and incorporate microorganism to soil and water and will benefit crop production. In this study, a total of forty bacteria were purified from a water and soil contaminated with heavy metal in Latacunga, Ecuador. By using gram staining and solid-liquid resistance assays, it was found that most of the strains are gram negative and resist to mercury (Hg), Chromium(Cr) and Zinc (Zn). Future studies include the evaluation of resistance to heavy metal with different concentration, pH variations and identification of bacteria applying biomolecular techniques.

References

1. Mark W. Rosegrant, Claudia Ringler, and Tingju Zhu. (2009). "Water for Agriculture: Maintaining Food Security under Growing Scarcity." *Annual Review of Environment and Resources* 205 – 222
2. Ayers R.S., and D.W. Westcot. (1994). "Water quality for agriculture ." *FAO Corporate Document Repository* 1 - 29.

3. Molden D. (2007). "Water for Food, Water for Life: A Comprehensive Assessment of Water Management in Agriculture." London/Colombo, Sri Lanka: Earthscan/Int. *Water Manag. Inst.* 1 - 20.
4. Abbas ST, Sarfraz M, Mehdi SM, Hassan G, Obaid Ur Rehman. n.d. (2007) "Trace elements accumulation in soil and rice plants irrigated with the contaminated water." *Soil and Tillage Research* 503 - 509.
5. Boothoathy R. (2000). "Factors Limiting bioremediation Technologies." *Biosource Technology* 63 - 67.
6. Vidali. M . (2001). "Bioremediation. An overview ." *Pure and Applied Chemistry* 1163 – 1172.
7. Gadd, Geoffrey Michael. (2010). "Metals, minerals and microbes: geomicrobiology and bioremediation ." *Microbiology* 609-643.
8. Ecuador, South America. (2016). *Central Intelligence agency.* september 16. Accessed september 24, (2016). https://www.cia.gov/library/publications/the-world-factbook/geos/ec.html.
9. Hucker G, and Conn H. 1(923). "Methods of Gram Staining". *New York Agricultural Experiment Station* (The Station) 93: 1 - 37.
10. Serrano Bueno G, Hernandez A, Lopez Lluch G, Perez Castiñeira J, Navas P, Serrano A. (2012). "Inorganic Pyrophosphatase Defects Lead to Cell Cycle Arrest and Autophagic Cell Death through NAD Depletion in Fermenting Yeast." *The Journal of Biological Chemistry* 13082–13092.
11. Congeevaram S, Dhanarani S, Park J, Dexilin M, Thamaraiselvi K. (2007). "Biosorption of chromium and nickel by heavy metal resistant fungal and bacterial isolates." *Journal of Hazardous Materials* 270 - 277.
12. Leung W.C, Wong M.F, Chua H, Lo W, Yu PHF, Leung C.K. (2000). "Removal and recovery of heavy metals by bacteria isolated from activated sludge treating industrial effluents and municipal wastewater." *Water Science and Technology* 233 - 240.

470271 Performance of a Multistage Fluidized Bed Adsorber Using Polymeric Adsorbent to Capture Volatile Organic Compounds

Abstract:

Fluidized bed adsorbers are particularly suited for treating large flow rates of dilute air streams such as volatile organic compounds (VOC) laden air from automotive and other industrial painting operations. This research investigates the effect of different operational parameters on the performance of a multistage fluidized bed adsorber for adsorption of a VOC (1,2,4-trimethylbenzene) on a mesoporous polymeric adsorbent (DOWEX OPTIPORE). A pilot-scale six-stage fluidized bed adsorber was tested to assess the effect of the adsorber operating conditions on VOC removal efficiency. The results showed that adsorbent feed rate, VOC inlet concentration, and air flow rate had the most effect on adsorber performance. On the other hand, the inlet stream humidity had negligible effect on the adsorber removal efficiency and enhanced particle fluidization due to reduced electrostatic charge buildup and associated particle agglomeration. The results help determine optimal conditions for operating a fluidized bed adsorber using this adsorbent.

470829 Integrated Environmental Assessment and Economic Optimization of Chemical Process Plants Based on Knowledge Management

Abstract:

Process industry is increasingly committed to decrease the environmental impact of its processes while, maximizing its economic performance. Industrial processes, encompassing production, marketing, sales and supply activities, usually interact with each other and share a large amount of data and information. Therefore, the integration of environmental assessment and economic performance optimization is a highly challenging objective. The effective integration of environmental and economic issues within the enterprise structure can only be achieved improving information sharing and communication. A promising solution approach stems from the application of knowledge management technologies. Several works in the literature consider environmental assessment within enterprise frameworks (1,2); however, none of them combines the potential of integrating environmental and economic issues with available enterprise data systems and analytical methods. Therefore, the representation of the life cycle inventory within a knowledge management model will provide decision makers a framework to assess and evaluate the environmental and economic impact of the enterprise activities. The aim of this work is to provide process analytical models with the necessary data and information related to the life cycle assessment, in order to optimize and evaluate them from an environmental and economic perspective thus supporting decision-making. The scope of these decisions encompasses the operational, tactical and strategic levels. Therefore, a semantic model approach representing an integrated enterprise framework and considering the different enterprise decision levels is proposed. A semantic model is developed using a systematic methodology for ontology development [3]. As a result, more environmentally conscious production processes can be obtained. The environmental performance is calculated using different environmental performance indicators. The assessment of these indicators is based on the resources consumption (energetic, raw material, human, etc.) and pollutants generation of products and processes.

Several examples related to waste-to-energy systems in the industry have been studied since it stands as an opportunity

for providing decision makers with new technologies to assess and evaluate the plant performance. The framework supports information management based on a semantic representation of the whole process and its waste treatment, namely its operational and logistics functions, and assesses the potential of energy generation and savings. It is demonstrated that the ontological framework can adapt and recognize the different elements found through the hierarchy levels associated with the enterprise functions and relate them to their environmental and economic impact. As a result, it allows assessing the economic and environmental performance of the enterprises by communicating process life cycle data to analytical models.

As a whole, the main contribution of this work lies in a higher efficiency in communication and coordination procedures within the enterprise in order to assess and optimize its environmental and economic performance.

471327 Analysis of Multivariate Statistical Methods Performance Applied to Tropospheric Ozone Level Prediction in the Metropolitan Area of SÃo Paulo

Abstract:

According to the World Health Organization (WHO), in 2012, around 7 million people died due to exposure to atmospheric pollution. That is equivalent to one in eight of total global deaths. Because of that, air pollution is now considered the largest single environmental health risk (WHO, 2014).

For this reason, the prediction of critical episodes of air pollution is a growing concern to authorities, especially in large urban centers, where there is high concentration of people and, therefore, high levels of air pollutants emission.

Tropospheric ozone is one of the most concerning pollutants. An acute exposure of humans to high levels of ozone may lead to biochemical and cellular changes in the lung (Devlin et al., 1991). Ozone is also capable of impacting forest trees in many ways, decreasing growth and productivity (Karnosky 2006).

The tropospheric ozone formation is a result of Volatile Organic Compounds (VOCs) oxidation and photolysis of oxides of nitrogen (NOx). The chemical complexity of atmospheric VOC oxidations is illustrated by the Harwell photochemical trajectory model of Derwent and Jenkin, which contains 384 chemical species and 684 chemical reactions (Finlayson-Pitts and Pitts Jr., 2012).

Additional complexity for ozone formation modeling is given by the fact that ozone precursors, VOCs and NOx, are commonly released to the atmosphere by mobile sources, such as cars and trucks, which cannot be monitored, and the dispersion of these reactants depends on the local wind velocity and direction.

The air quality in large urban centers is affected by a high amount of variables that represent the meteorological conditions and the emission sources. Therefore, model results are highly dependent on the local conditions for each specific region of interest. The existing deterministic models for ozone levels prediction in the lower atmosphere have limitations, given that all chemical and physical phenomena involved in the production and dispersion of pollutants in the urban atmosphere are not yet completely understood.

The proposed models are based on multivariate statistic methods, in the field of Statistical Machine Learning, such as Neural Networks, Random Forests and Discriminant Analysis (Pavón-Domingues, Jiménez-Hornero & Gutiérres de Ravé, 2014) (Burrows et al., 1994) (Guardani et al., 2003).

Neural Networks and Random Forests have been particularly effective in the capture of the complex and non linear relations between meteorological variables and air pollutant concentrations (Arhami, Kamali, & Rajabi, 2013) (Ochando et al., 2015) (Borges & Guardani, 2012).

This study is an extension of previous studies carried out at the University of São Paulo and CETESB, the São Paulo State Environmental Protection Agency (Guardani et al., 2003, Borges et al., 2012), and consists of a comparison of the performance of statistical models, such as Neural Networks and Random Forests, for the prediction of levels of tropospheric ozone in the São Paulo Metropolitan Area (SPMA), which is characterized by high concentration of inhabitants and intense economical activity, and where the air quality is particularly affected by episodes of high ozone levels.

The study focused on the prediction of daily maximum levels of ozone based on hourly values of meteorological variables such as Temperature, Pressure, Relative Humidity and Wind Velocity in two directions, North-South and East-West.

Two different sets of features were used during the tests: 1) the morning and afternoon averages for each of the variables; 2) The maximum daily values of Temperature, the minimum daily values of Pressure and Relative Humidity and the averages of Wind velocity in two directions. The selection of features affected the performance of the models.

Temporal curves of predicted values presented a good agreement with the observed values of maximum ozone levels, both for Neural Networks and Random Forests. However, the highest values of ozone concentration were underestimated by the predictors, indicating that different interactions among the factors exist in these scenarios.

REFERENCES

Arhami, M., Kamali, N., & Rajabi, M. M. (2013). Predicting hourly air pollutant levels usin artificial neural networks coupled with uncertainty analysis by Monte Carlo simulations

Borges, A. S., Andrade, M. F., & Guardani, R. (2012). Ground level ozone prediction using a neural network model based on meteorological variables and applied to the metropolitan area of São Paulo. *Int. J. Environment and Pollution* .

Burrows, W. R., Benjamin, M., Beauchamp, S., Lord, E. R., McCollor, D., & Thompson, B. (1994). CART Decision-Tree Statistical Analysis and Prediction of Summer Season Maximum Surface Ozone for the Vancouver, Montreal, and Atlantic Regions of Canada. *Journal of Applied Meteorology* , pp. 1848-1862.

Karnosky, D. F., Skelly, J. M., Percy, K. E., Chappelka, A. H. (2006). Perspectives regarding 50 years of research on effects of tropospheric ozone air pollution on US forests. *Elsevier. Environmental Pollution 147, pp. 489-506.*

Devlin, B. R., McDonnell, W. F., Mann, R., Becker, S., House, D. E., Schreinemachers, D., Hillel S. K. (1991). Exposure of Humans to Ambient Levels of Ozone for 6.6 Hours Causes Cellular and Biochemical Changes in the Lung. *American Journal of Respiratory Cell and Molecular Biology. Vol. 4. pp. 72-81.*

Finlayson-Pitts, B.J. , Pitts Jr, J.N. (1993). Atmospheric Chemistry of Tropospheric Ozone Formation: Scientific and Regulatory Implications. *Air & Waste, 43:8, 1091-1100.*

Guardani, R., Aguiar, J. L., Nascimento, C. A., Lacava, C. I., & Yanagi, Y. (2003). Ground-Level Ozone Mapping in Large Urban Areas Using Multivariate Statistica Analysis: Application to the São Paulo Metropolitan Area. *J. of Air & Waste Manage. Assoc.*

Guardani, R., Nascimento, C., Guardani, M., Martins, M., & Romano, J. (1999). Study of Atmospheric Ozone Formation by Means of a Neural Network-Based Model. *Journal of the Air & Waste Management Association* .

WHO. (2014). *World Health Organization*. Acesso em 31 de 05 de 2014, disponível em http://www.who.int/mediacentre/news/releases/2014/air-pollution/en/

Pavón-Domingues, P., Jiménez-Hornero, F. J., & Gutiérres de Ravé, E. (2014). Proposal for estimating ground-level ozone concentrations at urban areas based on multivariate statistical methods. *Elsevier.*

Ochando, L. C., Julián, C. I. F., Ochando, F. C., Ferri, C. (2015). Airvlc: An application for real-time forecasting urban air pollution.

471517 Sorptive Removal of Cationic – and Anionic Dyes By the Acid Activated Red Mud from Aqueous Solutions

Abstract:

Sorptive Removal of Cationic – and Anionic Dyes by the Acid Activated Red Mud from Aqueous Solutions

Sinem Ortaboy[1,2], Elif Turker Acar[1], Elif Tuzun[1], Gulten Atun[1]

[1]Chemistry, Istanbul University, Engineering Faculty, Istanbul, Turkey

[2]University of California at Berkeley, Berkeley, CA, United States of America

In recent years, many attempts have been made to use bauxite waste red mud (RM) for wastewater treatment to remove both cationic[1,2] and anionic colorants[3]. It has been shown that acid treatment improves surface properties of the RM sorbents.[1]

In this study, the RM was treated with HCl according to Shiao and Akashi's procedure before use in sorption experiments.[4] The diffusion coefficients of the MB and NB on the RM were estimated as 6.4×10^{-11} and 4.0×10^{-11} m^2/s, respectively, by using the Vermeulen approximation based on homogeneous surface diffusion.

Experimental equilibrium data obtained in the initial concentration range of 0.5 – 20 mmol/L were fitted to the Langmuir and Freundlich isotherm models. The monolayer capacities for MB increased from 0.019 to 0.056 mmol/g in the temperature range of 288–318 K. Thermodynamic calculations showed that spontaneity of the endothermic sorption process was entropy controlled. The sorption capacity of ~0.021

mmol/g for NB was not influenced by temperature but the sorption equilibrium constants were significantly decreased indicating that sorption process was exothermic in nature. The pH change during the sorption process showed that NB^- anions are adsorbed on positively charged surface hydroxyl groups while sorption of MB^+ cations takes place on the negatively charged surface sites. The diffuse reflectance infrared Fourier transforms (DRIFT) spectral analysis of the bare and dye loaded RMs were also used to elucidate sorption mechanisms.

REFERENCES

1. Shaobin Wang, Y. Boyjoo, A. Choueib and Z.H. Zhu. Removal of dyes from aqueous solution using fly ash and red mud. ***Water Research*** 39 (2005) 129–138.

2. V.K. Gupta, Suhas, Imran Ali, and V. K. Saini Removal of Rhodamine B, Fast Green, and Methylene Blue from Wastewater Using Red Mud, an Aluminum Industry Waste ***Ind. Eng. Chem. Res.***, ***43, (***2004) 1740-1747.

3. C. Namasivayam and D.J.S.E. Arasi. removal of congo red from wastewater by adsorption onto waste red mud. ***Chemosphere,*** 34 (2) (1997) 401-417.

4. S.J. Shiao and K. Akashi, "Phosphate Removal from Aqueous Solution from Activated Red Mud," J. Water Pollution Control Federation, 49 (2) (1997) 280-285.

471214 Effect of Surface Reactions in the Electrokinetic-Based Cleaning of Contaminated Soils

Abstract:

Electrokinetic-based coupled with hydrodynamic-based driving forces are used in the process of cleaning contaminated soils by electrokinetic-centered remediation. This type of remediation utilizes a relative low-voltage direct current (DC) electric field (EF) as its key cleaning-mechanism. Depending of operating and soil conditions, Joule heating, combined with heat of reaction are produced and, therefore, the temperature increases could impact the cleaning efficiency. In general, soils are non-uniform environments and thus we simulate this situation by a pore domain with non-symmetrical boundaries. The model uses a rectangular geometry (channel) that is resting on an incline position at a given angle. The channel is bounded by a surface reaction on one side while heat is being released from the opposite side at a different rate. In order to understand the possible impact of the temperature increase on the hydrodynamics of the system, a microscopic-scale model has been formulated. The mathematical model developed uses parameters such as the Nusselt number, the heat of reaction number, and Grashof number to describe the system behavior in the temperature/velocity profiles. The presentation will include details about the model formulation and illustrative results from the model predictions.

471972 Impact of Meteorology Datasets on Near Roadway Dispersion Model Estimates

Abstract:

More than 19% of the United States population lives near high traffic roadways where they are exposed to elevated near road pollutant concentrations. Vehicular emissions are one of the primary sources of air pollution in cities and associated with elevated morbidity and mortality rates in individuals living near roadways. Currently available near-roadway dispersion models use meteorological data from local weather stations, potentially limiting their implementation due to the unavailability of data in many areas. In order to overcome this limitation, process-based meteorological estimates from models such as the Weather Research and Forecasting (WRF) model are used as a substitute. This raises the question of how model estimates vary between process-based meteorological estimates and station data.

In this study we employ a Gaussian plume dispersion model, R-LINE, to simulate near road concentrations using both station data and Weather Research and Forecasting model (WRF) estimates. R-LINE simulates the dispersion from line sources by numerically integrating point source emissions along multiple road configurations. We evaluate the seasonal and diurnal variability of roadway dispersion for both of the meteorological input datasets and compare the model estimates. As we found fairly substantial differences in model estimates for the two input datasets, we also explore R-LINEâ€™s sensitivity to different meteorological parameters in both stable and unstable atmospheric conditions for multiple locations in Connecticut. This study illustrates and quantifies how R-LINEâ€™s estimates vary based on the use of different sources of meteorological inputs.

473018 Removal of Taste and Odor Compounds with Cactus Mucilage Beads Adsorption and Advance Oxidization: Scale up Applications for Continues Recirculating Aquaculture Systems

Abstract:

Recirculating aquaculture systems (RAS) are gaining popularity due to their inherited water savings compared conventional aquaculture. In RAS, the water is treated and recirculated, reducing fresh water inputs and volume of wastewater discharges. In addition, they provide more environmental control for fish growth, which in turn can result in higher production rates. However, one of the disadvantages of RAS is the production of off flavor compounds. These compounds, Geosmin (GSM) and 2-methylisoborneol (MIB), are secondary metabolites created by cyanobacteria and some actinomycetes. Such compounds induce an earthy musty flavor in fish that can be detected by the human senses at extremely low concentrations. One of the most common processes to remove such compounds involves purging them with fresh water. However, this technique requires large amounts of treated water or treating the water onsite. Removal of GSM and MIB can be achieved by conventional adsorption treatments with activated carbon.

Cactus mucilage is a biomaterial that used as food source in some countries and it is renewable, biodegradable, easy access and low cost. Previous research has found mucilage extraction can be used as dispersant in different cases, such as oil spills. Different concentrations and types of mucilage extraction can be combined into Alginate beads that are proved to be effective for removal of GSM and MIB due to adsorption mechanism. Further, modified UVis-TiO_2 photo-catalysis oxidization can be combined with the mucilage beads column to keep enhancing the total removal effectiveness of such odor compounds. It is found this combined system can possibly achieve more than 90 percent total removal of MIB and GSM, which may significantly benefit the fish production and market.

In the end, the system will be scaled up and tested in the Mote Marine Laboratory & Aquarium in order to further demonstrate its efficiency in a pilot unit for treating the wastewater from their fish production.

472725 Ozonation of Solid Wastes from Vegetable Processing Plants for Biogas Production

Abstract:

Many industries produce tremendous amounts of waste and wastewaters containing organic materials (solids and soluble substrates) that require treatment prior to disposal (often as environmental discharges). Example facilities include food processors, confined animal raising operations, slaughter houses, breweries, and food preparation operations. These organic wastes, especially the liquid wastes, are amenable to microbial digestion producing a gaseous product containing mainly methane and carbon dioxide (aka biogas). The solid wastes, on the other hand, usually require pretreatments to enhance their biodegradation. This study will focus on the ozonation as a pretreatment strategy to enhance the digestion of solid wastes from vegetable processing facilities. Different variables including solid to liquid ratio, ozone dose, and ozonation time will be studied. Initial results indicated a 3-fold increase in sugar (glucose, xylose) content of the sample after ozonation. The pretreated solid wastes will be subjecte to anaerobic digestion to determine the treatment conditions that will result to an optimum biogas production.

472997 Cogeneration of Electricity, Hydrogen and Carbon-Containing Chemicals from Natural Gas Coalsupplies

Abstract:

A no-carbon dioxide-emitting process flowsheet is presented, in which electric power, hydrogen and formic acid are coproduced from natural gas and coal. The principal technologies employed in the process network include combustion, steam methane reforming (SMR), pressure swing adsorption, and formic acid production from CO2 and H2. The concept of thermodynamic self-sufficiency is employed to access a feasible envelope for the considered process.

Further thermodynamic analysis will be carried out, so that we can find operating limits for the proposed process. Here I employ UniSim to describe flowsheet and simulate the process. Heat and power integration studies show this flowsheet to be energetically self-sufficient through the use of heat engine and heat pump subnetworks.

473031 Simulation of Atmospheric Cr Speciation in Droplets

Abstract:

The two chromium oxidation states found in ambient atmospheric particulate matter are trivalent [Cr(III)] and hexavalent [Cr(VI)] chromium. Cr(III) is a trace element essential for the proper function of living organisms. However, Cr(VI) is toxic and exposure to Cr(VI) may lead to cancer, asthma and bronchitis. Therefore, it is important to accurately discriminate between these two species in atmospheric particulate matter (PM). Little is known regarding atmospheric Cr chemistry. It was found that atmospheric Cr(III) is oxidized to Cr(VI) in the presence of dissolved ozone in liquid coated particles or droplets or even at dry conditions, while Cr(VI) is reduced to Cr(III) in the presence of organic carbons and metal ions such as Fe(II), As (III) and V(II) at high humidity. This inter-conversion between Cr(III) and Cr(VI) results in some biases in measuring atmospheric Cr concentrations. Therefore, it is critical to estimate the interconversion rate of Cr(III) and Cr(VI) to be able to quantify Cr(VI) concentrations in the ambient air. This study focuses on the simulation of interconversion of atmospheric chromium in the atmospheric droplets in the presence of PM matrix. Recent field measurements of concentrations of atmospheric Cr and other reductants and oxidants were used as model input. The simulation results demonstrated that the major form of Cr(VI) in the atmosphere is insoluble Cr(VI) which is in agreement with the literature findings. Insoluble Cr(VI) forms quickly within few seconds while the system reach steady state after approximately 15 hours. No insoluble form of Cr(III) was found when initial Cr(III) concentration was zero, and small amount of soluble Cr(III) was formed. However, the final form of Cr(III) can be influenced by the initial Cr(III) concentrations that may vary with the type of he source and the measurement location. This study is being continued for different pH of the solutions and initial Cr(VI) and Cr(III) concentrations.

474068 Treatment of Deet in an Advanced Ozone Membrane Reactor

Abstract:

The risks posed by endocrine disrupting compounds (EDCs) on the developmental, behavioral and reproductive well-being of people and animals cannot be ignored considering the increasing occurrence of these compounds in the environment. The persistent and refractory nature of these pollutants is particularly worrying. This work investigates the conversion of DEET in an advanced ozone membrane reactor (AOMR). The reactor uses ozone-resistant polytetrafluoroethylene (PTFE) membrane for ozone distribution and polyvinylidene fluorine (PVDF) membrane acts as membrane contactor and separator. Activated carbon adsorbents were deposited within the membrane lumen to trap DEET and their byproducts to allow for more complete degradation by ozone. Mass measurements were used to determine the extent of DEET conversion and degradation, while bioassay kits were employed to monitor to acute toxicity, genotoxicity and estrogenic activity of the treated water. More complete mineralization of the DEET lead to lower bioactivity in the treated water compared to DEET treated by either ozonation or membrane separation.

474095 Performance of a Micro-Mini Pulsed Electric Field (PEF) Device for Drinking Water Disinfection

Abstract:

Clean drinking water supply is considered a major public health milestone of our generation. However, waterborne diseases are still the third leading cause of death. The Word Health Organization (WHO) reported that diarrhoeal diseases from unsafe water kills 1.5 million children each year and causes two billion cases of illness worldwide (WHO Diarrhoeal disease, fact sheet No.330). Recently, there was an outbreak of waterborne diseases in a hospital in US in 2012. Centers for Disease Control and Prevention (CDC) claimed the Legionella found in the hospital in 1982 is almost the same to the Legionella in this outbreak indicating the long viability of bacteria in contaminated water distribution system. Existing disinfection technologies relies on chemicals to safeguard water producing in many cases undesired byproducts and change in water quality in term of odor and taste. This work presents a design for a new micro-mini Pulsed Electric Field (PEF) device that utilizes a low voltage pulse electric field generator with micro-engineered electrodes to disinfect microorganisms found in drinking water. The device is designed to attain 99.9% bacterial reduction. The killing effect is achieved by a low-voltage input to the system to avoid electrolysis and has low energy consumption. The intense electric field perturbs the cell wall resulting in both irreversible and reversible enlargement of intracellular pores. The former results in rapid cell inactivation and death while the latter renders the cell nonviable by an increased ingress of residual chlorine in the tap water.

474362 Insights on Life Cycle Assesment of Ionic Liquid Based CO2 Capture Processes

Abstract:

We present results related to environmental life cycle assessment (LCA) of Ionic Liquid-MEA solution based CO2 capture process. Using a combination of process siumlation data and process life cycle assessment method we analyze three scenarios comparing two different variations of Ionic Liquid-MEA CO2 capture with the base case of aqueous MEA. Useful insights gained through this study will be presented in the context of LCA considerations during process design and optimization. Further, using the results, the utility and importance of performing LCA analysis during the design and devlopment stage of emerging technolgies will be emphasized.

462148 Effects of Surfactants on the Hygroscopicity and CCN Activity of Aerosols

Abstract:

Surface tension is a critical parameter affecting water interactions of aerosol particles. Surfactant compounds are those which alter the surface tension of solutions via interactions at the gas/liquid interface. In the atmosphere, compounds that may act as surfactants are produced from both biogenic and anthropogenic sources. When these compounds are combined with existing aerosol particles, they have the potential to alter the surface tension of liquid-phase interfaces which may change a particleâ€™s ability to take on water under sub-saturated or super-saturated conditions. In this work, we investigate the effects of various surfactants on both inorganic and organic aerosol particles on both hygroscopicity and CCN activity properties.

Dilute aqueous solutions were prepared using pure, single-component inorganic or organic compounds. These solutions were then mixed with various concentrations of different surfactants, atomized, and dried in diffusion dryers. The resulting aerosol stream was then measured for CCN activity using a DMT CCNC-100 instrument. In a second set of experiments, the same aerosols were passed through Nafion humidifier tubes exposing them to up to 97% relative humidity and their growth factors were measured using a TSI SMPS system. A final set of experiments exposed the aerosol streams first to the Nafion humidifier and then into the CCNC to determine if there was an effect of water vapor exposure time. Combined, these experiments investigated the effects of surfactant type, surfactant concentration, hygroscopicity changes, CCN changes, and kinetic limitations for water uptake by compounds that were already hygroscopic (such as ammonium sulfate) and organic compounds with much more limited water activity. The results of these experiments will be presented along with a comparison of experimental investigations available in the literature as well as some basic theory and modeling work on the subject.

465557 Chemical Characterization of Water Soluble Organic Matter in Rural and Contrasting Urban Environments in the Southeastern United States

Abstract:

Atmospheric particulate matter (PM) has substantial impacts on climate, air quality, and human health. One important component of atmospheric PM is water-soluble organic matter (WSOM), which can alter the hygroscopicity and surface tension of particles and further influence climate. However, the chemical composition and sources of WSOM are poorly understood. In order to continuously measure the chemical composition of WSOM, we developed a novel system by coupling a Particle-Into-Liquid-Sampler to a High-Resolution Time-of-Flight Mass Spectrometer (denoted as PILS-AMS). This novel system was deployed in two contrasting environments, rural Centreville in Alabama (part of the Southeast Oxidant and Aerosol Study) and urban Atlanta in Georgia in summer, 2013. We found that the majority of OM is water-soluble in both rural and urban sites, which is consistent with our previous finding. We performed Positive Matrix Factorization (PMF) analysis on both ambient total OM and WSOM for source apportionment. The comparison between ambient OM and WSOM provides insights about the water solubility of OM subtypes. We found that primary OM from vehicle emissions and cooking is not water-soluble. A large fraction of biomass burning OA is water soluble. Secondary OM from different sources show different water solubility, which will be discussed in detail in the presentation. Studying the water solubility of OM subtypes largely improves our understanding of the sources of PMF factors.

478291 New Findings Based on Airborne Measurements of Aerosol Particles and Stratocumulus Clouds off the California Coast

Abstract:

Recent aircraft field experiments with the Center for Interdisciplinary Remotely Piloted Aircraft Studies (CIRPAS) Twin Otter have yielded a rich set of data that have been analyzed to explore various aspects of aerosol-cloud interactions. These campaigns, based out of Marina, California in the July-August time frame, include the Eastern Pacific Emitted Aerosol Cloud Experiment (E-PEACE, 2011), Nucleation in California Experiment (NiCE, 2013), and the Biological Ocean Atmospheric Study (BOAS, 2015). This presentation will discuss the following: (i) differences between cloud chemical composition between coupled and decoupled marine stratocumulus clouds for dissolved non-water substances; and (ii) differences between cloud and aerosol composition during periods with and without wildfire influence.

464099 Elucidating the Role of Photons and Atmospheric Aerosols in the Reduction of Oxidized Mercury Species

Abstract:

Toxic mercury species are released to the atmosphere through both natural and anthropogenic emission sources, where coal-fired power plant emissions plumes constitute a primary source of anthropogenic mercury emissions.[1] The release of mercury to the atmosphere presents large concerns for both the environment and human health (e.g., bioaccumulation of mercury in aquatic organisms). The ultimate fate of mercury in the atmosphere is deposition to the Earth's surface, but the deposition rate of mercury is largely dependent on redox chemistry involving reactive mercury species (Hg(II)), due to high relative deposition rates of Hg(II) species as compared to elemental mercury (Hg^0).[2] Recent experimental and field studies have noted the reduction of Hg(II) to Hg^0, and that this chemistry involves interactions with photons and atmospheric aerosols; however, the mechanism for this reduction chemistry has remained elusive.[1,3]

In our work, we have employed first-principles density functional theory (DFT) calculations to probe the reduction of Hg(II) species on atmospheric aerosol surfaces.[4] Through this work, we will elucidate both the reduction mechanism for Hg(II) species over a range of aerosol surfaces and the combined role of atmospheric aerosols and photons in the reduction of Hg(II) species. We will also show the usefulness of DFT as a novel approach to study atmospheric aerosol chemistry.

References

1. Y. Tong el al. *Environ. Sci.: Processes Impacts*, **2013**, *15*, 1883–1888.

2. A.P. Rutter and J.J. Schauer. *Environ. Sci. Technol.*, **2007**, *41*, 3934–3939.

3. B. de Foy et al. *Atmos. Environ.*, **2016**, *134*, 27–39.

4. S.A. Tacey, L. Xu, M. Mavrikakis, and J.J. Schauer. *J. Phys. Chem. A*, **2016**, *120*, 2106–2113.

468217 Quantifying Atmospheric Nitrogen Deposition to U.S. Waterways

Abstract:

Increasing anthropogenic emissions of nitrogen released into the atmosphere contributes to water eutrophication. In this study we quantified seasonal dry and wet deposition of major nitrogen-containing species, including ammonia (NH_3), nitric acid (HNO3), nitrogen oxides (NOx), particulate ammonium (PNH_4) and particulate nitrate (PNO_3) onto watersheds in the contiguous United States. We used a regional chemical transport model, the Comprehensive Air Quality Model with eXtension (CAMx) to obtain deposition information in the contiguous United States.

Our results show that gaseous species dominant dry deposited mass. Ammonia (NH_3), nitric acid (HNO_3), and nitrogen dioxide (NO_2) each contribute 40% to 60%, 20% to 40%, and approximately 10% of the total nitrogen dry deposition, respectively. For wet deposition the major contributors are nitric acid (HNO_3) and particulate ammonium (PNH_4). HNO_3 contributes 40%-60% of the total nitrogen wet deposited mass, PNH_4 20%-40%, and NH_3 and particulate nitrate (PNO_3) together about 10%. Spring and summer consistently receive the more deposited nitrogen than fall or winter for both wet and dry deposition. The maximum deposition of spring is over twice that of fall due to more frequent rainfall, which removes the nitrogen-containing species from the atmosphere increasing the loading to watersheds. Increased vegetation in spring and summer increases dry deposition.

Over the different watersheds, the annual mass of nitrogen deposited through dry deposition ranges from around 110 to 910 kgN/km^2, over twice that of wet deposition, which ranges from about 40 to 390 kgN/km^2. The Eastern United States, including Ohio, Mid Atlantic, Great Lakes, Upper Mississippi, Lower Mississippi, Tennessee, and New England watersheds alongside the California, Texas-Gulf, and Arkansas-White-Red watersheds receive the most deposition during the year. These spatial trends exist because nitrogen sources are mainly vehicular emissions, power plants, and fertilizer use and these watersheds are highly populated.

455581 Air-Quality Conscious Study for Multiple Olefin Plants' Turnaround Operations

Abstract:

Ground-level ozone is a pervasive air pollutant, which can be potentially elevated by flaring emissions during turnaround operations (e.g., shut-down or start-up) of chemical plants. Especially in chemical plant concentrated regions, the local ozone concentration can be aggravated by simultaneous start-up and shut-down operations. Thus, it is environmentally and cost-effective important for study on multiple plant turnaround operations under allowable manufacturing constraints to minimize the induced potential adverse air-quality impact. In this paper, a systematic methodology on air-quality conscious study for multi-plant turnaround operations has been developed. Through case studies, it demonstrates that multi-plant turnarounds without any coordination could result in the worst air quality impact (i.e., 11.4 ppb of maximum 8-hour ozone increment); however, the optimal scheduling plan with several-hour difference tuning on their turnaround starting time would significantly reduce such impact (only 1.4 ppb of maximum 8-hour ozone increment). The study couples process-level dynamic simulation for flare emissions with regional-level air-quality modeling together; and explores cost-effective and environmentally benign air-quality control strategies. It provides valuable quantitative supports for relevant stakeholders, including environmental agencies, regional plants, and local communities.

463659 Jet Aircraft Non-Volatile Particulate Matter Characterization and Estimation

Abstract:

Jet engine aircraft exhaust contains combustion byproducts and particulate matter in the form of non-volatile particulate matter (nvPM), black carbon (BC) is used synonymously for nvPM throughout this paper. Aircraft cruise emissions are the only direct source of anthropogenic BC particles at altitudes above the tropopause.[1] Black carbon aerosols are strong solar radiation absorbers and have long atmospheric lifetimes.[2] Therefore, BC results in positive radiative forcing and is believed to be the second largest contributor to climate change.[3] Additionally, upper troposphere and lower stratosphere BC particles contribute to climate forcing indirectly by acting as ice nucleation sites and cloud activators.[4-6] With regards to human health, a link between cardiopulmonary diseases and carbonaceous black particulate matter has recently been suggested.[7] As concern for human health risks and environmental impacts caused by aviation BC emissions increases, emission reduction strategies will need to be implemented. An ambitious carbon, solid and gaseous, emission reduction goal of 50% reduction by 2050 as compared to 2000-2005 levels have already been defined by the International Air Transport Association and Advisory Council for Aviation Research and Innovation in Europe.[8] Meeting these goals will require significant engineering advancements requiring a long implementation period. However, in the near term, alternative jet fuels with reduced aromatic content are an attractive solution for reducing BC emission.[9-12] Alternative aviation fuels containing synthetic blend components with near zero aromatic content (synthetic paraffinic kerosenes, SPKs) such as those synthesized via the Fischer-Tropsch (FT-SPK) process and hydrotreated esters and fatty acids (HEFA-SPK) overall contain highly reduced aromatic content compared to conventional fuel and thus significantly reduce aircraft engine BC emissions.[9-12]

Currently there is not a direct regulation on BC emissions from jet engines. Rather, BC emissions during the landing and take-off (LTO) cycle are limited by the International Civil Aviation Organization (ICAO) through regulations on smoke number (SN).[13] The smoke number regulation introduced in 1981 was put in place with the purpose of reducing plume visibility and no engines have failed this regulation since 1990.[14] With increasing concern on both human health and environmental impacts caused by jet engine BC emissions the EPA is expected to place regulations on such emissions.[15] The ICAO's Committee on Aviation Environmental Protection is currently developing a regulatory standard for BC emissions. The pending regulation will require BC emissions from new jet engines to be measured by a standard procedure. A standardized measurement methodology was defined in the Aerospace Information Report 6241[16], with much of the research effort led by Missouri University of Science and Technology[17]. Such a regulation would likely apply to new engines with the existing fleet grandfathered in. However, in-service engine lifetimes can be in excess of 20 years and current engine designs will continue to be manufactured for several more years. Therefore, predictive tools capable of accurately estimating BC emissions from the current in-service fleet will be needed for the next couple decades to quantify atmospheric BC inventory from aviation.

Current models do not accurately predict BC emissions. The First Oder Approximation-3 (FOA3) methodology is used worldwide for estimating BC emissions within the vicinity of airports.[15] The FOA3 was endorsed by the (ICAO)[18] in February 2007 and relies on a measured SN to predict BC emission. Black carbon is most often reported as an emission index of black carbon (EI_{BC}), reported as milligrams of BC emitted per kilogram of fuel combusted. Due to inaccuracies in measuring low SNs produced by modern high bypass ratio engines, the FOA3 and its modifications are unreliable. Recently a kinetic model based on formation and oxidation rates termed the FOX method was reported.[19] The FOX does not require input of a SN, instead the input variables are engine conditions. Hence, the FOX avoids the measurement error built into the FOA3. However, the FOX is fuel independent and cannot be applied to predict EI_{BC} from alternative fuels and alternative fuels blended with conventional jet fuels. Both the FOA3 and the FOX methods are designed to predict EI_{BC} at ground level, which is important for assessing human health concerns at and in the vicinity of airports, however, it is the cruise EI_{BC} that is of the most importance in determining the role aviation BC plays on the Earth's radiative balance. The current practice to arrive at a predicted cruise EI_{BC}

and Lecht relation was developed there were limited cruise BC emission measurements. The available data was not representative of real aviation emissions because the aircraft operated at reduced weight and velocities compared to regular operation.

In this work current predictive methods are evaluated for accuracy by comparison to over a decade's worth of field campaign data collected by the National Aeronautics and Space Administration's (NASA) Langley Aerosol Research Group with inclusion of cruise data.[9] An improved semi-empirical method is developed. Accurate engine condition relations are developed based on proprietary engine cycle data for a common rich-quench-lean (RQL) style combustor. Alternative fuels and fuel blend predictive relations are developed as well as a direct cruise prediction. The intent is to provide an improved method to calculate EI_{BC} from in-service aircraft and account for EI_{BC} reductions from the use of alternative fuels.

References

1. Peck, J.; Oluwayemisi, O; Wong, H; Miake-Lye, R. An algorithm to estimate cruise black carbon emissions for use in developing a cruise emissions inventory. *J. Air Waste Manage. Assoc.* **2013,** *63*, 367-375.

2. Lee, D. S.; Fahey, D. W.; Forster, P. M.; Newton, P. J.; Wit, R. C. N.; Lim, L. L.; Owen, B.; Sausen, R. Aviation and global climate change in the 21st century. *Atmos. Environ.* **2009**, *43*, 3520-3537.

3. Bond, T.; Doherty, S.; Fahey, D.; Forster, P.; Berntsen, T.; DeAngelo, B.; Flanner, M.; Ghan, S.; Karcher, B.; Koch, D.; Kinne, S.; Kondo, Y.; Quinn, P.; Sarofim, M.; Schultz, M.; Schulz, M.; Venkataraman, C.; Zhang, H.; Zhang, S.; Bellouin, N.; Guttikunda, S.; Hopke, P.; Jacobson, M.; Kaiser, J.; Klimont, Z.;

Lohmann, U.; Schwarz, J.; Shindell, D.; Storelvmo, T.; Warren, S.; Zender, C. Bounding the role of black carbon in the climate system: A scientific assessment. *J. Geophys. Res.: Atmos.* **2013,** *118,* 5380-5552.

4. Haywood, J. M.; Shine, K. P. The Effect of Anthropogenic Sulfate and Soot Aerosol on the Clear-Sky Planetary Radiation Budget. *Geophys. Res. Lett.* **1995**, 22, 603-606

5. Karcher, B.; Peter, T.; Biermann, U. M. Schumann, U. The Initial Composition of Jet Condensation Trails. *J. Atmos. Sci.***1996**, *53*, 3066-3083.

6. Heymslield, A. J.; Lawson, R. P.; Sachse, G. W. Growth of Ice Crystals in Precipitating Contrail. *Geophys. Res. Lett.* **1998**, *25*, 1335-1338.

7. Pope, C. A.; Dockery, D. W. Health effects of fine particulate air pollution: Lines that connect. *J. Air Waste Manage. Assoc.* **2006**, *56*, 709-742.

8. Realising Europe's Vision for Aviation: Strategic Research & Innovation Agenda (Executive Summary); Advisory Council for Aviation Research and Innovation in Europe (ACARE): Brussels, Belgium, 2012.

9. Moore, R.; Shook, M.; Beyersdorf, A.; Corr, C.; Herndon, S.; Knighton, W.; Miake-Lye, R.; Winstead, S.; Yu, Z.; Ziemba, L.; Anderson, B. Influence of Jet Fuel Composition on Aircraft Engine Emissions: A synthesis of aerosol emissions data from the NASA APEX, AAFEX, and ACCESS missions. *Energy Fuels* **2015,** *29,* 2591-2600.

10. Timko, M. T.; Herndon, S. C.; Blanco, E. R.; Wood, E. C.; Yu, Z.; Miake-Lye, R. C.; Knighton, W. B.; Shafer, L.; DeWitt, M. J.; Corporan, E. Combustion products of petroleum jet fuel, a Fischer–Tropsch synthetic fuel, and a biomass fatty acid methyl ester fuel for a gas turbine engine. *Combust. Sci. Technol.* **2011**, *183*, 1039-1068.

11. Corporan, E.; Dewitt, M. J.; Belovich, V.; Pawlik, R.; Lynch, A. C.; Gord J. R.; Meyer, T. R. Emissions characteristics of a turbine engine and research combustor burning a Fischer–Tropsch jet fuel. *Energy Fuels* **2007**, *21*, 2615–2626.

12. Cain, J.; DeWitt, M.J.; Blunck, D.; Corporan, E.; Striebich, R.; Anneken, D.; Klingshirn, C.; Roquemore, W.; Vander Wal, R. Characterization of gaseous and particulate emissions from a turboshaft engine burning conventional, alternative, and surrogate fuels. *Energy Fuels* **2013**, *27,* 2290-2302.

13. ICAO. *ICAO Annex 16: Environmental Protection, Volume II – Aircraft Engine Emissions;* International Civil Aviation Organization: Montreal, Canada, 2008.

14. EASA ICAO Engine Emissions Databank [online]. Available at: http://easa.europa.eu/document-library/icao-aircraft-engine-emissions [Accessed 4 Jan 2016].

15. Wayson, R. L.; Fleming, G. G.; Lovinelli, R. Methodology to estimate particulate matter emissions from certified commercial aircraft engines. *J. Air Waste Manage. Assoc.* **2009**, *59*, 91-100.

16. SAE. Aerospace Information Report AIR6241 *Procedure for the Continuous Sampling and Measurement of Non-Volatile Particle Emissions from Aircraft Turbine Engines*; SAE International: Warrendale, PA, 2013.

17. Lobo, P.; Durdina, L.; Smallwood, G. J.; Rindlisbacher, T.' Siegerist, F.; Black, E. A.; Wang, J. Measurement of aircraft engine non-volatile PM emissions: Results of the aviation-particle regulatory instrumentation demonstration experiment (A-PRIDE) 4 campaign. *Aerosol Sci. Technol.* **2015**, *49*, 472-484.

18. ICAO. *Airport Air Quality Guidance Manual;* International Civil Aviation Organization: Montreal, Canada, 2011.

19. Stettler, M. E. J.; Boise, A. M.; Petzold, A.; Barrett, S. R. H. Global civil aviation black carbon emissions. *Environ. Sci. Technol.* **2013a**, *47*, 10397-10404.

472983 Simulation of Particulate Matter Formation during Heating Different Cooking Oils Using Aspen Plus

Abstract:

Evaporation of organic constitutes of cooking elements such as cooking oils, with subsequent cooling and supersaturation of vapor phase results in particle formation. Experimental studies can help us to understand source emission rates and size distributions of formed PM. However, experimental measurements of some physical properties such as supersaturation level of produced organic vapor, viscosity and surface tension of formed particles and partitioning of chemicals among the gas and particulate phases might be hard or even impossible. In this study, we tried to build a simulation tool using Aspen Plus simulation software which is extensively used in chemical processes to estimate such physical properties. This software works at steady state conditions, with ability to predict physical properties of thousands of chemicals. Aspen Plus also performs mass and energy balances, and includes advanced thermodynamic models to predict phase equilibria of complex chemical mixtures. It is also used to simulate chemical reactions with provided kinetics data by the users. In this study, heating six cooking oils including canola, soybean, safflower, sunflower, olive and peanut were simulated. The approach was to first predict the evaporation of triglycerides from heated oils, mixing with air containing VOCs and ozone, and then minimizing the Gibbs energy of the resulting supersaturated vapor at constant temperature and pressure. Then, PM emission rate due to homogeneous nucleation can be estimated. The experimental data in the literature were used to tune simulation parameters. The effects of indoor RH and temperature were also studied on PM emission rates of the heated oils. The current simulation model is being developed to account for cooking oil smoking as well as heterogeneous nucleation. The current approach in simulating cooking PM emissions can be utilized for simulating indoor air chemistry involving phase equilibria and reactions.

469060 Estimating the Life Cycle Impact of Chemicals from Molecular Descriptors and Thermodynamic Properties Via Mixed-Integer Linear Programming

Abstract:

The chemical industry is at present striving to decrease its environmental footprint during the transition towards a more sustainable manufacturing sector. In this context, there is a clear need to develop systematic methods to assess and optimize the environmental impact of chemicals, identify critical hotspots across their supply chains and define the best strategies to implement so as to enhance their sustainability level.

In the recent past, it has become clear that the environmental impact should be assessed considering all the stages in the life cycle of the products/services being assessed. As a result, Life Cycle Assessment (LCA) has become (arguably) the prevalent approach to quantify the environmental burdens of products, finding applications in a wide variety of areas, including the evaluation of the environmental footprint of chemicals from cradle to grave.

One of the main limitations of LCA is that it is highly data intensive, that is, it requires large amounts of data from several echelons of the product supply chain, which are typically owned by different companies that might be reluctant to share this environmental information or even lack the corresponding measurements. This is critical in the chemical industry, where chemicals are produced by complex interrelated chemical networks that exchange energy, mass and water and consume a wide range of intermediate products and feedstocks.

One possible way to simplify the LCA calculations is to resort to Streamlined LCA (SLCA) approaches, which attempt to reduce the upstream and downstream information required in a standard LCA by using proxy data, qualitative models and/or regression equations. These methods are typically tailored to a particular sector, as otherwise the underlying simplifications and assumptions would not hold, thereby leading to poor estimates and large errors.

SLCA methods have been developed for a wide range of production processes, including the building sector (where the size of the bill of materials often makes unpractical the use of standard approaches), urban water treatments and hydroelectric power plants, among others. In the context of chemical processes, SLCA has been applied in oil refining as well as to general chemical processes.

In this work we introduce a novel approach for evaluating the life cycle impact of chemicals that provides better estimates and reduces the complexity of previous models, thereby opening new avenues for their use within the framework of computer aided molecular design tools. In essence, building on the seminal paper by Wernet et Al. (2008)[1] that predicted the life cycle impact of chemicals from their molecular structure using neural networks, we present here an alternative method based on rigorous mixed-integer optimization techniques that uses key thermodynamic properties in addition to molecular descriptors in order to produce more accurate estimates of impact. Our methodology was applied to estimate 9 environmental impact categories of 88 chemicals using 17 structural descriptors and 15 thermodynamic properties. Numerical results show that the life cycle impacts can be predicted with relative errors ranging from 18 to 45% depending on the specific indicator. Furthermore, the inclusion of thermodynamic properties in addition to molecular descriptors improves greatly the quality of the estimate, with relative errors dropping from 22 to 18% when moving from a molecular-based to a combined-based estimation.

1 G. Wernet, S. Hellweg, U. Fischer, S. Papadokonstantakis and K. Hungerbuìˆhler, *Environ. Sci. Technol.*, 2008, **42**, 6717â€"6722.

462465 Optimal Production of Light Olefins from Lignocellulosic Biomass (BTO): Process Synthesis and Global Optimization

Abstract:

The growing interest toward alternative carbonaceous energy sources stems from concerns surrounding energy independency, energy affordability, and the generation of lower carbon energy, as stated in a recent perspective [1]. Specifically, there is a need for the development of sustainable and carbon neutral energy processes that have the potential to reduce global energy related carbon dioxide emissions. Lignocellulosic biomass is a potential feedstock that can serve as an alternative to fossil-fuels and offset the growing energy demand while simultaneously reducing the net carbon emissions due to the absorption of CO_2 during photosynthesis. The U.S. has the potential of sustainably supplying more than one billion tons of biomass annually [2,3]. Biomass has already been shown to be a viable feedstock for the production of liquid transportation fuels at low break-even oil prices [4]. However, the transportation fuel demand is expected to stall in the near future while the chemicals demand is expected to raise by 65% [5]. This increase in the chemicals demand is expected to be the major driver of the crude oil demand growth [5].

This work investigates the optimal production of light olefins from biomass within a process synthesis and global optimization framework. Light olefins are the most demanded class of petrochemicals and command two-thirds of the petrochemical commodities market [6]. These chemicals can be produced from carbonaceous feedstocks through several novel/competing process alternatives [7]. We propose a process superstructure with alternatives regarding (i) biomass conversion, (ii) synthesis gas cleaning, (iii) methanol production and conversion, (iv) olefins purification, (v) light gas recycle, and (vi) hydrogen and oxygen production among others. Simultaneous heat, power, and water integration is included to ensure minimum utility usage. The objective function of the proposed mathematical model maximizes the profit of the refinery. The resulting large-scale mixed integer nonlinear optimization model (MINLP) is solved to global optimality with a branch-and-bound global optimization framework. Several case studies are

investigated to observe the effect of biomass type, plant scale, and topological decisions on the overall plant profitability. Significant reductions in the lifecycle GHG emissions are observed through the utilization of biomass.

[1]: Floudas, C. A.; Niziolek, A. M.; Onel, O.; Matthews, L. R. Multi-scale systems engineering for energy and the environment: Challenges and opportunities. AIChE Journal 2016, 62 (3), 602-623.

[2]: Department of Energy, Biomass as Feedstock for a Bioenergy and Bioproducts Industry: The Technical Feasibility of a Billion-Ton Annual Supply.

[3]: Department of Energy, US Billion-Ton Update: Biomass Supply for a Bioenergy and Bioproducts Industry.

[4]: Baliban, R. C.; Elia, J. A.; Floudas, C. A. Biomass to liquid transportation fuels (BTL) systems: process synthesis and global optimization framework. Energy Environ. Sci. 2013, 6 (1), 267-287.

[5]: ExxonMobil, The Outlook for Energy, A View to 2040

[6]: de Klerk, A. Fischer-Tropsch Refining; Wiley-VCH Verlag & Co. KGaA, 2011.

[7]: Onel, O.; Niziolek, A. M; Floudas, C. A Optimal Production of Light Olefins from Natural Gas via the Methanol Intermediate. Industrial & Engineering Chemistry Research 2016, 55 (11), 3043-3063.

455402 Sub and Supercritical Conversion of Kirchneriella Algal Biomass into Biocrude Oil

Abstract:

Sub and supercritical water technology utilizes wet biomass as feedstock to produce liquid fuels. Water at elevated temperature and pressure exhibit extraordinary solvent properties. Here, water acts as reactant as well as reaction medium in performing various complex chemical reactions. Algal biomass was an interesting feedstock for producing biofuel via sub and supercritical technology due to their high water content. We investigated the potential of microalgae (Kirchneriella sp.) for biofuel production. Several process parameters such as reaction temperature, pressure, time and solid loading were studied. Biocrude oil yield of 45.5% was achieved at 300°C, 30 min, 100 bar and 10% solids loading. Lower biocrude oil was produced at supercritical conditions (374°C and 220 bar) compared to subcritical conditions due to intensive gasification at higher temperatures. The high heating values (HHV) of biocrude oil and bio-char were determined with a bomb calorimeter. The biocrude oils and water soluble organics were Analyzed using gas chromatography mass spectrometry (GC-MS) and calorimetrically. CHNS/O analyzer was used to determine the elemental composition of algae, biocrude and biochar samples.

471483 Selective Aromatic Ring-Opening of Biorefinery Lignin Towards Dicarboxylic Acids: Influence of Reaction Temperature and pH

Abstract:

Lignin is the worldâ€™s only sustainable source of aromatic compounds, which has resulted in a significant quantity of research being conducted on the depolymerization of this complex biopolymer for the production of low molecular weight aromatic compounds. These aromatic products can then be further broken down via selective ring-opening to form dicarboxylic acids.

Dicarboxylic acids are a valuable class of organic compounds which are utilized in numerous commercial industries, including the production of biopolymers, pharmaceuticals, and food products. Like phenolic compounds, dicarboxylic acids are currently dependent on crude oil as a feedstock. However, lignin has the potential to act as a sustainable alternative.

This work investigates the selective production of dicarboxylic acids from several different types of lignins, via thermochemical breakdown. The treatment conditions employed a temperature range of 60 to 250 Â°C, and a reaction mixture pH of 4 to 13. The major products obtained in this work were oxalic, malonic and succinic acids. Yields of up to 10 wt.% oxalic acid were achieved, based on the mass of lignin treated.

467791 Lignin in Ethylene Glycol and Poly(Ethylene Glycol): Fortified Lubricants with Internal Hydrogen Bonding

Abstract:

Lignin, one of the most naturally abundant polymers, has been successfully incorporated into ethylene glycol (EG) and polyethylene glycol (PEG) in this work and fortified lubricating properties were achieved in EG/lignin and PEG/lignin. The molecular interaction between lignin and EG (or PEG) has been revealed as hydrogen bonding, which serves as the dominating factor that determines the thermal, rheological and tribological properties of the mixed systems of EG/lignin and PEG/lignin. The physicochemical properties of the mixed lubricants are tightly related to the state of internal hydrogen bonding (EG-EG, PEG-PEG, EG-lignin, PEG-lignin and lignin-lignin) and are well correlated to their lubrication properties. Generally, larger lignin fraction leads to better lubricating performance in both EG and PEG systems. Lignin liquefaction in PEG has been addressed by catalytic degradation with the presence of sulfuric acid, which was then neutralized by triethanolamine for lubricant development. Lignin in PEG significantly improves the lubricating property at higher pressure conditions, where a wear reduction of 94.6% was observed. Lignin fortified EG and PEG based lubricants show outstanding non-corrosive characteristic to the mostly used metal materials such as aluminum and iron.

469693 Enzyme Selection for Hydrolysis of Lignocellulosic Biomass Coupled with Fermentation

Abstract:

The India - Norway collaborative project EcoLodge aims at developing a new, integrated biotechnological production process for butyl butyrate, a promising supplement/substitute for diesel and jet engine fuels. The feedstock, lignocellulosic biomass is enzymatically hydrolyzed to produce C5 and C6 sugars. Further, the sugars are fermented by the bacteria to produce butanol and butyric acid, which are the building blocks of butyl butyrate. Butanol and butyric acid fermentations are developed and optimized separately to achieve high yield.

Major challenges for fermentation of biomass derived sugars are low product yield and high hydrolysis cost (Sun and Cheng, 2002). Exploitation of all the C5 and C6 sugar polymers locked in the lignocellulosic biomass can provide a solution to the low yield problem. However, the hydrolysis of all the polymers requires an optimal enzyme mixture, which is usually need to be tailor made or carefully selected for the specific substrate (Jørgensen et. al., 2007). The challenge of high enzymatic hydrolysis cost is mainly due to the enzyme since utility costs are low owing to mild operating conditions (pH 4.8 and temperature 45 - 50 °C) (Duff and Murray, 1996). Therefore, this paper suggests an optimal enzyme mixture, which yields the most favorable composition of C5 and C6 sugars necessary for high yield butanol and butyric acid fermentations.

Mathematical models and kinetic parameters for bacterial growth on mixed substrates from the literature are employed for simulation of fermentations with varied sugar compositions (Lendemann and Egli, 1998). Various composition values of C5 and C6 sugar mixtures are taken from the literature and used in the model simulations as well for comparison purposes (Van Dyk and Pletschke, 2012). The simulation results illustrate the effect of sugar composition on the fermentation yield; eventually provide the optimal value of the sugar composition and corresponding enzyme mixtures. The approach enables backward knowledge flow, from fermentation to enzymatic hydrolysis, to investigate the optimal process conditions. Therefore, it suggests the optimal enzyme mixture after evaluating the possibilities in terms of the cost, composition and the operating conditions.

References

Sun, Y., & Cheng, J. (2002). Hydrolysis of lignocellulosic materials for ethanol production: a review. *Bioresource technology*, *83*(1), 1-11.

Jørgensen, H., Kristensen, J. B., & Felby, C. (2007). Enzymatic conversion of lignocellulose into fermentable sugars: challenges and opportunities.*Biofuels, Bioproducts and Biorefining*, *1*(2), 119-134.

Duff, S. J., & Murray, W. D. (1996). Bioconversion of forest products industry waste cellulosics to fuel ethanol: a review. *Bioresource technology*,*55*(1), 1-33.

Lendenmann, U., & Egli, T. (1998). Kinetic models for the growth of Escherichia coli with mixtures of sugars under carbon-limited conditions.*Biotechnology and bioengineering*, *59*(1), 99-107.

Van Dyk, J. S., & Pletschke, B. I. (2012). A review of lignocellulose bioconversion using enzymatic hydrolysis and synergistic cooperation between enzymesâ€"factors affecting enzymes, conversion and synergy.*Biotechnology advances*, *30*(6), 1458-1480.

462847 Janus-like Meso-Porous Hybrid Frameworks for Super-Efficient and Cost-Competitive Water Desalination By Membrane Evaporation

Abstract:

Membrane evaporation (ME) is a novel desalination process, whereby a heat-conductive membrane is used to evaporate water in contact with the material. Similarly to membrne distillation, the difference of liquid vapour pressure across the pores of the membrane leads to vapour transport from the feed side to the permeate side, where vapours are thereafter condensed.
In contrast to other desalination techniques such as nanofiltration and reverse osmosis, ME offers a potentially low energy and high rejection route to the desalination of highly dirty or salty waters. However, to become competitive with other desalination technologies, specific membrane materials must be developed and tailored to allow for high flux, low fouling tendency and long term stability.
Specifically, in order to achieve efficient vapour transport from the hot to the cold side, a ME membrane must be highly porous, as thin as possible and not wettable by the process liquids. The pores need therefore to be large enough to facilitate vapour transport, while having sufficiently small dimensions to avoid liquid wetting and formation of a direct liquid bridge between the feed and the permeate sides. In addition, the membrane material must be heat conductive to allow for direct heating and evaporation of water in direct contact with the material surface. Most previous studies have investigated low porosity membrane materials formed from sintered stainless steel powders. While these materials are highly hydrophobic, they are also expensive and difficult to process. It is consequently important to investigate other alternatives as well as techniques for improving the process efficiency by modifying the membrane properties and structure.
Here, novel metal based scaffolds were decorated for the first time with ultra-thin layers of graphene materials by a direct electroplating. The micron-sized pores across the metal scaffolds were found by micro-Raman analysis to be entirely covred with controllable thicknesses of graphene, which latter was used for seeding specific functionalization and provide a strong interface as well as a corrosion protection barrier against chloride anions present in water. Both sides of the membranes were then specifically modified to alter the water evaporation and vapour diffusion and condensation characteristics. A Janus structure was prodcued by (i) grafting on one side of the membrane fluorosilane chains, used to generate a super-hydrophobic interface, while (ii) the second side of the membrane was coated with a nanoscale but yet thermally insulating and semi-hydrophilic alumina layer by Atomic Layer Deposition (ALD). The macro-properties of the materials, including their wettability, thermal conductivity, mechanical strength, as well as their surface charge and morphologies were systematically investigated and correlated to the electroplating conditions. The performance of the membranes for desalting model 3.5 wt% NaCl solutions were also linked to process parameters, including the liquid velocity within the module and the temperature difference between both sides of the membranes, as well as to the materials properties, fluorosilane functional groups densities and morphology of the alumina layer deposited.
In short, the membranes were found to exhibit Janus behaviours in nearly all aspects of their behaviours. The water contact angle values ranged from 160° to 60°, while the thermal conductivity of the alumina-coated side was found to be up to 25 times lower than that of the fluoro-sialne grafter graphene side. In addition, >99.99% desalination performance were achieved for flux as high as 145 LMH for a feed to permeate temperature difference of only 50°C. The hybrid Janus membranes were also found to be extremely stable and desalination tests for up to 5 consecutive days were performed without loss of performance. This work demonstrates the feasabiity of hybrid materials with controlled porosities and pore distributions for the competitive desalination of brines.

469836 Electrospun Nanocarbon Fibers for the Chromium Removal in Water

Abstract:

Uniform electrospun polyacrylonitrile / iron (III) nitrate nanofibers with small diameters were successfully prepared by systematically optimizing the electrospinning parameters including polymer concentration, salt loading, applied voltage, tip to collector distance, and flow rate. These parameters are shown to affect the fiber morphology and the addition of iron (III) nitrate salt was found to be able to reduce the fiber size and required electrospinning voltage. The improved thermal properties of the composite fibers were investigated and shown by thermal gravimetric analysis (TGA) and differential scanning calorimetry (DSC). The composite nanofibers were further carbonized to carbon nanofibers. Applications of the nanofibers for chromium removal have shown promise.

469583 Chemically Crosslinking Graphene Oxide and Chitosan for Scalable Water Treatment Membranes

Abstract:

Graphene Oxide (GO) has emerged as a next generation material for water treatment. Although numerous studies have focused on the use of GO as an adsorbent material, recent models proposed in the literature suggest that graphene oxide can be effectively applied toward membrane technologies. However fabrication of GO membranes for separation processes has been limited to small-scale vacuum filtration with an anodized aluminum oxide Anodisc®. This is due to the need for aluminum (Al) ions to crosslink the stacked GO layers preventing the membranes from breaking down from electrostatic repulsion in water. The acidic nature of GO causes Al ions to leach into the GO solution and crosslink oxygen functionalities between the GO lamellar layers keeping the membrane intact when submerged. The vacuum-assisted self-assembly (VASA) method is effective but limits the scale and scope in which these membranes can be studied and used.

To remedy this problem GO was crosslinked with chitosan (CSGO) to form a composite material that can be wet cast and fabricated at any size via convective deposition. The CSGO membraneâ€™s performance was tested in a crossflow reactor and challenged with a positively charged dye (methylene blue). The CSGO membranes were able to remove 100% of methylene blue from solution at concentrations ranging from 1-100 ppm. Mass balances obtained from the concentrate and permeate are indicative of size exclusion for the mechanism of removal. The flux rates for these solutions ranged from 2-4.5 LMH with a transmembrane pressure of 50psi. Future work involves increasing the flux across the membrane, while maintaining removal rates, and challenging CSGO composites with negatively charged dyes, radionuclides, divalent and monovalent salts.

453499 Metal-Organic Framework/Î±-Alumina Composite Withnovel Geometry for Enhanced Adsorptive Separation

Abstract:

UiO-66, as a prototypical zirconium-based metal-organic framework (MOF), provides a rapid uptake of arsenic from water when compared to other typical adsorbents with the same order-of-magnitude particle size. This fast kinetics allows an efficient adsorptive separation to be realized within a micro-space. Moreover, α-alumina can be specifically structured into a novel hollow fiber geometry: containing a plurality of open radial micro-channels on the shell side and a very thin barrier layer at the lumen. Through a facile vacuum filtration method, a MOF/α-alumina composite with novel geometry is developed and optimized in this study. The composite leads to a new concept for enhanced adsorptive separation: efficient adsorption occurs within numerous conical micro-channels with no loss of the active adsorbents during the process. As a proof of concept, this composite can effectively remediate arsenic contaminated water producing potable water recovery, whereas the conventional fixed bed requires eight times the amount of active adsorbents to achieve a similar performance. Looking forward, different functional composites can be prepared based on specific adsorptive applications, as a wide range of adsorbents can be loaded into the α-alumina hollow fiber, of which the micro-channel size and barrier layer pore size can be easily manipulated during fabrication.

462597 Multifunctional Epoxy Nanocomposites

Abstract:

In this work, the multifunctional epoxy nanocomposites filled with different nanofillers including magnetite, silica, and multi-walled carbon nanotubes were prepared. The conductive polymer polyaniline was introduced on the surface of nanofillers through surface initiated polymerization method to improve the interfacial interaction between nanofillers and epoxy matrix. The improved mechanical property of magnetite/epoxy nanocomposites was obtained due to the well-dispersed magnetite nanoparticles coverd by polyaniline. The tensile strength in polyaniline stabilized multi-walled carbon nanotubes/epoxy nanocomposites was enhanced 85% compared with that of pure epoxy. The introduction of phosphoric acid doped polyaniline and silica decreased the heat release rate peak of epoxy, providing the flame retardant property to epoxy.

468250 Graphene/Aramid Nanofibers Composite Electrodes for Structural Energy and Power

Abstract:

Flexible electronics, such as roll-up displays and wearable systems have been widely studied with the fast increasing demand for potable electronics. For these flexible, portable electronics, structural elements which can bear mechanical load and dissipate mechanical stress are required in the energy storage systems such as lithium ion batteries and supercapacitors. Flexible graphene or graphene oxide papers have attracted much attention as structural electrodes due to their extraordinary mechanical and electrical properties. The mechanical properties and the electrochemical performance of the graphene paper electrodes can be manipulated when combined with guest materials. In this presentation, we report on the composite paper electrodes containing graphene oxide and aramid nanofibers fabricated by flow-directed assembly. Recently developed aramid nanofibers, well-known for its high mechanical strength, have been used as the guest material for creating composites. The addition of the aramid nanofibers produced flexible, mechanically robust electrodes, where aramid nanofibers act as the guest material and interfacially interact with graphene sheets. In addition, the incorporation of the guest material increased porosity and alleviated the diffusion limitation for application as electrodes for supercapacitors. The composition, morphology, and the structure of the paper were characterized using scanning electron microscopy, Raman spectroscopy, and X-ray photoelectron spectroscopy. Mechanical and electrochemical performances of the composite electrodes were explored as structural electrodes.

453890 Composites Based on Biomimicry of the *Prunus* Spp. Seed Cyanide Defense System as Alternative to Pesticides

Abstract:

Many crops are ill-protected against insect pests during storage. Globally, there is still a significant loss of stored grains due to pest infestation especially in developing countries. To protect cereal grains from herbivores during storage, pesticides are often applied. While pesticides have an undoubtable functionality, increasing concerns are arising about their application.

Recently, we were able to mimic the principle of cyanogenic defense from plants such as bitter almonds (*Prunus dulcis*) and apply it onto wheat grains that naturally do not possess any cyanogenic defense.[1] Our motivation was to engineer a protective biomimetic coating material that defends the grain only upon triggering, is cost-effective, and free of conventional pesticides commonly used for grain protection during storage. To achieve an artificial cyanogenic defense mechanism similar to that found in seeds of *Prunus*species, we developed a novel onion-type grain coating consisting of separate layers containing either pure biodegradable polylactic acid (PLA), PLA mixed with the cyanogenic precursor mandelonitrile or the HCN releasing enzyme hydroxynitrile lyase. While the grains showed a cyanogenic potential high enough to theoretically have a effect on herbivores, their germination rate was reduced to approximately 25% compared to untreated grains, which was a clear drawback.

Here we would like to present the results from our recent study where we have significantly improved the cyanogenic grain coating design in terms of its effect on germination efficiency and have tested its actual performance against herbivore insects in laboratory and field experiments.[2] The optimized cyanogenic multi-layer coating consists of biodegradable polylactic acid with individual layers containing either the cyanogenic precursor amygdalin or Î²-glucosidase which is capable of degrading amygdalin to HCN. Cyanogenesis occurred only when the layers were ruptured *e.g.* by a herbivore attack. We could show that upon feeding coated cyanogenic wheat grains to the insect pest

species *Tenebrio molitor* (mealworm beetle), *Rhizopertha dominica* (lesser grain borer) and *Plodia interpunctella* (Indianmeal moth), their reproduction as well as consumption rate were significantly reduced, whereas germination ability increased compared to non-coated grains. In field experiments, we observed an initial growth delay compared to uncoated grains which became negligible at later growth stages. Our proposed and experimentally tested strategy to artificially mimic and apply a naturally occurring defense mechanisms could be expanded to other crops than wheat and has the potential to replace certain pesticides with the benefit of complete biodegradability and increased safety during storage.

[1] Halter, J. G.; Chen, W. D.; Hild, N.; Mora, C. A.; Stoessel, P. R.; Koehler, F. M.; Grass, R. N.; Stark, W. J. Induced cyanogenesis from hydroxynitrile lyase and mandelonitrile on wheat with polylactic acid multilayer-coating produces self-defending seeds. *Journal of Materials Chemistry A* **2014**, 2, 853-858.

[2] Mora, C. A.; Halter, J. G.; Adler, C.; Hund, A.; Anders, H.; Yu, K.; Stark, W. J. Application of the *Prunus* spp. cyanide seed defense system onto wheat: Reduced insect feeding and field growth tests. *Journal of Agricultural and Food Chemistry* **2016** (in print).

451359 Preparation and Photocatalytic Properties of g-C_3N_4/TiO_2/$BiVO_4$

Abstract:

Preparation and Photocatalytic Properties of g-C_3N_4/TiO_2/$BiVO_4$

Shen Xinlin, Wang Youliang, Li Kebin, Wang Fengyun

The Institute of Industrial Chemistry, Nanjing University of Science & Technology, Nanjing 210094, P.R.China

Keyword: g-C_3N_4/TiO_2/$BiVO_4$; preparation; MB degradation

Over the last few decades, the semiconductor-based photocatalysis has emerged as an environmentally benign technology for wastewater treatment due to its destructive ability towards a wide range of inorganic and organic pollutants[1-3] . TiO_2 is by far the most commonly being researched photocatalytic material owing to its many advantages for photocatalytic applications, e.g. high redox potentials, long-term thermodynamic stability, cost-effective and non-toxicity[1,4,5]. Whereas the bare TiO_2 suffers from several disadvantages including the limit to ultraviolet (UV) light absorption due to its wide band gap, as well as low quantum efficiency resulting from the rapid recombination of photogenerated electron¨Chole pairs. To overcome these drawbacks and enhance the photocatalytic activity of TiO_2, we constructed and prepared a ternary catalyst, i.e., mpg-C_3N_4/TiO_2/$BiVO_4$.

The mpg-C_3N_4 was obtained in a typical synthesis: dicyandiamide was used as raw material and SBA-15 as template agent and calcined under nitrogen atmospheres. The binary and ternary catalysts were prepared by hydrothermal. The XRD patterns of catalysts are shown in Figure 1. There are two peaks found in mpg-C_3N_4 sample at 12.28¡ãand 27.51¡ã, which can be indexed to (100) and (002) peak. The peaks at 25.5¡ã (101), 38.2¡ã (004), 48.2¡ã (200) and 54.9¡ã (105) indicate the existence of anatase TiO_2 phase (JCPDS No. 21-1272). The other diffraction peaks are agreement with monoclinic $BiVO_4$ phase (JCPDS No. 14-0688). No characteristic diffraction peaks of mpg-C_3N_4 can be observed for the ternary composite because of its low content and poor crystallization. But, its existence can be evidenced by FTIR spectra in figure 2. We have prepared the ternary mpg-C_3N_4/ $BiVO_4$/TiO_2 composite successfully.

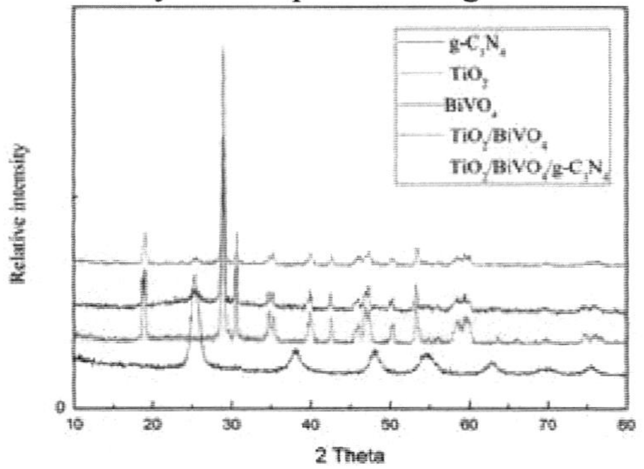

Figure 1 : XRD patterns of mpg-C_3N_4; $BiVO_4$; TiO_2; $BiVO_4$/TiO_2; mpg-C_3N_4/ $BiVO_4$/TiO_2

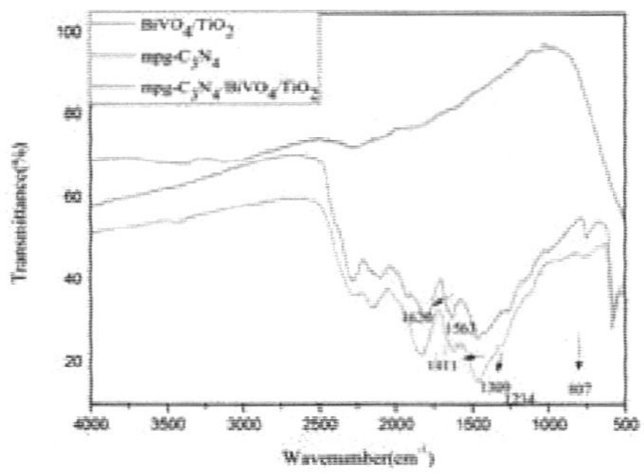

Figure 2 : FTIR spectra of mpg-C_3N_4; $BiVO_4$/TiO_2;mpg-C_3N_4/ $BiVO_4$/TiO_2

The TEM images of catalysts are shown in figure 3, we can see the binary and ternary catalysts are the formation of heterojunction. Although the size of the structure reach micron grade, the minimum structure of the hierarchical structure unit is still the nanoscale.

The effect of photocatalytic degradation for methylene blue (MB) under visible light radiation (¦Ë ¡Ý 420 nm) is shown in figure 4. It shows that the methylene blue is almost completely degraded by the ternary mpg-C_3N_4/ $BiVO_4$/TiO_2 composite after 2 h. It proves that the mpg-C_3N_4/ $BiVO_4$/TiO_2 composite is more efficiently than the binary and pure catalysts.

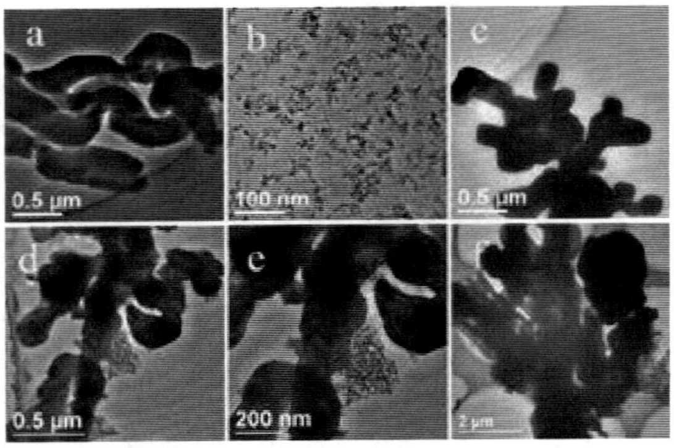

Figure 3 : The TEM image of catalysts : (a) mpg-C_3N_4 ; (b) TiO_2 ; (c) $BiVO_4$; (d-e)$BiVO_4$/TiO_2 ; (f)mpg-C_3N_4/ $BiVO_4$/TiO_2

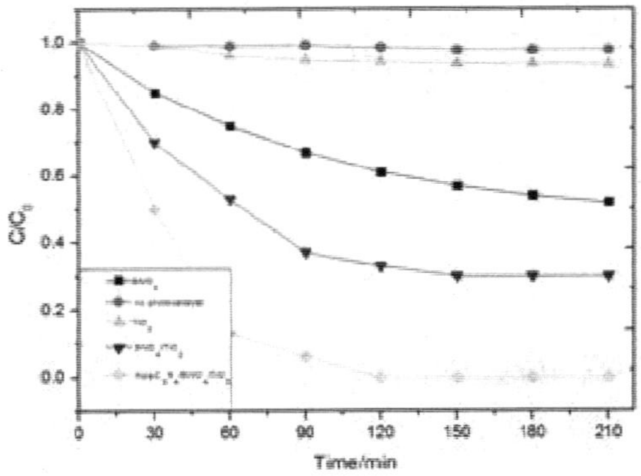

Figure 4 : The effect of photocatalytic degradation of methylene blue under visible radiation (¦Ë ¡Ý 420 nm)

The ternary mpg-C_3N_4/$BiVO_4$/TiO_2 catalysts have been prepared by calcination- hydrothermal successfully. It has been comfirmed of the formation of heterojunction by XRD, FTIR and TEM. The titled photocatalyst is efficiency for degradation of methylene blue under visible light radiation (¦Ë ¡Ý 420 nm).

References

[1] Bai X, Lv L, Zhang X, et al. Synthesis and photocatalytic properties of Palladium-loaded three dimensional flower-like anatase TiO_2 with dominant {001} facets[J]. J Colloid Interface Sci, 2016. 467: 1-9.

[2] Zhao W, Zhang Z, Zhang J, et al. Synthesis of Ag/TiO_2/graphene and its photocatalytic properties under visible light[J]. Materials Letters, 2016. 171: 182-186.

[3] Yu J, Liu Z, Zhai L, et al. Reduced graphene oxide supported TiO_2 as high performance catalysts for oxygen reduction reaction[J]. International Journal of Hydrogen Energy, 2016. 41(5): 3436-3445.

[4] Tan L-L, Ong W-J, Chai S-P, et al. Visible-light-activated oxygen-rich TiO_2 as next generation photocatalyst: Importance of annealing temperature on the photoactivity toward reduction of carbon dioxide[J]. Chemical Engineering Journal, 2016. 283: 1254-1263.

[5] Xu J, Wang G, Fan J, et al. g-C_3N_4 modified TiO_2 nanosheets with enhanced photoelectric conversion efficiency in dye-sensitized solar cells[J]. Journal of Power Sources, 2015. 274: 77-84.

473001 *Sustainable System Synthesis and Analysis using a Novel Sustainability Concept*

Abstract:

Sustainable system analysis and synthesis

using a novel sustainability concept

One of the first definitions for sustainability has been introduced by United Nations in 1987, in a report entitled, Our Common Future (also known as the Brundtland report) which define sustainable developments as "...development that meets the needs of the present without compromising the ability of future generations to meet their own needs." Although this definition captured the essence of the word sustainability but it is very vague in order to be applicable for other organizations. Efforts have been made by researchers to quantify the abstract definitions of sustainability and give it an applicable view. Previous attempts include the ecological footprint (1992, 1996), the use of Fisher information theory (2002), the pressure-state-response model of environmental sustainability (2003), the environmental sustainability index (2005), the sustainability assessment by fuzzy evaluation (SAFE) model (2001-2009), and the AIChE Sustainability Index (2007).

In the current work a novel sustainability concept is put forward, which allows a rigorous quantification of sustainability for a variety of system descriptions. The new concept allows societal and personal input to be incorporated in the sustainability assessment process. Rigorous mathematical criteria are provided for its assessment, involving the global solution of finite dimensional optimization problems. In addition, the proposed concept lends itself to the synthesis of sustainable systems. Case studies for 2-dimensional and 3-dimensional dynamical systems are used to illustrate the introduced new concept, and its applicability to both sustainability analysis and synthesis.

473967 Systematic Approach Towards Establishing Thermodynamic Principles of Sustainable Coupled Industrial-Natural Systems (CINS)

Abstract:

Fundamentals of thermodynamics apply to wide variety of systems at multiple scales such as materials and biogeochemical cycles [1, 2] - ranging from quantum to planetary scale. It will not be an exaggeration to say that systems that violate physical laws of thermodynamics cannot exist as stable systems. This raises two very critical questions for the field of sustainability: a) What are the thermodynamic principles of sustainability science and b) How do we approach hypothesizing, establishing and validation of these principles in a scientifically rigorous way? For the first question, I will qualitatively present some plausible theories of thermodynamics in context of sustainability such as Maximum Entropy principle [3] and non-equilibrium thermodynamics [4]. For the second question, I will present a framework that can be adopted to study thermodynamic properties of coupled industrial-natural systems (CINS) and relating it to approaches for testing the applicability of thermodynamic principles for CINS. The quantitative study will focus on the thermodynamic properties of nutrient flows (Nitrogen) in an industrial ecosystems coupled with regional-ecological nitrogen cycle. Following the course of establishment of thermodynamic laws in other fields, it is suggested that quantitative modeling framework to observe thermodynamic properties over a long span may provide clues to answer question 1 whereas more validation approaches may need to be developed for drawing some general principles.

References

1. D Panagiotaras et al., 2015, "Biogeochemical Cycling of Nutrients and Thermodynamic Aspects", J. Thermodynamic Catalysis, 6:2
2. J.J. Vallino and C.K. Algar, 2016, "The Thermodynamics of Marine Biogeochemical Cycles : Lotka Revisited", Annu. Rev. Mar. Science , 8:333-356
3. R. E. Ulanowicz and B.M. Hannon, 1987, "Life and the production of entropy", Proc. R. Soc. Lond., B 232, 181-192
4. C.G. Chakrabarti and K. Ghosh, 2009, "Non-equilibrium thermodynamics of ecosystems : Entropic analysis of stability and diversity", Ecological Modelling, 220, 1950-1956

474368 Analysis of United Nations Clean Development Mechanism Carbon Emission Reduction Projects from a Life Cycle Assessment Perspective

Abstract:

We analyze, from a life cycle perspective, the effectiveness of the United Nations Clean Development Mechanism (CDM) emission reduction projects implemented in developing countries. Typically, the carbon offset credit allocated to these projects account for avoided direct emissions of GHGs but do not include indirect emissions that occur during upstream and downstream steps (for both avoided as well as the project emissions). We explore, through a case study approach, how these projects perform when life cycle calcualtions are used. Further, we also explore if these proejcts lead to lower (or higher) health and ecosystem impacts (co-benefits). A small but diverse set of renewable energy generation CDM projects are compared on an equivalent bassis to assess any potential problem shifting issues. Results from this study can help policy makers assess effectivenss of such projects in a more holistic manner and fine tune their emission calculation methodologies.

468174 Evolution and Robustness of the Global Agricultural-Phosphorus Trade Network

Abstract:

Containing one of the three essential nutrients for plant growth, phosphatic fertilizers are an integral input to modern agriculture, and will always have a global demand. As the world population continues to grow through mid-century, a significant increase in food production must follow, potentially furthering our current dependence on synthetic fertilizers. While the global resource-criticality of phosphate rock reserves – from which the majority of these fertilizers are derived – may be uncertain, the trade network upon which many countries depend for access to these fertilizers (and their precursors) is a well-documented complex system, subject to an array of physical, economic, and geopolitical constraints. If we are to collectively achieve global food security by 2030, as per the U.N. Sustainable Development Goals, the global security of agricultural nutrients must also be ensured. To formalize questions of access to these nutrients, network analysis provides an attractive quantitative framework with which to examine features of individual relationships and of the underlying structures of complex systems.

The present work focuses on characterizing the structure and understanding the systemic vulnerabilities of the global agricultural phosphorus trade network (GAPTN). We accomplish this by modeling a directed, weighted global network of annual phosphorus-containing commodity exchanges, including all synthetic phosphatic fertilizers and their chemical precursors. All commodities are further normalized to their weight content of elemental phosphorus. As such, nodes are taken to be countries participating in phosphorus trade, and edges are taken to be the aggregate of the directed trade flows of elemental phosphorus between each pair of countries, resulting in the construction of a global agricultural phosphorus trade network (GAPTN). The primary objectives of this study are to (1) characterize the network topology and evolutionary trends of the GAPTN. (2) Elucidate the sets of nodes and edges that are critical to the overall function of the network. (3) Identify those countries most dependent on the network's functioning, and outline engineering and policy measures such that they may ensure their consistent access to this essential resource.

Edges within the GAPTN display mild heterophily, preferentially connecting nodes of dissimilar degree. Based on statistical testing, the distribution of GAPTN node degrees follows a heavy-tail distribution, as does the distribution of in-, out-, and total-strength for 2014. Such distributions indicate that a small subset of all of the participating countries exchange most of the resources that comprise the system, leaving the network robust to random disruptions and highly vulnerable to inopportune ones. The potential for inopportune disruptions stems from both the price volatility of phosphatic fertilizers with respect to energy prices, and the geopolitical tension that surrounds Western Sahara & Morocco, which owns some 75% of the estimated global phosphate rock reserves. Further confirming the implications of the degree distributions, a robustness analysis of randomized versus targeted removal of nodes from the network indicates that some 80% of the network's overall activity halts when some 5% of the nodes do not participate. Results from node-level analyses reveal a small subset of countries to be particularly vulnerable to supply disruptions, even as the overall network has become more redundant and dense from 1990 to 2014. In light of these findings, both international cooperation on phosphorus management and country-level efforts to close local nutrient cycles can aid in improving food security while also minimizing the ecological impacts associated with fertilizer run-off, especially eutrophication.

471802 A Framework for Considering Synergies Between Nature and Engineering from Process to Planetary Scales

Abstract:

Two major requirements for designing sustainable systems include considering all the systems through the entire life cycle to prevent the shifting of impacts to different scales; and to account for the role played by nature while making decisions in order to stay within the carrying capacity of ecosystems to prevent unintended harm or ecological degradation.

The recently developed Process-to-Planet (P2P) modeling framework [1,2] integrates fundamental engineering models and key processes in the upstream supply chain with macroeconomic systems. It improves upon conventional Sustainable Process Design (SPD) methods by incorporating the entire upstream life cycle while solving a process design problem, thus reducing the chances of shifting of environmental impacts across narrow system boundaries. The Techno-Ecological Synergy (TES) framework [3], on the other hand takes into account the carrying capacity of ecological systems by quantifying the demand and supply of ecosystem services, accounting for all the interactions between technological and ecological systems at multiple scales. Including ecosystems within the boundary of traditional engineering design decisions is expected to identify design solutions that aim to keep the impact of engineering activities within the regenerative capacity of ecosystems, depending on the scale of relevance of ecological systems. While the TES framework is the first of its kind to account for interactions between engineered systems and ecological systems, the framework lacks connections with life cycle models and while the P2P considers a comprehensive analysis boundary, it does not consider ecological systems at any scale.

This work introduces the methodology of "Process-to-Planet Techno-Ecological Synergy" (P2P-TES) for designing sustainable systems. The objective behind developing this framework is to make design decisions at the local scale by capturing the entire upstream life cycle boundary. The Structural Path Analysis algorithm (SPA) embedded within the P2P framework helps to distribute the processes or sectors in the system among the different scales. Inclusion of ecosystem services through TES within the P2P framework will expand the solution space and help to discover optimal solutions which enhance both the environmental and economic profitability of the system. In other words, P2P-TES framework is expected to provide "win-win" solutions to the process design problem compared to one where ecosystem services are not included. In addition, accounting for the imbalance between the demand and supply of ecosystem services will enable making design decisions that keep human activities within the carrying capacity of ecological systems. Ecosystem services that will be included in the framework are carbon sequestration, within the national boundary and air quality regulation service within a regional serviceshed.

P2P-TES framework will be applied to two separate case studies which present their own different set of challenges. Successful application to both of them will prove the robustness of the framework as well expose existent deficiencies.

References

[1] Hanes, Rebecca J., and Bhavik R. Bakshi. "Process to planet: A multiscale modeling framework toward sustainable engineering." AIChE Journal 61.10 (2015): 3332-3352.

[2] Hanes, Rebecca J., and Bhavik R. Bakshi. "Sustainable process design by the process to planet framework." AIChE Journal 61.10 (2015): 3320-3331.

[3] Bakshi, Bhavik R., Guy Ziv, and Michael D. Lepech. "Techno-ecological synergy: A framework for sustainable engineering." Environmental science & technology 49.3 (2015): 1752-1760.

452566 Impact of the Surface Energy of Particulate Foulants on Membrane Fouling

Abstract:

Foulant-foulant and foulant-membrane interfacial interactions play an important role in dictating the extent of fouling. In order to understand the impact of the surface energy of particulate foulants on the fouling extent, the direct observation through the membrane (DOTM) technique was used to characterize critical flux and also assess the initial evolution of foulant deposition. Polystyrene and glass with diameters approximating 10 μm were used as foulants because (i) the similar particle diameter eliminated the differences due to particle back-transport, and (ii) the Gibbs free energies (ΔG) of foulant-membrane and foulant-foulant interactions were both negative (i.e., attractive) and both positive (i.e., repulsive) respectively for polystyrene and glass. Results indicate that: (1) because of the attractive and repulsive *foulant-membrane* interactions of the polystyrene and glass, respectively, (i) critical flux was lower for polystyrene than glass, (ii) the phenomenon of a flowing particle layer was observed at a lower cross-flow velocity (CFV) for glass than polystyrene, and (iii) relaxation was more effective for glass than polystyrene; (2) because of the attractive and repulsive *foulant-foulant* interactions of the polystyrene and glass, respectively, clustering was observed for the former but not latter; (3) out of the four particle deposition mechanisms (namely, convective, entrapment, dead zone and clustering) distinguished, the first three occurred for both attractive and repulsive interfacial interactions, but the forth (i.e., clustering) only occurred when foulant-foulant interactions were attractive.

452791 Operating Strategy for the Energy-Efficient Reverse Osmosis (EERO) Desalination Process

Abstract:

Operating Strategy for the Energy-Efficient Reverse Osmosis (EERO) Desalination Process

William B. Krantz[a,c], Tzyy Haur Chong[b,c], Siew-Leng Loo[b,c]

[a] Dept. of Chemical & Biological Engineering, University of Colorado, Boulder CO 80309-0424

[b] School of Civil & Environmental Engineering, Nanyang Technological University, Singapore 639798

[c] Singapore Membrane Technology Center, Nanyang Technological University, Singapore 637141

The energy-efficient reverse osmosis (EERO) process, patented in 2014 by the Singapore Membrane Technology Center, reduces the osmotic pressure differential (OPD) and the cost for potable water production from a saline water feed by combining Single-Stage Reverse Osmosis (SSRO) with a countercurrent membrane cascade with recycle (CMCR). The retentate from the SSRO stage is the feed to the CMCR and its permeate is blended with that from the CMCR to produce the potable water product. Key features of the EERO process are: (1) optimum injection of the SSRO retentate into the CMCR; (2) operating all stages at the same OPD; (3) countercurrent retentate and permeate flow; (4) permeate recycle to the retentate side in the CMCR; and (5) retentate recycle to the permeate side by employing one or more NF stages in the CMCR. These features reduce the concentration difference across the membrane in each stage, thereby reducing the OPD and net specific energy consumption (SEC_{net}).
Since the permeate from an NF stage has a reduced divalent salt concentration and is the feed to the terminal CMCR stage, the latter can be operated at a higher recovery. Operating the EERO stages at the same OPD requires that the recoveries of adjacent CMCR stages add to one. Hence, fouling in the NF stage whose feed has a high divalent salt concentration can be mitigated by operating it at a lower recovery (higher safety factor).

Optimal operation of the EERO process can require using NF membranes with very low salt rejections in one or more of the CMCR stages. If a membrane with a higher than optimal rejection is used, near optimum EERO performance can be achieved at the cost of a small increase in the SEC_{net} by boosting the pressure in the NF stage.

EERO process performance for different operating strategies is compared to that for SSRO and two SSROs in series based on the OPD and SEC_{net}. The annualized total cost of water production for EERO operation at 75% recovery is shown to be less than that for SSRO operation at 50% recovery.

At 75% recovery EERO results in a two-fold increase in the retentate concentration that not only reduces the brine disposal, but also translates to a four-fold increase in the energy density possible via a hybrid EERO-PRO (pressure-retarded osmosis) process for recovering the osmotic potential energy in the brine. As such, this hybrid process would cross the 10 W/m^2 threshold for economic PRO operation.

Acknowledgment

This research was supported by the Singapore National Research Foundation under its Environmental & Water Technologies Strategic Research Program and administered by the Environment & Water Industry Program office (EWI) of the PUB. The Singapore Membrane Technology Center, Nanyang Environment and Water Research Institute. Nanyang Technological University is supported by the Economic Development Board of Singapore.

456121 Separation and Modeling of Oil-in-Water Emulsions Stabilized By Different Types of Surfactants Using Electrospun Fiber Membranes

Abstract:

In this work, we examine the influence of surfactants on the microfiltration of oil-in-water emulsions using electrospun fibrous membranes of poly(trimethyl hexamethylene terephthalamide) (PA6(3)T) with fiber diameters of [121 ± 22 nm]. Four types of surfactants (anionic, cationic, non-ionic and zwitterionic) were used to create monodisperse emulsions of dodecane in water, with average droplet sizes of [475 ± 14 nm]. Both dead-end and cross-flow filtration configurations were employed to study the interaction (i.e. fouling and cleaning) between the emulsion and membrane.

In both dead-end and cross-flow configurations, the emulsions prepared with anionic and non-ionic surfactants exhibited the highest fluxes, around five-fold higher than that for the emulsion stabilized by cationic surfactant. The emulsion with zwitterionic surfactant led to intermediate flux. The oil rejections were generally as high as 90%, with the exception of the emulsion stabilized by non-ionic surfactant, whose rejection was closer to 60%. The observed fluxes using the cross-flow system were about one order of magnitude higher than those observed with the dead-end system, due to a mechanism for foulant removal (cleaning) attributed to hydrodynamic forces. The differences in the fluxes and rejections for emulstions stabilized by different surfactants is traced to electrostatic interactions between the emulsified droplets and the membrane, based on zeta potential measurements and flux recovery after backflushing.

To study the fouling and cleaning mechanism, several models were also applied in this work. Blocking filtration models suggest that the membranes fouled initially in the regime of complete blocking, but then transitioned to cake filtration as the flux continued to decline. Foulant resistivity modeling supported a series resistance model, consistent with cake filtration. The emulsion stabilized by anionic surfactant showed the lowest resistance, while that stabilized by cationic surfactant showed the highest resistance. Using the surface renewal model, the emulsion stabilized by anionic surfactant was found to be most readily removed by hydrodynamic cleaning during cross-flow filtration.

Lastly, the decay of flux was well described by the conformally coated fiber model, wherein the fibrous structure of the membrane is considered explicitly, and the flow through the membrane is modified as a foulant layer builds up on the fibers.

461223 Development of Cordierite Honeycomb Membrane Filters for Produced Water Treatment

Abstract:

Oil and gas exploration and production processes produce large volume of waste as produced water that needs to be treated before reuse or discharge due to fresh water scarcity and upcoming tighter regulation. In this talk, we will present Corningâ€™s effort to develop chemically durable cordierite honeycomb microfiltration membranes for oil-water filtration. Two types of cordierite membrane have been developed with the pore size of 0.2 and 0.5 micron, respectively. The clean water permeance is in the range of 1500-2000 L/m2/h/bar. The chemical durability test demonstrated that they are durable in the solutions from pH3 to pH13 at the temperature up to 80 degree C. The oil-water filtration performance will be discussed.

464875 Influence of Graphene Quantum Dot Surface and Pore Modification on Membrane Fouling

Abstract:

Influence of Graphene Quantum Dot Surface and Pore Modification on Membrane Fouling

Membranes have been widely utilized for many applications including water softening, pharmaceutical purification, and dairy processing. In systems with organic foulants or sparingly soluble salts present, fouling of the membrane surface leads to reduced performance and shortened membrane lifespans. Graphene oxide has been of great interest as an antifouling material for membrane modification due itâ€™s inherit hydrophilicity and the presence of negatively charged carboxyl groups. Graphene oxide quantum dots are highly hydrophilic particles that maximize functionality in a size (1-3 nm) that makes them a particularly interesting candidate for membrane functionalization.

In this work, the benefits of GQD modification toward surface fouling behavior and pore blockage by organic foulants will be studied. Through EDC/NHS coupling, GQDs were covalently bonded to the surface of nanofiltration membranes.Â Zeta potential analysis shows higher negative surface charge properties after GQD coupling and increased rejection of Na_2SO_4 (97% compared to 76% for the standard membrane). The GQD coupled membrane exhibits double the normalized flux after 2 Â½ hours of recovery of solution containing 400 ppm BSA when compared to the standard membrane.

GQDâ€™s were integrated throughout the pore structure of cellulose ultrafiltration membranes by means of phase inversion. A common ionic liquid solvent between GQDs and cellulose membranes was used to deter aggregation and promote the selective entrapment of GQD in the membrane pore network. The effect of water flux, selectivity and antifouling behavior for BSA will be studied.

Acknowledgements

The author acknowledges the support of the Southern Company in funding and collaboration, and additional funding from the National Science Foundation EPSCoR for funding. Nanostone Membrane of Oceanside, CA has provided support and collaboration in the development of membranes and spiral wound membrane modules.

465388 Water Desalination through Molecular Layer-By-Layer Membranes: Insights from Molecular Dynamics Simulations

Abstract:

Reverse osmosis (RO) membranes based on polyamide (PA) membranes synthesized via interfacial polymerization (IP) of trimesoyl chloride (TMC) and m-phenylenediamine (MPD) represent one of the most common membranes used for producing potable water from seawater. However, typical RO membranes synthesized via IP can have several drawbacks which can impact the efficiency of the membrane. In particular, IP membranes lack control over (1) film thickness, (2) surface roughness, (3) cross-linkage, and (4) local chemical composition, which influence the membrane performance and efficiency. As an alternative to these challenges, ultraâ€'thin PA membranes synthesized via molecular layer-by-layer (mLbL) deposition have been introduced.

Although PA-mLbL have been studied in previous work, there are still challenges in both experimental characterization and understanding the relationship between membrane performance and membrane variables (grafting density and monomer ratio). In this work, molecular dynamics simulations were employed to characterize and to investigate membrane performance with respect to these membrane variables. We first applied a general simulated polymerization algorithm, Polymatic, developed in our group [Theor. Chem. Acc. 2013, 132, 1334] to virtually synthesize RO membranes via mLbL deposition on silicon substrates. The resulting simulated structures for twenty cycles are characterized by analyzing densities, membrane thickness, free volume, swelling, molecular orientations and local chemical compositions. Membrane performance was evaluated by analyzing water and salt dynamics in mLbL membranes with respect to grafting density and monomer ratio. This simulation study provided molecular level insight into water desalination using PA-mLbL membranes and illustrates that molecular simulations can play a crucial role in the understanding of next generation RO membranes synthesized via mLbL deposition.

468420 Aquathermolysis of Waste Triglycerides in a Continuous Flow Reactor for Jet Fuels Production

Abstract:

The unavoidable dwindling fossil fuel and its attendant climate change in our time have given impetus to researchers in the energy sector to delve into different sustainable energy sources. One of such energy sources is the abundant waste triglycerides whose long chain hydrocarbons can be separated from their short oxygen chain via hydrodeoxygenation, decarbonylation, and decarboxylation in the presence of a heterogeneous catalyst and hydrogen. However, the use of gaseous hydrogen is limited by a diffusion process since it is not readily soluble in liquid (polar) components and requires an excess use of hydrogen. To address these challenges, a single-pot â€˜Formic Acid Assisted Autotheromolysis to Jet Fuel (FAT2JF)â€™ process is studied. The use of aqueous phase (formic acid and water mixture) serves as an *in situ* source of H_2 (hydrogen precursor and hydrogen transfer agent) and organic solvent where the reactions in aqueous media (aquathermolysis) helps in reducing mass transfer resistances that results in significant increase in reaction rates in this FAT2JF process. The phase diagram of water, formic acid and some typical triglycerides will be presented.

The present study on aquathermolysis is focused on developing a metal oxide-based low cost catalyst (e.g. Fe-Ni/activated carbon, Fe-Ni-Co/Al_2O_3) system that is stable in aqueous media. The study is conducted in a packed bed continuous flow tubular reactor in the range of 250 to 400âµ'C under hydrothermal conditions. In order to alleviate coke formation and evaluate the stability of the catalyst, the effects of reactants flowrates, feedstock composition, pressure, temperature, and formic acid concentration are studied. GC-MS and elemental analyzer are used to analyze the product compositions. Thermal gravimetric analysis (TGA), energy dispersive X-ray, SEM, BET method, FT-IR, and XRD are used to characterize both fresh and used catalyst. The proposed novel approach has certain advantages over conventional commercial hydrotreating processes including no handling of hydrogen gas, faster reaction times, and use of renewable formic acid as hydrogen source.

469513 Rheology of Biomass Slurries to Determine Pumpability for Hydrothermal Liquefaction

Abstract:

The pursuit of renewable fuels and chemicals has led to the development of many different biomass conversion processes. One of these processes, hydrothermal liquefaction (HTL), has the advantage of high energy recovery, no drying step, and utilization of a green aqueous reaction media.[1] However, transport of biomass and water slurries during HTL has proven to be a challenge.[2,3] Typical reactors can only process biomass between 10 and 15 wt% solids because higher solids loadings plug reactor systems. To improve the efficiency of HTL, and other processes reliant on pumping biomass slurries, it is essential to increase solids loading. Increasing solids loading reduces the volume of waste streams and increases process throughput. In an effort to increase biomass solids loadings many people have studied the rheological properties of biomass slurries to improve pumpability.[4] Despite these efforts there has yet to be a strong correlation developed between typical rheological properties, like viscosity, and biomass pumpability.

This work takes a unique approach to studying biomass slurry rheology by combining traditional rheometric measurements, such as viscosity and storage/loss moduli, with a novel flow characterization system. Traditional rheometric measurements are made on a rheometer capable of analysis at temperatures up to 300 °C and 2000 psi, which mimics conditions found in some traditional HTL systems. Novel flow studies performed at Idaho National Laboratory are made on an in-house designed syringe attached to an Instron load frame. This system measures the force required to push biomass through a reducing orifice and can study plugging phenomena in biomass slurry transport systems. These studies have shown how plugging can be related to biomass dewatering, followed by bridging across the flow channel. Pumpability studies have been performed on a wide variety of materials from herbaceous to woody crops and residues. In addition to characterizing pumpability in biomass slurries this work has also focused on improving the pumpability of such slurries through preprocessing techniques including: size reduction, blending, and torrefaction. This work will ultimately lead to improvements in feed handling related to hydrothermal liquefaction and other biomass slurry transport systems.

1 Toor, S. S., Rosendahl, L. & Rudolf, A. Hydrothermal liquefaction of biomass: A review of subcritical water technologies. *Energy* **36**, 2328-2342, doi:10.1016/j.energy.2011.03.013 (2011).

2 Elliott, D. C., Biller, P., Ross, A. B., Schmidt, A. J. & Jones, S. B. Hydrothermal liquefaction of biomass: Developments from batch to continuous process. *Bioresource technology* **178C**, 147-156, doi:10.1016/j.biortech.2014.09.132 (2015).

3 Berglin, E. J., Enderlin, C. W. & Schmidt, A. J. *Review and Assessment of Commercial Vendors/Options for Feeding and Pumping Biomass Slurries for Hydrothermal Liquefaction*. (Pacific Northwest National Laboratory, 2012).

4 Stickel, J. J. *et al.* Rheology measurements of a biomass slurry: an inter-laboratory study. *Rheologica Acta* **48**, 1005-1015, doi:10.1007/s00397-009-0382-8 (2009).

469372 The Upgrading of Biomass-Derived Dimethyl Ether to High-Octane Hydrocarbons: The Effect of Process Conditions on Catalyst Performance

Abstract:

Dimethyl ether (DME) and/or methanol can be converted into C_{5+} branched hydrocarbons at low temperatures and pressures over large-pore acidic zeolites. Recently, we have demonstrated that Cu-modification of the existing commercial catalyst (H-BEA) resulted in a 2-fold increase in total hydrocarbon productivity at 200 °C with co-fed H_2.[1] The Cu/H-BEA catalyst incorporated H_2 with DME into the desired branched hydrocarbons, while maintaining the high C_{5+} selectivity of the parent H-BEA. Here, we compare the performance of Cu/H-BEA and the parent H-BEA at 175, 200 and 225 °C, and at weight-hourly space velocities (WHSV) for DME ranging from 0.43-1.4 h^{-1}. The productivity of C_{5+} products increased with temperature over H-BEA, with no significant changes over Cu/H-BEA due to a decrease in C_{5+} selectivity at elevated temperature, which was observed for both catalysts. Operation at 175 °C resulted in a significant increase in C_{5+} selectivity (75-85% for Cu/H-BEA), but lower productivity than at 200 °C due to lower reactant conversion. Lower space velocities increased reactant conversion over both catalysts, but significantly decreased the C_{5+} productivity over Cu/H-BEA. The Cu/H-BEA catalyst was stable for 100 h at 200 °C after an initial 40 h break-in period. In previous research, we demonstrated that C_4-C_8 olefins produced in the DME homologation reaction can be oligomerized to C_{12}-C_{22} branched hydrocarbons with properties that align with a synthetic paraffinic kerosene product.[2] Following a simple vacuum distillation, the fuel properties of the distillate-range product align well with a jet fuel blend when compared with 5 ASTM methods. This DME-to-hydrocarbons reaction represents a promising route to transform renewable biomass sources into commercially viable, high-performance biofuels.[3]

[1]Schaidle, J. A.; Ruddy, D. A.; Habas, S. E.; Pan, M.; Zhang, G.; Miller, J. T.; Hensley, J. E. *ACS Catal.* **2015**, *5*, 1794.

[2]Behl, M.; Schaidle, J. A.; Christensen, E.; Hensley, J. E. *Energy & Fuels* **2015**, *29*, 6078.

[3]Tan, E. C.; Talmadge, M.; Dutta, A.; Hensley, J.; Snowden-Swan, L. J.; Humbird, D.; Schaidle, J.; Biddy, M. *Biofuels, Bioproducts and Biorefining* **2016**, *10*, 17.

462619 A System-Level Analysis on Biomass Thermal Fractionation and Catalytic Upgrading Processes

Abstract:

Biomass presents a promising source for sustainable liquid fuels production. There are three principle routes for converting biomass into liquid fuels, classified by the initial treatment of the biomass: gasification, hydrolysis, or pyrolysis and liquefaction. The pyrolysis route involves thermal decomposition of the biomass at high temperature (500-800°C), and it has demonstrated high yields. Unfortunately, the wide range of components in the resulting pyrolysis vapor (oil) makes it unstable, corrosive, and challenging to catalytically upgrade. Furthermore, the high content of light (<C6) oxygenates in the pyrolysis are unsuitable for liquid fuels and are therefore of limited utility. Multi-stage torrefaction of biomass with catalytic upgrading presents a promising method of overcoming these challenges. Importantly, the decomposition of the biomass over several stages can reduce the number of product species within each fraction, allowing one to mold the catalytic upgrading strategy to target a subset of the chemical functionalities.

Although staged thermal decomposition of biomass coupled with targeted upgrading has a number of advantages over single stage pyrolysis with hydro-treating, there are many unknowns regarding the best way to design such a system. Importantly, the optimization of the number and conditions of each thermal decomposition stage is a complex problem that cannot be solved without knowing the influence on the final yields and process complexity resulting from the subsequent catalytic upgrading. Furthermore, there are a wide variety of chemistries available to perform upgrading. In addition, the reaction sequence must also be configured to optimize the overall process. In order to improve the impact of such an approach, one must also determine if general practices from this strategy may be applicable to different types of biomass feeds.

Accordingly, the ultimate goal of this work is to design a broad roadmap that can be used to identify the key parameters and trade-offs in a bio-refinery process based on multi-stage torrefaction of biomass with catalytic upgrading. In order to investigate the effect of the number of torrefaction stages and the specific chemistries on the yield of fuel-range grade product and the cost, we first derived a series of process alternatives based on experimental torrefaction yields obtained for a 3-stage torrefaction process (270 °C, 360 °C, and 500 °C). In designing the alternatives, we carefully selected the chemistries that are used to upgrade the thermal decomposition fractions, to maximize the yield of fuel-grade product and to minimize the hydrogen consumption, considering the abundance of various chemical functionalities within the biomass fractionation product. We suggested twelve cases where there is 1, 2, or 3 thermal decomposition stages and where each fraction is upgraded 1, 2, 3, or 4 times. Through the analysis, we found that there exists a trade-off between fuel-range carbon yields versus the number of upgrading steps. In general, increasing the complexity of the fraction upgrading systems increases the ultimate yield of C6+ product, thought there are diminishing returns on the increase in product yield versus the complexity of the catalytic upgrading sequences. In contrast, however, the choice of the number of thermal decomposition stages is found to be not simple, requiring careful consideration of the chemistries available to upgrade different components and the relative abundances of these different components. More thermal decomposition stages allow for more fine-tuning of the upgrading processes but lead to more complex and potentially more expensive processes. Thus, the design of the decomposition conditions and number of stages must be done simultaneously with the design of the fraction upgrading system. Finally, we also suggest some hybrid processes by eliminating several upgrading steps that have small marginal effects on the fuel-range carbon yield and discuss about separations that could enhance the process performance via prevention of unwanted side reactions or protection a functionality from conversion.

474875 Lawrence K. Cecil Award Lecture: Atmospheric Organic Nanoparticles, Air Quality and Climate Change

Abstract:

Spyros Pandis is Research Professor of Chemical Engineering and Engineering and Public Policy in Carnegie Mellon University and Professor in the Chemical Engineering Department of the University of Patras in Greece. He received his PhD from the California Institute of Technology in 1991 and joined the faculty of Carnegie Mellon University in 1993 and of the University of Patras in 2004. His research applies principles of chemical engineering to the control of air pollution and climate change. Prof. Pandis has published 230 peer-reviewer papers which have received approximately 14000 citations. He has served as mentor to 30 PhD graduates. The work of his group has contributed to successful regulation and improvements in air quality in areas around the world.

Lecture title: Atmospheric Organic Nanoparticles, Air Quality and Climate Change

Abstract: The human development of our planet has a variety of negative impacts on the composition of its atmosphere at every scale â€" locally, regionally, and even globally. One of these dramatic changes has been the increase in the mass concentrations of sub-micrometer particles by one to sometimes two orders of magnitude over populated areas in the Northern Hemisphere. These atmospheric aerosols can cause serious health problems, reduce visibility, contribute to acidic deposition and material damage, but are also cooling the planet by reflecting sunlight back to space.

Organic particulate material has been traditional classified as either primary or secondary with the primary component being treated as non-volatile and inert. Laboratory and field studies during the last decade, demonstrate that primary combustion aerosol is highly dynamic, consisting of mostlysemi-volatile material that moves between the gas and particulate phases in the atmosphere and at the same time is oxidized forming a variety of oxygenated products. A unifying framework for the description of all organic components based on their volatility distribution and oxygen content (the two-dimensional volatility basis set) can be used for the treatment of a wide range of processes affecting organic aerosol loadings and composition in the atmosphere. Thismodeling framework is combined with emission characterization studies, laboratory smog chamber studies, and field measurements to simulate the atmospheric evolution of these organic emissions. Applications of this framework to areas around the world are used to provide insights about the sources of these particles and ways to reduce their corresponding negative impacts.

474876 Early Career Award: Incorporating Sustainability in Process Systems Engineering

Abstract:

Process design for sustainability using various available techniques is still limited to computer-aided design featuring process optimization of energy and material flow plus minimizing greenhouse gas emission and water conservation. Sustainable process demands more, such as minimizing the impacts from other harmful emissions, discharges, waste creation, economic, and societal impacts. The author will demonstrate her research and educational interests in integrating sustainability principles for process engineering.

471609 Challenging Separations in the Power Generation Industry

Abstract:

The power generation industry has diverse challenging physical and chemical separations, including many in which membranes are important. For coal-fired steam generators, physical separations of ash (particulate matter) from the flue gas using electrostatic precipitation, wet scrubbing, or fabric filters pose challenges on various fuel types (interference from other constituents). Chemical separations include adsorption of contaminants such as mercury, reactive separations targeting sulur dioxide or sulfuric acid, and removal of carbon dioxide. Environmental contaminants (such as mercury, selenium, and nitrates) must also be removed from water used in the flue gas cleanup and ash handling.

However, other power generators include challenging separations. Energy storage technologies such as batteries and flow batteries rely on membranes to maintain charge separation and respond quickly when electrical demand is high. Advanced nuclear power generators need process contaminants removed to maintain high efficiency.

Poly-generation (such as hydrogen and power production) relies on separation technologies. Produced hydrogen must be separated from co-products using hydrogen separation membranes, and utilized in fuel cells that requires different membranes. Advanced hydrogen generation technologies such as electrolysis and photolysis require advanced membranes, and thermochemically produced hydrogen has challenging separations in each of the unit operations.

This talk will survey the key progress and remaining challenges in each area of power generation: coal-fired, storage, nuclear, and hydrogen.

473791 Self-Adaptive Water Treatment and Desalination of Source Water of Variable Feed Quality

Abstract:

Membrane based water treatment and desalination technologies have been gaining momentum in potable water production from seawater and brackish groundwater, municipal and industrial water reuse, as well as for management and reuse of agricultural drainage water. Considerable progress has been made since the early days of Reverse Osmosis (RO) water treatment/desalination regarding reduction in energy consumption, development of high performance fouling and chlorine resistant membranes, as well as improvements of in RO feed pretreatment. Given the increased level of water scarcity in various regions of the world and in particular in the Western US there has been growing efforts to expand the use of otherwise unutilized inland groundwater resources. A major challenge, however, is to implement a strategy that produces treated water of the right quality for the right use at the desired location. Temporal variability of source water quality, however, is a major challenge. Also, many of the various locations/communities, where water purification/desalination is needed, are remote and are not part of a centralized water infrastructure. In such locations, it is critical that water purification is accomplished at high recovery. Addressing the above challenges requires the development of a robust strategy for self-adaptive operation of membrane based filtration and desalination operations that will enable the deployment of distributed water systems. Such water systems have to readily adapt to changing water capacity demand and water quality fluctuations. Accordingly, an approach will be described whereby the concept of self-adaptive distributed water systems is developed, following a rapid process field evaluation approach that then serves to optimize process configuration. The above is enabled via direct real-time membrane monitoring and considerations of high recovery in a small plant footprint with salt harvesting via novel accelerated process of chemical demineralization. Examples of recent advances in the above areas will be discussed along with presentation of successful field demonstrations of seawater, brackish and industrial water treatment and reuse, in addition to purification of impaired groundwater.

473771 Desalination Performances of Large Hollow Fiber-Based DCMD Devices

Abstract:

Membrane distillation (MD) can be utilized to desalt highly saline waters that have considerable osmotic pressure. Current reverse osmosis desalination technology is not designed for saline waters having high osmotic pressure. Realization of this goal requires MD membranes to be highly fouling resistant. Fouling is a major obstacle that MD technique should potentially eliminate or control by selecting appropriate membrane material/surface, flow manipulation, additional force field and cleaning procedure. In direct contact membrane distillation (DCMD), Song et al. (2008) have shown that a rectangular cross-flow hollow fiber membrane configuration with a specific type of plasma polymerized superhydrophobic coating on the membrane outside surface and a novel method of operation ensured no precipitation based fouling from $CaCO_3$ and $CaSO_4$. We dwell on this novel aspect compared to conventional methods of fouling control. The simulation model of Song et al. (2008) assumed the membrane mass transfer coefficient, k_m, as an adjustable parameter for large hollow fiber modules with crossflow in a given range guided by estimated values and Zukauskas equation for predicting heat transfer coefficients in the two boundary layers. Here k_m was predicted from an enhanced model from Li and Sirkar (2016) without any adjustable parameters. To develop efficient designs, the model further simulated the performance of hollow fiber modules where the fiber ID was varied. Effects of fiber length on water vapor flux and water production rate were also modeled. This analysis will help design optimum hollow fiber membrane modules for desalination by DCMD for brines with low as well as high salinity.

472716 Development of a Multi-Stage Membrane Distillation System for Treatment of Hydraulic Fracturing Flow Back Waters

Abstract:

The reuse of wastewater for beneficial uses has become increasingly important in recent years. There is an urgent need to develop innovative and more effective technologies for treatment of wastewaters. Many of these wastewaters such as hydraulic fracturing flow back waters, contain very high dissolved solids (TDS). For example, treatment of hydraulic fracturing flowback waters can be very challenging as they often contain very high TDS, in excess of 200,000 ppm as well as dissolved surfactants and small organic compounds. Treatment of these highly impaired wastewaters is therefore very challenging. Pressure driven membrane processes such as reverse osmosis are impractical for treating very high salinity wastewaters due to the high osmotic back pressure that must be overcome. Membrane distillation has been proposed as a new unit operation for treatment of very high TDS wastewaters. One of the advantages of membrane distillation is the fact that low grade waste heat may be used.

Here we have developed a multi-staged membrane distillation system for maximizing water recovery from actual hydraulic fracturing flowback waters. The multi-stage system has been designed to different on-site applications for wastewaters with specific TDS levels. We show that ideal membrane properties vary as the TDS of the feed increases above 300,000. At lower TDS values membranes should be selected that maximize flux. However at TDS values above 300,000, membrane surface properties such as roughness becomes important. At theses higher TDS values leakage of the feed through the membrane pores and into the permeate side is a major concern. In addition appropriate pretreatment of the feed to remove dissolved organic compounds is essential prior to membrane distillation.

Here different membranes have been used in the different stages and performance of the system in terms of water recovery potential and specific energy consumption has been evaluated. Significant concentration of the feed has been conducted and over 60% water recovery has been achieved for actual hydraulic fracturing flowback waters from

Marcellus shale containing over 100,000 ppm TDS.

465789 Role of Membrane in Iemb Treatment of Groundwater Highly Contaminated with Oxyanions

Abstract:

Role of Membrane in IEMB treatment of groundwater highly contaminated with oxyanions

Perchlorate contamination of ground water is a worldwide concern. Several sites in Israel's coastal aquifer are contaminated with perchlorate, with hundreds of ppm found in one area accompanied with significant concentrations of nitrate and chlorate as well. This has prevented water production from wells in the area.

The ion-exchange membrane bioreactor (IEMB) [1] is a hybrid process for safe treatment of groundwater highly contaminated with oxyanions (perchlorate, nitrate, and chlorate). By Donnan dialysis anionic contaminants migrate across an anion exchange membrane (AEM) from a feed-water compartment to a bio-compartment. Once in the bio-compartment, the anions undergo microbial reduction to safer species such as chloride ions and nitrogen. The AEM acts as a barrier and keeps both compartments completely separate. Glycerol is used as an exogenous carbon and electron source for the biodegradation process [2]. This arrangement keeps the carbon source, reaction byproducts and bacteria confined in the bio-reactor thus preventing secondary contamination of the treated water. This contrasts with standard anaerobic bioreactors for treating such contaminations, for which the Ministry of Health requires additional barriers including MF, UV and GAC.

The present study examined the role of membrane type in performance of the IEMB in removing perchlorate and other anions (nitrate and chlorate) at levels of hundreds mg L^{-1} from ground water with a mixture of high levels of oxyanions. This was based on Donnan dialysis studies and was studied initially for synthetic and actual ground water fed to the water side while feeding 0.1 N of NaCl to the bio-compartment. To understand the results, conductivity and selectivity measurements were carried out on the different membranes. The anion exchange membranes examined included ACS (Tokuyama Soda), PCA-100 (PC-Cell GmBH), and AMI-7001 (Membrane International).

For ACS membrane and all experiment conditions studied here, perchlorate dominates the flux across the AEM and fluxes deviated from linear dependence on the driving force at less than 0.5 mM. Even though perchlorate concentration is considerably lower than nitrate and chlorate, its flux is greater than the sum of the other anion fluxes. At an effective perchlorate driving force (EDF) value above 0.7 [mM] perchlorate had a negative effect on the other anions resulting in a decrease in the flux for both nitrate and chlorate while its own flux only increased.

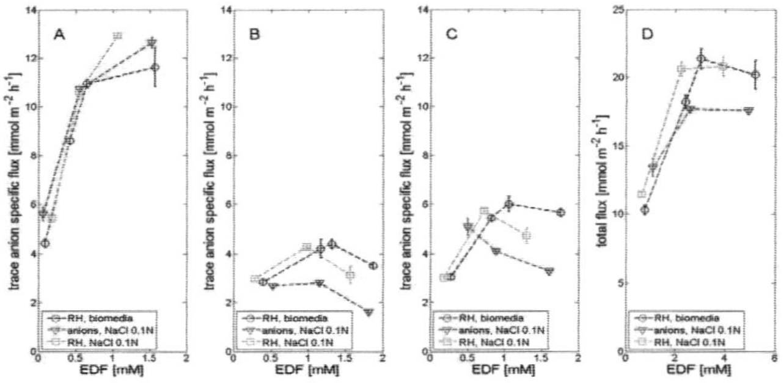

Figure 1 – Results of Donnan dialysis on ACS membrane. Oxyanions specific flux vs. the Effective Driving Force (EDF) of (A) perchlorate, (B) chlorate, (C) nitrate and (D) anions total flux vs. the total EDF, for IEMB Donnan dialysis experiments fed with (○) RHGW and bio media, (▼) synthetic anions solution and NaCl 0.1N and RHGW and NaCl 0.1N (■) into the water compartment and bio-compartment, respectively.

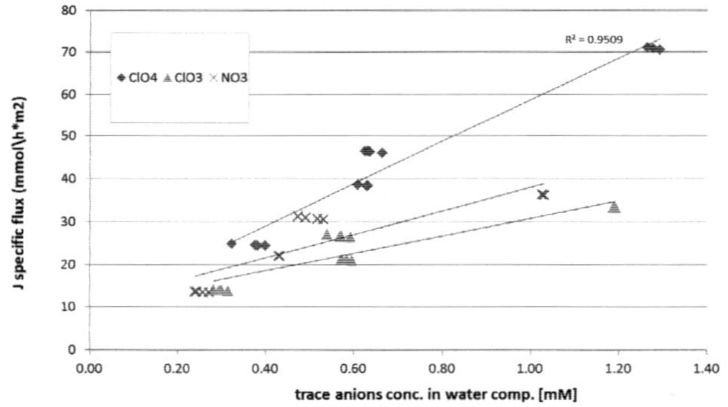

Figure 2: Specific flux in Donnan dialysis on synthetic mixture of oxyanions using PCA-100 anions exchange membrane. Bio-side contained 0.1 N NaCl.

On the other hand, experiments with PCA-100 membranes showed a linear dependence of flux on driving force to much higher levels and much less intereference of the perchlorate on the transport of nitrate and chlorate (Figure 2). Tests of membrane conductivity and regenerability supported the improved results with the PCA-100 membranes. Another type (AMI-7001) also showed promise based on conducitivity but may be problematic in terms of polymer structure. The membrane selectivity for perchlorate and the other oxyanions over chloride was clearly lower for the PCA-100 membranes than ACS. A qualitative model will be proposed to explain these findings and relate it to the differences in ion exchange membrane structure.

If time allows, the implications of these findings for operating the IEMB as a plug flow reactor will be outlined.

[1] A.D. Fonseca, J.G. Crespo, J.S. Almeida, M.A. Reis, Drinking water denitrification using a novel ion-exchange membrane bioreactor, Environ. Sci. Technol. 34 (2000) 1557–1562.

[2] S. Fox, Y. Oren, Z. Ronen, J. Gilron, Ion exchange membrane bioreactor for treating groundwater contaminated with high perchlorate concentrations., J. Hazard. Mater. 264 (2014) 552–9.

470031 Economic Comparison of Pressure Driven Membrane Processes to Electrically Driven Processes for Use in Hydraulic Fracturing

Abstract:

Hydraulic fracturing has provided many new resources for the United States in oil and gas. These resources continue to be an economic driver going forward in many areas of the country. However, significant challenges remain with the environmental sustainability of this technology. One of the areas that is of particular concern is the re-use of the water that is used in fracking operations. The current method of dealing with this water, deep well injenction, takes water out of the eco-system and has been shown to cause earthquakes. This water has high levels of suspeneded and disolved organics as well as high levels of dissolved salts. Because of this the osmotic pressures of these streams are significantly more than seawater reverse osmosis in several instanced. Our group has looked at several possible alternatives of dealing with this water including reverse osmosis, nanofiltration, electrodialysis, and electrodeionization. All of these technologies have different advantages depending on the concentration of the water and ultimate re-use objective. This study has shown that electrodeionization is the most flexible technology, due to selective ion removal and high water recoveries, with the potential to work for both re-use and drinking production as well as acceptable economics at high and low ion concentrations. Experimental results as well as economic modeling will be presented in this talk.

472208 Evaluation of the Cactus Based-Mucilage As an Alternative Natural Dispersant on Toxicity, Effectiveness, and Surface Tension of Oil-in-Water Emulsion

Abstract:

Cactus based-mucilage has the advantage of being biodegradable and can be obtained through sustainable agriculture. This functional natural surfactant was extracted from the *Opuntia ficus indica* cactus plant, and the extraction process yielded two types of mucilage: the non-gelling extract (NE) and the gelling extract (GE). The properties of the cactus plant-based mucilage were studied on the toxicity, surface tension, droplet size and stability of dispersed crude oil in water emulsions (O/W emulsions). The toxicity of the mucilage and conventional dispersants (Corexit 9500A) were evaluated by a standard EPA toxicity test using Daphnia magna colonies exposed to both NE and GE mucilage extracts in concentrations ranging from 0 to 2000 mg/L for 48 hours. It has shown that mucilage can be classified as practically non-toxic to the Daphnia magna colonies with a LC_{50} above 2000 mg/L. The LC_{50} of Corexit 9500A was below 2 mg/L, which is moderately toxic to the Daphnia magna colonies. Cactus mucilage extracts and conventional dispersants has been evaluated by testing the surface tension and measuring the droplet size of O/W emulsions under different concentrations, including three concentrations of oil (3, 6, 30% v/v) and several dispersant to oil ratios (1:1, 1:3, 1:10). Synthetic seawater and surrogate oil from BP were used. The results of the dispersion obtained using mucilage extracts were compared with using Corexit 9500A. The surface tensions of emulsions with cactus mucilage were similar as with Corexit 9500A. The average droplet size in the systems with cactus mucilage (1.5% of NE has generated 5 μm droplets) were smaller when compared with the systems using the Corexit 9500A (1.5% of Corixit 9500A has formed 6.2 μm droplets). Smaller droplet size implies higher emulsion stability, and may improve the biodegradation of oil. An improved laboratory dispersant testing protocol developed by EPA, baffled flask test, was performed to measure the dispersion effectiveness at DOR 1:20, 1:50, and 1:100. The effectiveness of DOR at 1:20 and 1:50 were around 35%. Cactus mucilage is readily available through sustainable agriculture, it is inexpensive to

extract, and has exhibited a long shelf live. Therefore, cactus mucilage can be an alternative technology to mitigate the damage that oil may cause to the aquatic ecosystem and minimize undesired effects associated with the use of synthetic dispersants in oil spills.

462488 Optimization of UF Backwash with Real-Time Control of Coagulant Dosing

Abstract:

Over the past two decades, reverse osmosis (RO) has emerged as the leading technology for seawater and brackish water desalination. However, membrane fouling remains a major challenge for robust RO operation. Hence, effective RO feed pre-treatment is vital to long-term operation of RO desalination plants. Amongst existing RO pre-treatment technologies, ultrafiltration (UF) has proven to be effective in producing consistently high quality filtrate for RO feed. One of the major advantages of UF membrane pre-filtration is the ability to reverse flow through an UF module (a process known as backwash) and thereby removing the foulant layer. After UF backwash, a portion of the foulant may remain on the membrane (i.e., not removed by backwash) as irreversible foulant layer. Irreversible fouling is typically removed through chemical washing which can significantly increase RO feed pretreatment. UF filtration performance can be increased via coagulant dosing of the raw feed water. Studies have also shown that UF backwash effectiveness improves as coagulant dose increases up to a certain threshold, beyond which UF backwash effectiveness no longer improves via additional coagulant dosing. Optimization of coagulant dosing is therefore critical in order to reach both optimal UF and backwash effectiveness while reducing coagulant consumption. However, real-time coagulant dose optimization is a major challenge when feed water quality is temporally variable both in the short term as well as seasonally. Under such conditions coagulant dose optimization via static jar tests is inappropriate. Therefore, in the present work, a novel method is introduced to allow real-time adjustment of coagulant dose so as to ensure effective UF backwash even under conditions of changing water feed quality. In the present method, the change in initial UF filtration resistance (post backwash) is used as a metric of UF backwash effectiveness. The above information is utilized by a coagulant dose controller designed to utilize the above metric to determine whether the coagulant is being underdosed or overdosed, and correspondingly adjust the coagulant dose to the UF feed. The controller was implemented in an integrated UF-RO seawater desalination pilot plant (capable of 18,000 GPD production of fresh water). The controller performance was evaluated, in a field study in California, with respect to its performance,

including during a storm event in which raw seawater feed quality rapidly worsened, and its ability to reduce coagulant consumption. Results of the study demonstrated that the present approach enabled system operation that was able to adapt to worsening feed water quality conditions while reducing coagulant dose by up to ~29% relative to operation without real-time coagulant dose optimization.

451292 Performance Analysis of Texas Eagle Ford Shale Oil Hydro Fracturing Produced Water Treatment Process

Abstract:

Performance Analysis of Texas Eagle Ford Shale Oil Hydro Fracturing Produced Water Treatment Process

Chris Skrivanos, Nico Dwarica, Cameron McKay, John Allen Floyd and Mahbub Uddin

Trinity University, San Antonio, Texas

Extended Abstract

With the rise of horizontal drilling, hydraulic fracturing has become increasingly common and it is estimated that 90 percent of new wells drilled in the United States are hydraulically fractured. Hydraulic fracturing involves the injection of a very large quantity of high pressure liquid to fracture the rock formation and pry open the fracture. This fracturing allows natural gas/oil to flow freely from the formation into the well for collection. The components used in the hydraulic fracturing process consist primarily of water, sand, salts, gel, diluted acid, surfactant corrosion inhibitor, scale inhibitor, and other additives to aid in the hydro fracturing process. During the hydro fracturing operation when the pressure used to inject the fracturing fluid into the well is released, some of the fluid returns to the surface; this fluid is called "flowback" water. Once gas production begins at the well, all wastewater emerging from the well is called "produced water". Both types of wastewater – flowback and production phase water – contain potentially harmful constituents. These constituents can be broadly grouped into several principle categories: salts, organic hydrocarbons, metals, chemical additives and some naturally occurring radioactive material. Because of these constituents, shale oil/gas wastewater should not be released into the environment without adequate treatment. The purpose of this paper is to report the performance analysis of a water treatment process of Texas Eagle Ford Shale oil hydro fracturing produced. The treatment process consists of four primary unit operations: Adsorption column, Reverse osmosis, Ion-Exchange and Pump. The produced water is first pumped through the adsorption column, then a second high-pressure pump which then forces the water through the reverse osmosis membrane. The permeate of the membrane then flows through two ion exchange columns and received in a collection bin.

Adsorption is the adhesion of atoms, ions, or molecules from a fluid or dissolved solid to an insoluble surface and is used to remove suspended organics like oil and grease as well as dissolved organics like benzene and toluene. Our adsorption column consists of a cartridge and housing unit rated for operating pressures of 90 psi and temperatures no higher than 125 °F. The cartridge was 10" tall and accepted 10" x 4.5" cartridges. Reverse osmosis filters work on the principle of membrane divergence. An extremely high pressure (800 psi in our process) is applied at the housing entrance that forces the fluid through the fine membrane. Due to the pressure spike, particles that are small enough pass through the fine holes in the filter transfer, while the larger particles are rejected and thus removed from the membrane. In our application, the reverse osmosis membrane rejected the ions and other larger particles like metals, minerals, and organics, allowing only water and smaller particles to remain. The membrane we used is a FilmTec Sea Water Membrane designed to handle increased salinity (greater than 10,000 ppm). The reverse osmosis unit operation used had a flow ratio of permeate to retetante of 73 to 27, which was determined by a series of mass balances across the unit operation. We used a mixed ion exchange in order to remove any remaining positive or negative ions. Our process consists of two ion exchange columns in parallel, both of which were 20" vertical columns housed in separate units. The process uses two pumps. The first pump transferred the water from the tank to the adsorption column at a low pressure. The second pump primed the system to the operating pressure, 800 psi, to transfer the produced water through the reverse osmosis membrane and the ion exchange.

Each water sample analyzed was obtained from one of 4 major positions. Position 1: the untreated water, Position 2: after the activated carbon operation, Position 3: after the reverse osmosis operation and Position 4: after the ion exchange operation. Produced water processed by the system once is defined as first iteration treated water. Produced water processed by the system twice is defined as second iteration treated water. The first iteration sample collection began after flowing untreated produced water through the system for 60 seconds. The second iteration was run identically to the first iteration except produced water that had previously been treated by the system was used instead of untreated produced water. Once collected, samples were labeled according to the iteration, position and the trial number of extraction.

In order to assess the effectiveness of each unit operation and iteration in the wastewater treatment system, a series of experiments were designed to monitor the flow ratio, total dissolved solids present in solution, pH, and change in organic content. Inductively Coupled Plasma Optical Emission Spectrometry (ICP-OES) was implemented as a means to measure the metallic and salt content present in the water at various purification steps. The initial metallic content was determined to be 184.45 ppm, and the initial salt concentration was determined to be 2262.26 ppm for the produced water. A simple pH test using a probe was conducted on samples to monitor the change in acidity of the water as it passed through each unit operation, and the initial pH of the produced water was determined to be 4.25. Gas chromatography measurements were implemented in order to monitor the changing levels of total organic compounds present in the water.

Analysis of Results

First iteration ICP-OES results showed that the initial salt content of the water was 2262.26 ppm and the heavy metal content was 184.45 ppm. The carbon adsorption column removed 99.01 percent of the initial salt content and 98.5 percent of the initial heavy metallic content. The reverse osmosis membrane removed an additional 0.14 percent and 0.16 percent of the initial salt and heavy metal content respectively. After the ion exchange column approximately 100 percent of the total salt content was removed and 99.84 percent of the total heavy metals were removed, leaving a heavy metal concentration of 0.29 ppm and the salt concentration of essentially 0 ppm. Analysis of the second iteration treatment indicated the carbon adsorption column added 0.37 ppm of heavy metal and 2.09 ppm of salt to the water. The reverse osmosis membrane removed 0.29 ppm of heavy metals and added 0.85 ppm of salt content. The ion exchange column removed 0.04 ppm of heavy metals and 1.55 ppm of salts, bringing the final concentration of the second iteration treated water to 0.32 ppm of heavy metals and 1.39 ppm of salts. The first iteration pH results showed that the produced water had an initial pH of 4.25. The carbon adsorption column increased the pH to 8.26, the reverse osmosis membrane lowered the pH to 7.24, and the ion exchange column lowered the pH to the final value of 5.31. Second iteration treated water pH results showed that the carbon adsorption column increased the pH from 5.31 to 8.27. The reverse osmosis membrane decreased the pH to 7.24. The ion exchange column brought the pH to a final value of 6.49. The produced water contained a plethora of unknown organic constituents and it would be unpractical to identify each individual constituent to generate a calibration curve. So instead of measuring the changing concentrations of each constituent, the GC measurements were used to observe the change in the overall organic content in the water, relative to position 1 of the first iteration which is representative of the initial organic content of the produced water. The first iteration GC results showed that the carbon adsorption column removed 48.4 percent of the initial organic content, the reverse osmosis membrane removed an additional 28.7 percent of the initial organic content, and the ion exchange column removed 6.4 percent of the total organic content; the system overall removed 83.5 percent of the initial organic content.

Conclusions and Recommendations

The produced water treated in our process in a single iteration met the EPA standards for acceptable levels of metallic and salt content for reclaimed water, however the pH is at the lower end of the standard. The first iteration treatment reduced the metallic content by 99.84 percent, reduced the salt content by essentially 100 percent, raised the pH from 4.25 to 5.31 and reduced organic content by 83.5 percent. With a second iteration, the treated water pH is within the acceptable range for reclaimed water. The second iteration was shown to be unnecessary since the metallic and salt concentration changed negligibly, and the organic content increased. The pH was raised to acceptable levels, but the addition of a controlled injection pH adjustment unit operation would allow the pH to be raised without the need for a second iteration, making the produced water treatment process more efficient. This change to the process would generate treated water that meets the EPA drinking water standard. Our analysis indicates that the wastewater treatment processes has applications in commercial and residential areas where hydraulic fracturing is common, to treat wastewater that meets EPA standards.

461844 Macroalgae and Chitosan-Macroalgae Biocomposite As Potential Adsorbents of Water-Soluble Hydrocarbons: Effect of pH, Organic Matter and Ionic Strength

Abstract:

Macroalgae and Chitosan-macroalgae biocomposite as potential adsorbents of water-soluble hydrocarbons: Effect of pH, organic matter and ionic strength.

Abstract

Nowadays the recovery of low-molecular aromatic hydrocarbons (HC) released into natural bodies of water continues to be a challenging task. These contaminants bring sever consequences to the environment and to the human health. The oil spill cleanup strategies are primarily design to deal with the heavy fractions accumulated in the water surface. Unfortunately, very limited information is available regarding the treatment of dissolved fractions. Biosorption on macroalgae (Ma) biomass seem to be a potential alternative to overcome disadvantages associated to high-priced costs and complexity of activated carbon production. Three seaweeds of a representative sample of brown, green and red macroalgae (*Macrocystis pyrifera*, *Ulva expansa* and *Acanthophora spicifera*, respectively) were evaluated to remove benzene and toluene from water, which are the most soluble hydrocarbons. Toward this objective, the influences of temperature and ionic strength were also determined. Raw biomasses were characterized by different physical and chemical techniques to assess their potential as biosorbents and the mechanisms involved. The organic and inorganic dissolved fractions were also considered. The characterization techniques mainly identified carbohydrates, lignin and proteins. *M. pyrifera* biomass registered the highest removal capacities: about 112 and 28 $mg \cdot g^{-1}$ for benzene and toluene, respectively, probably attributed to alginate and lignin content. It was found that the hydrocarbons biosorption affinity was not affect up to an ionic strength of 0.6 M, due to the biosorbents pore occlusion by water clusters. For

benzene and toluene nonpolar solutes, the biosorption mechanism is an addition of simultaneous interactions between the sorbents and the chemical constituents of the cell wall, presumably by London (dispersion) forces onto â€“OH and â€“NH active sites from polysaccharides and proteins and hydrophobic interactions from lignin and lipid fractions.

Despite these advantageous properties, the physical characteristics (small particle size, low strength and density) of such biomaterials are not viable for a continuous process operation and make biomass difficult to apply. Therefore the development of innovative low-cost techniques of immobilized biosorbents with special attention to increase their effectiveness in the biosorption process is mandatory. The biopolymer-biopolymer interactions between positive and negative macromolecules can enhance the compatibility and stability of biosorbents, and these new structures are called polyelectrolytes complexes (PEC). The linear amino polycationic compound of chitosan (Ch) possess remarkable quantity of amine and hydroxyl groups that interacts with negatively charged substances with excellent chelation behavior and high affinity toward pollutants like metals and dyes. On the other hand pectin (Pe) is a negatively charged polysaccharide composed mostly by esterified D-galacturonic acid residues linked together by Î±-(1-4) chain. Pectine molecule is also used in the food industry for its well-known thickening, stabilization and gelling properties.

Â The main objective of this research was to explore the performance of Ma and PEC complexes of different proportions of Ma/Ch/Pe on biosorption capacity of the two main soluble hydrocarbons (benzene and toluene) in water. The biocomposite synthesis was optimized by application of the factorial design and response surface methodology. Moreover, a through chemical and physical characterization analysis, by textural properties, potentiometric titrations, elemental content, chemical stability, KBr-FT-IR and TGA analysis, were performed to explain the adsorption mechanisms.

458236 Concentrating 17 α-Ethinyl Estradiol (EE2) Using PES Hollow-Fibre Membranes: Experimental Exploration and Molecular Modelling

Abstract:

The 17Î±-ethinyl estradiol (EE2) is a synthetic estrogen derived from 17Î²-estradiol (E2), the major endogenous estrogen in humans. EE2 is widely used as an active ingredient for birth control and other hormonal treatments. EE2 is excreted into the sewage and the environment mainly via urine. Wastewater treatment plants are usually ineffective in removing EE2 from water which results in the release of EE2 into the environment. The concentration of EE2 in surface water bodies are typically at sub-ppb levels. Even at such low concentration, several studies have reported negative environmental effects of EE2 including feminization of aquatic animals. At present, it is not known if EE2, at such low concentration, has any adverse effect on human or other mammals over short and long term exposure. Developing novel strategies for the efficient removal and treatment of EE2 from water has become an imperative in water reclamation and reuse. Here, we report a novel approach whereas a widely used polyethersulfone (PES) membrane was used to concentrate EE2 through adsorption and solvent elution. The concentrated EE2 can then be further treated with high efficiency using approaches such as biodegradation. Molecular mechanisms underlying the interaction between EE2 and PES membrane was elucidated using molecular dynamic simulation.

454263 Natural Organic Matter Fouling Behaviors on Superwetting Nanofiltration Membranes

Abstract:

Natural organic matter fouling behaviors on superwetting nanofiltration membranes

Linglong Shan, Hongwei Fan, Hongxia Guo, Shulan Ji and Guojun Zhang

Beijing Key Laboratory for Green Catalysis and Separation, College of Environmental and Energy Engineering, Beijing University of Technology, Beijing, China. Email: shanlinglong@126.com

In natural water, natural organic matter (NOM) components have been identified as precursors of disinfection by-products. Nanofiltration has been widely recognized as a promising technology to remove NOM from natural water. Currently, the most challenging aspect is the often inevitable occurrence of membrane fouling, which is caused by NOM accumulation on the membrane surface, as in NOM removal in drinking water treatment and production. NOM fouling is considered to be governed by the interface relationship between NOM and the membrane surface in aqueous solution. Much work has been carried out, but the effect of hydrophilicity/-phobicity of NOM and the membrane surface on fouling behavior is still a controversial issue. [1-7] The controversy originates from the unilateral consideration of membrane wettability or foulant hydrophilicity/-phobicity, when in fact both need to be considered. In our study, both superhydrophilic and superhydrophobic membranes with stable structure have been successfully obtained using two facile hybrid approaches (see Fig. 1). The fouling behavior of NOM on superwetting nanofiltration membranes has been extensively investigated using humic acid (HA: model foulant) and real natural water. The extended Derjaguin-Landau-Verwey-Overbeek (XDLVO) approach and nanoindentor scratch tests suggested that the superhydrophilic membrane had the strongest repulsion force to HA due to the highest positive ¦¤G^{TOT} value and the lowest critical load. Excitation emission matrix analyses of natural water also indicated that the superhydrophilic membrane showed resistance to fouling by hydrophobic substances and therefore high removal thereof. Conversely, the superhydrophobic membrane showed resistance to fouling by hydrophilic substances and therefore high removal thereof (see Figs. 2 and 3). Long-term operation suggested that the superhydrophilic membrane had high stability due to its anti-NOM fouling capacity, which is crucial for practical application. Based on the different anti-fouling properties of the studied superwetting membranes, a combination of superhydrophilic and superhydrophobic membranes was examined to further improve the removal of both hydrophobic and hydrophilic pollutants. The R_{UV254} and R_{DOC} could be increased to 83.6% and 73.3%, respectively, by a combination of superhydrophilic and superhydrophobic membranes. In view of the hydrophilic and hydrophobic NOM fouling behaviors on superwetting membrane surfaces, the superwetting strategy can be regarded as a very important approach for obtaining a membrane with both high NOM rejection and good anti-NOM fouling properties.

Keywords: Superhydrophilic, Superhydrophilic, Nanofiltration, NOM, Fouling

References

[1] L. Shan, H. Fan, H. Guo, S. Ji, G. Zhang, Natural organic matter fouling behaviors on superwetting nanofiltration membranes, Water res., 93 (2016) 121-132.

[2] L. Shan, H. Guo, Z. Qin, N. Wang, S. Ji, G. Zhang, Z. Zhang, Covalent crosslinked polyelectrolyte complex membrane with high negative charges towards anti-natural organic matter fouling nanofiltration, RSC Adv., 5 (2015) 11515-11523.

[3] M. Ng, S. Liu, C.W.K. Chow, M. Drikas, R. Amal, M. Lim, Understanding effects of water characteristics on natural organic matter treatability by PACI and a novel PACI-chitosan soagulants, J. Hazard. Mater., 263 (2013) 718-725.

[4] J.A. Nilson, F.A. DiGiano, Influence of NOM composition on nanofiltration, J. Am. Water Works Assoc., 88(5) (1996) 53-66.

[5] H. Yamamura, K. Okimoto, K. Kinura, W. Yoshimasa, Hydrophilic fraction of natural organic matter causing irreversible fouling of microfiltration and ultrafiltration membranes, Water Res., 54 (2014) 123-136.

[6] L. Zhang, G. Chowdhury, C. Feng, T. Matsuura, R. Narbaitz, Effect of surface-modifying macromolecules and membrane morphology on fouling of polyethersulfone ultrafiltration membranes, J. Appl. Polym. Sci., 88 (2003), 3132-3138.

[7] A. Tiraferri, Y. Kang, E.P. Giannelis, M. Elimelech, Superhydrophilic thin-film composite forward osmosis membranes for organic fouling control: fouling behavior and antifouling mechanisms, Environ. Sci. Technol., 46 (2012) 11135-11144.

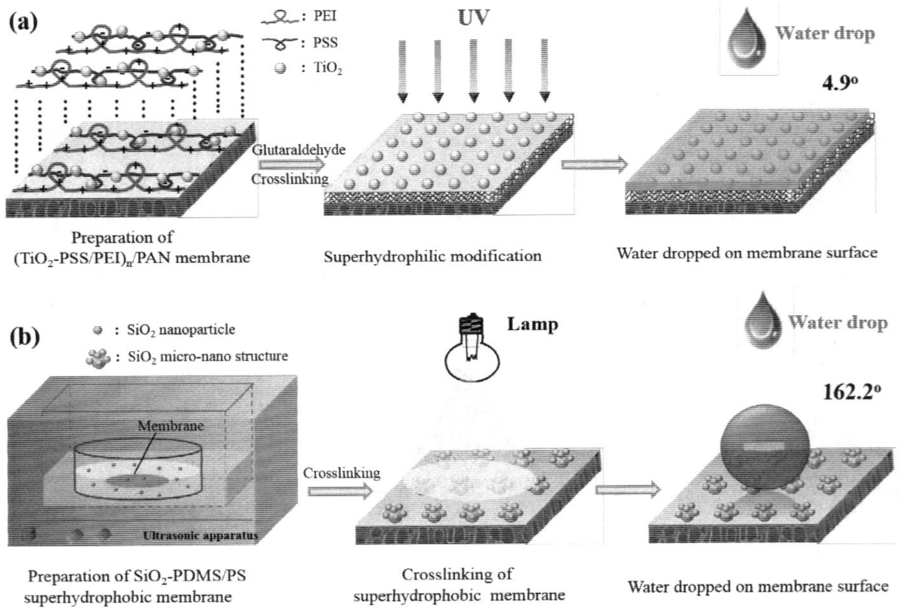

Figure 1. (a) Schematic illustration of the preparation of the superhydrophilic membrane and its surface wettability; (b) schematic illustration of the preparation of the superhydrophobic membrane and its surface wettability.

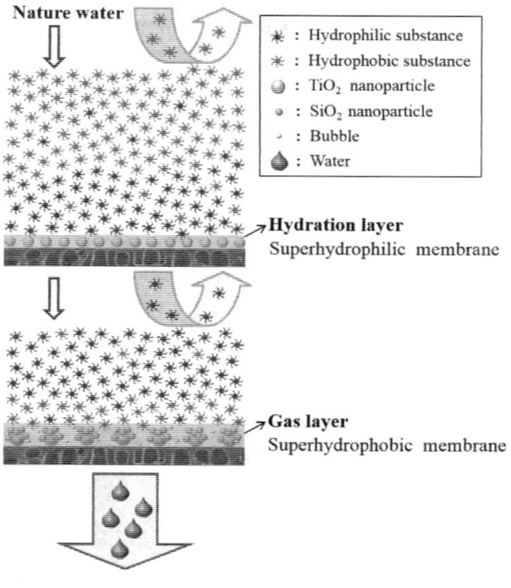

Figure 2. Schematic illustration of different anti-fouling properties of the studied superwetting membranes. Superhydrophilic membrane showed resistance to fouling by hydrophobic substances, while the superhydrophobic membrane showed resistance to fouling by hydrophilic substances.

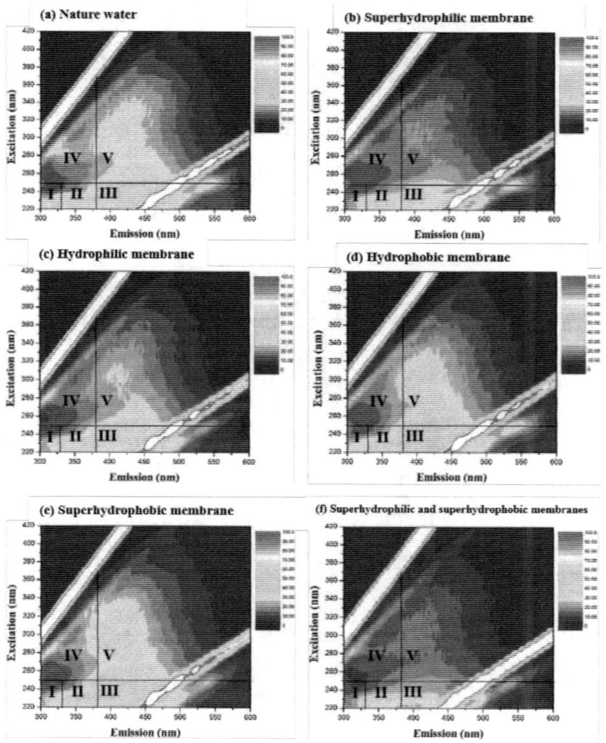

Figure 3. EEMs for (a) natural water; (b) natural water nanofiltrated by superhydrophilic membrane; (c) natural water nanofiltrated by hydrophilic membrane; (d) natural water nanofiltrated by hydrophobic membrane; (e) natural water nanofiltrated by superhydrophobic membrane; (f) natural water nanofiltrated by superhydrophilic membrane-superhydrophobic membrane.

Experimental and Simulation Analysis for Optimization of Solar Desalination System Using Humidification-Dehumidification Method: Open and Close Air Cycle

Nickyar Ghadirinejad [1], *Farshad Farshchi Tabrizi* [2], *Mohammmad Mehdi Zarei* [3], *Taher Chegini* [4], *Roberto Moreno-Atanasio* [5]

[1] Chemical Engineering Department, University of Sistan and Baluchestan, Zahedn, Iran
[2] Chemical Engineering Department, Shiraz University, Shiraz, Iran
[3] Chemical Engineering Department, Khazar University, Baku, Azerbaijan
[4] Civil Engineering Department, Khazar University, Baku, Azerbaijan
[5] Chemical Engineering Department, Newcastle University, Callaghan, Australia

Abstract:

Over the last decades, desalination of brackish water and seawater has been studied extensively for producing fresh water. Desalination technologies are divided into two major categories: Thermal and Membrane. Humidification-Dehumidification (HDH) which is the subject of the present research, is one of the thermal technologies. It is inspired by the rain cycle and is based on evaporation and consecutive condensation.

In a series of experiments, the processes of production of fresh water has been investigated under various operational parameters such as temperature and flowrate of cooling water, humidifier intake from water boiler, air flowrate and height of packed bed column. Subsequently, an appropriate mathematical model of the system has been developed and validated using the collected data. Furthermore, based on the developed model a series of simulations has been carried out in ASPEN Plus in order to obtain optimum physical parameters of various parts of the desalination system.

The experimental results showed that under constant heat flux and optimum air flowrate, three parameters increase the production of distilled water the most: increasing flowrate of cooling water, decreasing flowrate of water inlet and increasing temperature. Moreover, although increasing flowrate of cooling water leads to an augmentation in the distilled water flowrate, but this effect is mitigated at high rates.

Furthermore, based on the obtained overall heat transfer coefficients of the humidifier and condenser, the most reliable Height Equivalent to a Theoretical Plate (HETP) model of random packing is found to be Mangers & Ponter. The estimation error of the model in predicting the height of packed bed was less than 4%. Based on the results of this model, closed air system increases the efficiency of the production of distilled water by 15% in comparison to open air. This increase is the result of prevention of wasting of air humidity in the condenser outlet.

Finally, optimum parameters of operational units are obtained by applying a Differential Evolution (DE) strategy. The optimum values of flowrate and temperature of cooling water are 0.0298 m^3/hr, 13 °C, respectively. The best value of flowrate of closed air is found to be 0.008075 m^3/hr and of water intake to be 0.021497 m^3/hr. These values produced fresh water with flowrate of 0.3777 m^3/hr.

Keywords: Solar Desalination, Dehumidification, Humidification, Heat Transfer

1- Introduction:

The rapid growth of industries and population, rivers and lakes pollution problems caused by industrial wastewater and flowing municipal wastewater into freshwater resources have caused a shortage of fresh water [1]. However, it should be noted that the only sources of never-ending water reservoir are oceans; water salinity reduction to overcome the shortage of fresh water, is a more applicable solution [2]. Desalination includes many different methods and required energy for this procedure also encompasses a wide range. The heat produced by fossil fuels and/or the excess heat-released of nuclear reactions, renewable energies such as ocean waves or solar energy, all can supply heat power for water desalinisation plants. Desalination technologies are divided into two categories: Thermal and membrane methods. Water is heated in thermal processes and after evaporation, fresh water is produced via condensation. In membrane methods, thin permeable membranes are used to separate feed into both fresh water flow and more saline flow [3].

In recent years, Humidification-Dehumidification (HDH) system, as a new method has been investigated; through this system, air flows as the intemediate fluid. In fact, this method is a combination of membrane processes and thermal distillation. It operates based on the mass diffusion principles and uses cooling air due to water evaporation and humidification. Fresh water is resulted via dehumidification (condensation of humid air). The obvious benefit of HDH is to be one of the most cost-effective desalination mechanism regarding to terms of low temperature and pressure [4]. The investigated method includes solar collector/s, humidification tower/s (evaporation unit), dehumidification tower (condensation unit) which even to achieve higher efficiency, it could include heat recovery unit as well. In order to increase the efficiency of heat transfer, at first, water is passed through the condenser coil, by heat absorption from hot air, water is pre-heated. Afterward, the pre-heated water is led into a solar heater and an additional heater (auxiliary heater) subsequetly to reach a higher temperature. In the next step, hot water is showered into the humidification tower.

2- Closed-air, open-water (CAOW) water-heated systems:

Alluded humidification tower including random packings/ Bubble Columns/ Wetted-wall Columns [5], showered with hot water, gets totally moist using counter-current air flow from the botton; subsequently, after heat absorption, hot air is guided through the condenser to produce condnesate-water (fresh water). In closed air cycle, after passing the condenser, the air is recycled into the system. HDH method has been studied by Al-Hallaj et al. [6,7], Ben-Bacha et al. [8], Garg et al. [9] ،Nafey et al. [10] ،Nawayseh et al. [11], [12] Muller-Holst.

In order to achieve optimum operational conditions and achieve the best yield, many research projects have been studied [13,14,15,16,17] and various mathematical models have been proposed considering measurement of mass and heat transfer coefficients [13], impacts of input parameters into the HDH system, mass and energy balance equations, Consideration of waste heat during HDH process and finding a solution due to heat recovery [18]. To simulate a system including both a flat-plate water-heater and a double-pass solar air heater, assuming some simplified assumptions, Yildirim and Solmus [19] proposed an optimized mathematical model by using fourth order Runge-Kutta solution method. Cost-estimation of produced fresh water via an elaborated, multiple solar desalination system [20] and via a novel solar HDH system [21] has been investigated.

3-HDH Apparatus:

In the study, the humidifier consisted of emedded random packing (Pall ring type with 25 mm nominal diameter). Water was showered from the top and air wsa flown counter-currently. The coil-condensor was a single 8-meter-long cupper pipe. For circulation of air in side the HDH system, two fans were used, without any direct contact between motors and humid air.

As it shown in figure 1, intake salt water in low temperature (as feed water into the HDH system) entered into the condenser's shell, and after relative pre-heating subsequently entered into the solar heat collector. During experimental research, to achieve higher temperatures an electrically additional heater was used. Heated saline water was showered inside the humidifier. Water flow rate was regulated by passing through some valves and rotameters to reach desired amounts. Air-flow rates adjusted through entire HDH system using fans' regulators. By directly counter-current contact between hot saline water and air in random-packed bed, the air temperature got augmented. Consequently, it tended to embrace greater capacity to absorb more moisture. Passing through the dehumidifier, fresh water was by-passed and collected; outlet air was recycled in closed air system and was blowed from the bottom into the humidifier.

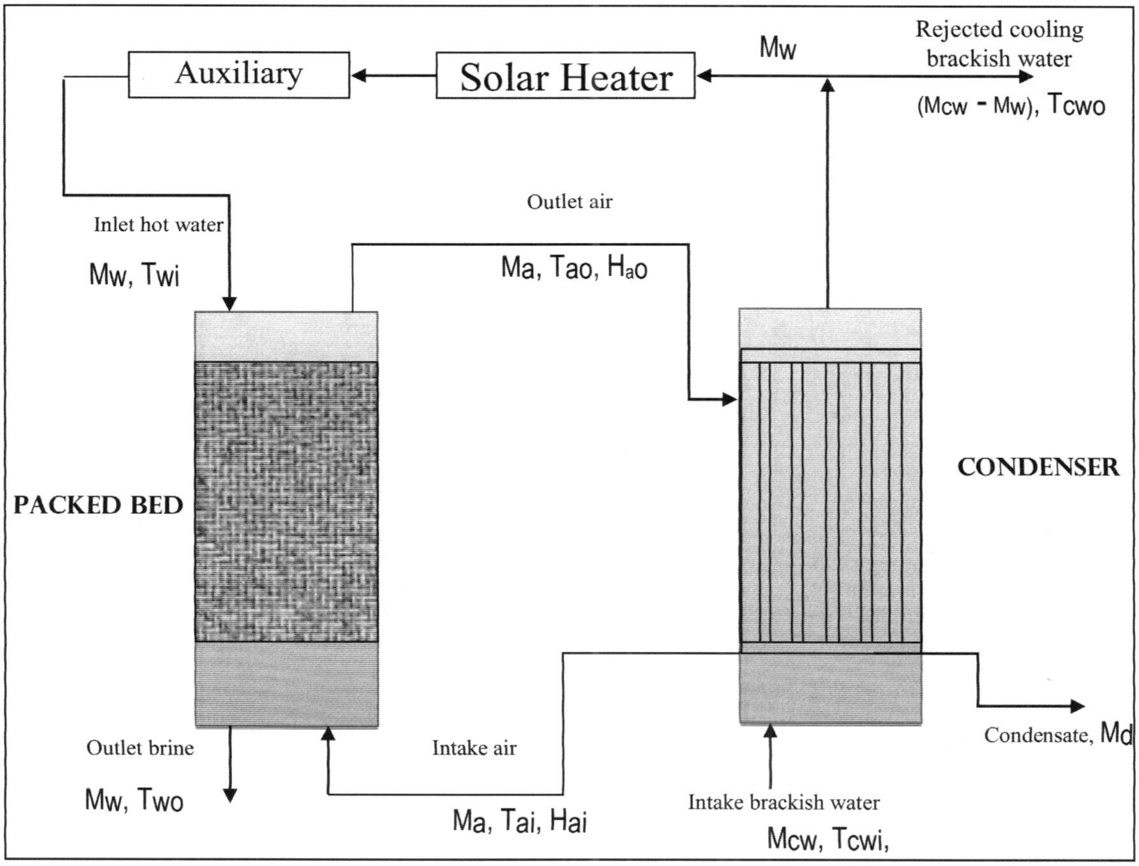

Fig. 1: Layout of HDH system

For experimental data collection, the system was ran continuously for 15 to 20 minutes in the same mode, reaching considered steady-state temperatures and humidity amount in each test; then tests were performed. Each of the experiments were repeated at least three times to ensure their correction.

4- Modeling of humidification-dehumidification system:

In order to determination of diameter and Height Equivalent to a Theoretical Plate (HETP), based on the obtained overall heat transfer coefficients of for the random-packed-bed humidifier and condenser [22], Considering given data about Rasching ring and Pall ring packings [5] following models have been investigated: Shulman et al. [23], Shulman & de Grouff [24], Van Krevelen & Hoftijzer [25,26], Mangers & Ponter [27], Shi & Mersmann [28], Billet and Schultes [29], Onda et al. [30].

The estimation HTEP model of **Mangers & Ponter**:

$$k_L = \frac{(3.22\times10^{8}) Sc_L^{0.5} D_L}{a_s(1-\cos\theta)} \left(\frac{\rho_L u_L}{\mu_L}\right)^{\alpha} \left(\frac{\rho_L^2\, g\, h_{pe}}{\mu_L^2}\right)^{0.27} \qquad (1)$$

As α in the above equation is calculated by:

$$\alpha = 0.49 \left[(1-\cos\theta)^{0.6} \left(\frac{\rho_L \sigma_L^3}{\mu_L^4 g}\right)^{0.2}\right]^{0.108} \qquad (2)$$

$$k_G = 0.2\, \frac{D_G}{d_c} \left[\frac{\rho_L u_L}{u_L a_p}\right]^{0.8} Sc_G^{\frac{1}{3}} \qquad (3)$$

Additionaly, for comparison the results obtained from the mathematical modeling and practical experiments, both physical and chemical operational behaviors of HDH units -at steady state conditions- have been investigated by modeling in ASPEN Plus. Initial feed's conditions to HDH system, constituted 3884.8 ppm salinity, Ca^{2+}, Mg^{2+}, K^+, Cl^-, HCO_3^-, SO_4^{2-}, NO_3^- Na^+ aqueous solutions were imported to ASPEN Plus model. Enick's Redlich-Kwong-UNIFAC equations [31] and Pall ring packings were selected for the mentioned processes.

In preliminary calculation, the capacity of existing packing in the dumidifier was assumed 10 HETP. Regrading experimental results, output data from run of each stage in ASPEN Plus were compared.

4-1- Modeling of HDH system in open and close air cycle:

Effects of intake water flow rate and its temperature entering the humidifier, air and cooling water flow rates, and effect of packed-bed height on the yield of desalinated water amounts have been studied by developing mass and energy transfer balances for overally and even on every sinlge unit in the HDH system.

$$Mcw.Cp.Tcwi - (Mcw - Mw).Cp.Tcwo + Q_{Heater} = Mw.Cp.Two + Md.Cp.Ta \quad (4)$$

Saturated-air, as a main simplifying assumption was considered to solve formulated equations. On the contrary, unlike precedent models, the term of air solubility in outlet saline water, as well as unsteady air and water flow rates recycling the entire HDH system was considered [32].

To improve derived equations and to reach more realistic modeling in the range of $0 \leq S_p \leq 16000$ ppm and $10 \leq T \leq 180$ °C, correlations of seawater density (ρ_{sw}), specific heat (C_p), seawater thermal conductivity (K_{sw}) [32] were applied:

$$\rho_{sw} = 10^3(A_1.F_1 + A_2.F_2 + A_3.F_3 + A_4.F_4) \quad (5)$$

Which A_n and F_n are specific variables proposed by Isdale and Morris [33].

Evaporation latent heat (λ) (which is T in K degree)

$$\lambda = 2501.89 - 2.407 \times T + 1.1922 \times 10^{-3} \times T^2 - 1.586 \times 10^{-5} \times T^3 \quad (6)$$

Specific heat coefficient of air (T in K degree)

$$C_{pa} = 1.03409 - 2.84887 \times 10^{-4}.T + 7.816818 \times 10^{-7}.T^2 - 4.970786 \times 10^{-10}.T^3 + 1.077024 \times 10^{-13}.T^4 \quad (7)$$

To calculate air density with respect to Al-Sahali & Ettouney's proposal [13] for HDH desalination plant, air assumed as an ideal gas.

$$\rho_a = \frac{P.M_{wei}}{R.T} \quad (8)$$

Thermal conductivity coefficient of air:

$$K_a = -2.276501 \times 10^{-3} + 1.2598485 \times 10^{-4}.T - 1.4815235 \times 10^{-7}.T^2 + 1.73550646 \times 10^{-10}.T^3 - 1.066657 \times 10^{-13}.T^4 + 2.47663035 \times 10^{-17}.T^5 \quad (9)$$

In order to calculate the amount of dissolved air in the water (dependent on its pressure and temerature) in the humidifier [34] Henry's constants for air solubility in water (K_{sol}) were considered too.

Newton's method was used to solve mentioned non-linear equations, using some most probable intial guesses via experimental data. For optimization of outlet fresh water dependent on Q, Mw, Ma, Tcwi and Mcw values, Differential Evolution (DE) strategy was applied to find optimum values. DE strategy [35], one of genetic algorithm methods, via successive tries to generate correct interpolations was coded in MATLAB software.

In the same method, calculations of open-air cycle in the HDH system were performed by rewriting similar equations as above.

5- Results and discussion:

Keeping constant given thermal energy to the inlet water (including solar heater and additional heater), several sets of experimental data were collected in various conditions. By changing the initial independent parameters, trends of intermediate variables and the final variable (amount of produced fresh water) were evaluated.

5-1 Effect of inlet water flowrate and its temperature on deslination production rate:

At constant thermal energy, inlet water flow rate and its temperature were realized as two dependent vaiables. In this series of tests, parameters of air flow, cooling water flow and packed-bed height were kept constant; and consequently, effects of two mentioned parameters of inlet water entering to the humidifier on the yield of HDH system were investigated (Fig.2). As it is shown, at constant conditions, by increasing the temperature of inlet water, production rate of fresh water increases. It is noticible that during experiments, relative humidty of circulating air was checked continuously and it was near to saturation state (97.33%).

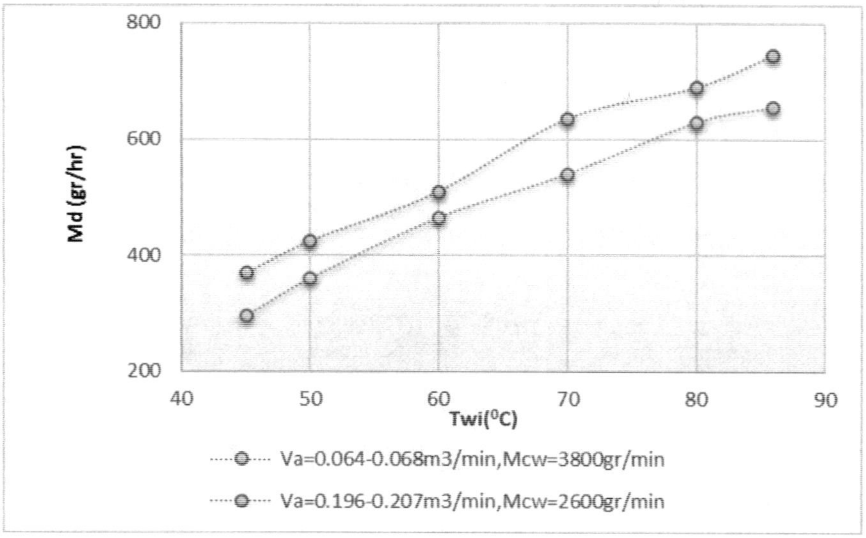

Fig. 2: Effect of temperature of inlet water on produced fresh water

5-2 Survey on air flow rate, considering effects of cooling water flowrate on yield of the HDH system:

As it presented in Figure 3 for three flow rates of circulated-air, the temperature of the exhausted air from the tower is a dependent variable of cooling water flow rate. At constant air flow rate, increasing in water temperature leads to augmentation of air temperature and consequently fresh water production.

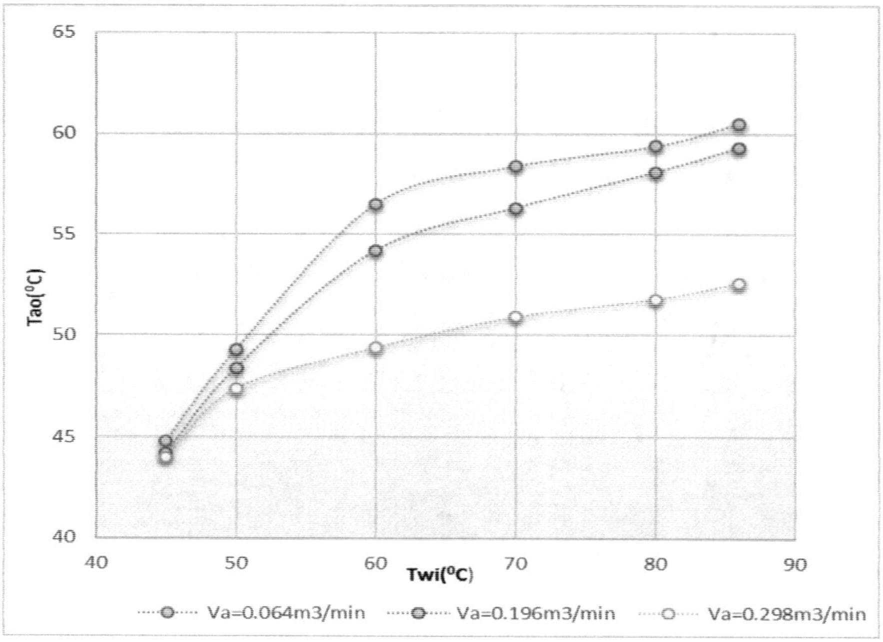

Fig. 3: effects of changes in Twi on Tao

In the present HDH design, to avoid wasting moisture, transfered by exhaust air flow from the condenser, and to spend less energy, close air cycle was contrived. It is noteworthy that the validity of this assumption has been proved by comparing output-results of close and open air models. Experimental tests were conducted in three different ranges (0.064-0.068 $\frac{m^3}{min}$, 0.196-0.207 $\frac{m^3}{min}$ and 0.298-0.310 $\frac{m^3}{min}$). Regarding Fig. 3, 4 and 5, range of 0.196-0.207 $\frac{m^3}{min}$ was determined as the optimal one for the air flow. Obviously, increase of air velocity causes decline of outlet air temperature in the HDH system. Besides, it leads to reduction in air settelment time in the humidifier and finally shrinkage of distilled water.

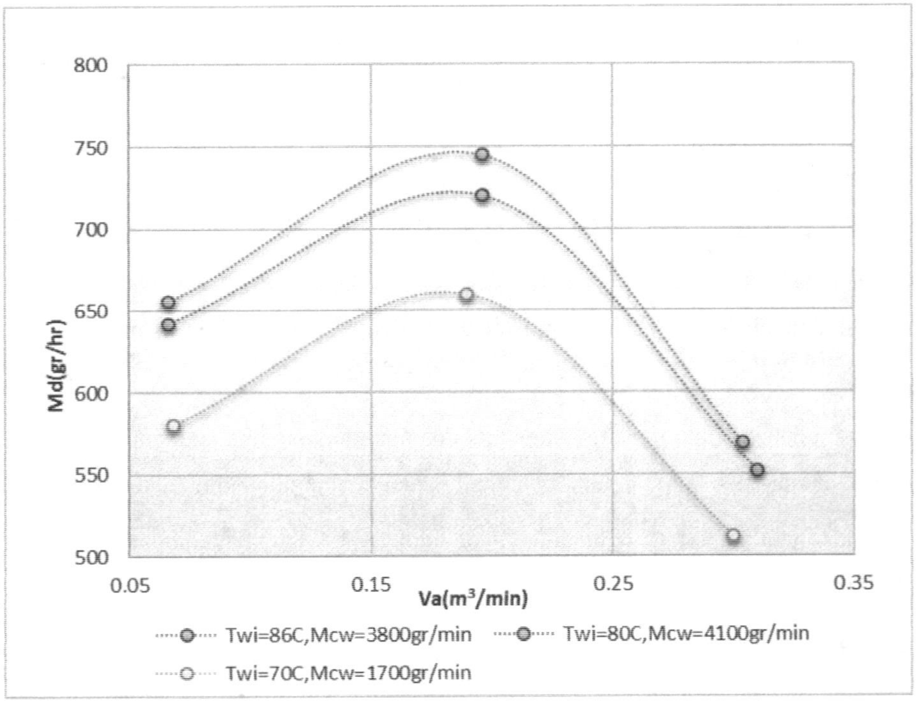

Fig. 4: Trends of M_d in three different air velocities at 70, 80 and 86 °C.

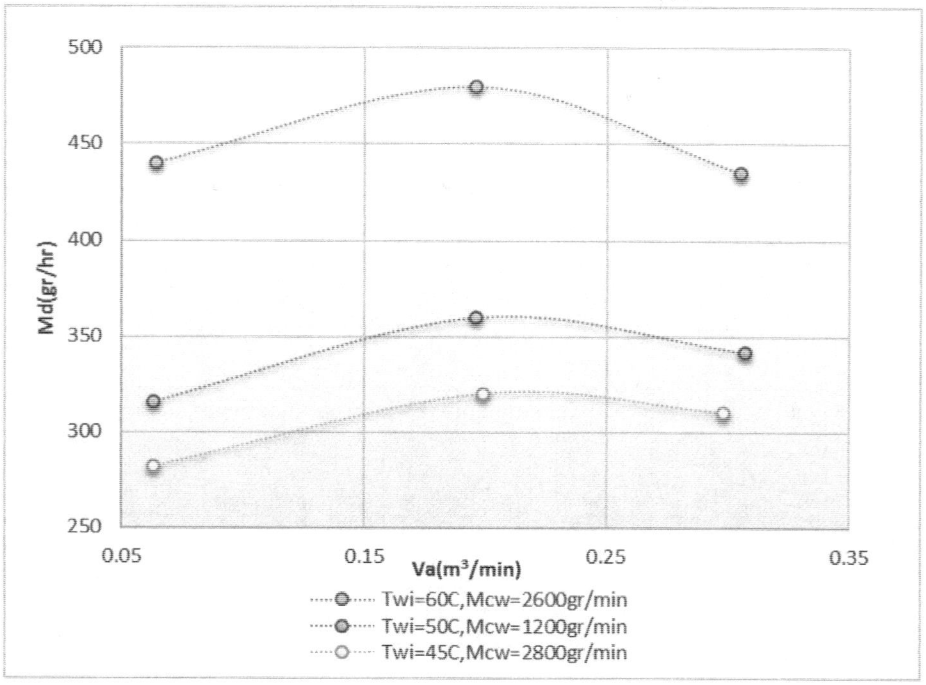

Fig. 5: Trends of M_d in three different air velocities at 45, 50 and 60 °C.

5-1 Distinguishing between results of open and clos air cycle in the investigated HDH system:

This comparibility is shown in table 1. It shows 15% higher productivity of fresh water in close air cycle than production rate in open air cycle. The most probalbe justification for this conlcusion could be waste-heat recovery and lack of moisture loss in the exhaust air from the condenser and recycling to the processes.

Table 1: comparability of 4 samples for open and closed air cycle

Closed cycle				**Open cycle**			
Twi (°C)	Tao (°C)	Tcwo (°C)	Md (m3/hr)	Twi (°C)	Tao (°C)	Tcwo (°C)	Md (m3/hr)
71.6	60.4	32.3	1.70E-04	70.4	56.2	32.1	1.01E-04
91.3	66.5	32.6	1.95E-04	88.4	61.4	31.1	1.50E-04
40.4	41.6	24.9	1.28E-05	37.6	35.1	25.6	1.23E-05
80.2	44.6	33.8	1.71E-04	79.4	39.0	33.2	1.27E-04

5-2 HETP investigation and HDH optimization:

To achieve optimal HETP for humidification tower, several experiments have been executed (Fig. 6). At constant practice conditions, using k_L and k_G equations, the most reliable HETP model has been proposed; Mangers & Ponter's model was determined as the most precise one (Table 2).

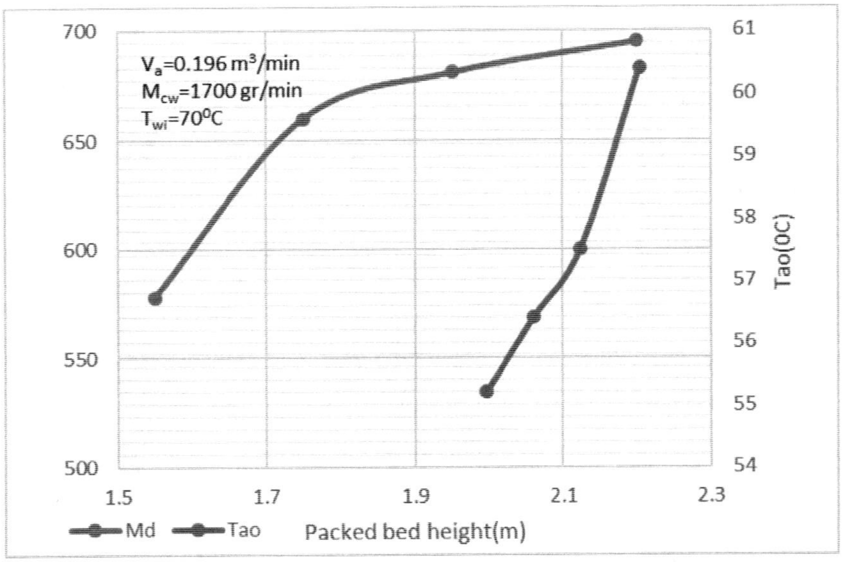

Fig. 6: effects of Packed bed height on M_d ($\frac{gr}{hr}$)

Table 2: estimating HETP models

Model	HETP (m)
Shulman et al.	0.042601
Shulman & de Grouff	0.047809
Van Krevelen & Hoftijzer	0.394546
Mangers & Ponter	0.511026
Shi & Mersmann	0.437404
Billet and Schultes	0.407311
Onda et al.	0.660333

Furthermore, due to admit validity of collected data modeling and simulation in ASPEN Plus have been done. At first, 10-equivalent-step was presumed. After every run of functions implemented in ASPEN Plus and comparison between outcome data and experimental values, it was observed that a tower including two or three steps can help us to meet our expected moisture balance (one single run is shown in table 3).

Table 3: ASPEN Plus simulation for 5-equivalent-setps of HETP

Exp. Cond.	Stages	Twi (°C)	Two (°C)	Tao (°C)	Tcwo (°C)	Md (mol/s)	Mw (mol/s)	Md/Mw
Mw=0.00305 (m3/hr) Ma=0.00108 (m3/hr) Tcwi=28(°C) Tai=32.9(°C)	5	82.31	42.23	66.48	32.9	0.04489	0.6095	0.07365
	4	82.17	43.77	65.9	32.9	0.04334	0.6095	0.07112
	3	81.95	46.18	64.92	32.9	0.04089	0.6095	0.06709
	2	81.56	50.26	62.97	32.9	0.03647	0.6095	0.05985
	REAL	80.4	36.2	58.1	30.2			0.06011

According to the Eckert's proposed coefficients [36] and the use of Pall ring packing with 25 mm nominal diameter, the effective height of humidifier was suggested about 170 cm.

Using Differential Evolution (DE) strategy and via differing the number of dependent variables (such as air flow rate, T_{cwi}, M_w and finally M_{cw}) individually and also simultaneously, oprimized model was correlated successfully. Practical values which were resulted optimum M_d=0.3777 $\frac{m^3}{hr}$ were M_{cw}=0.0298 $\frac{m^3}{hr}$ at T_{cwi}=13 °C, wiht M_a=0.008075 $\frac{m^3}{hr}$ and M_w= 0.021497 $\frac{m^3}{hr}$.

6- Conclusion:

One of the most applicable methods for desalination brackish water is humidification-dehumidification system. This thermal method has been studied on a pilot system. After experimental data collecetion due to data validity proof and also finding an optimum fresh water rate, two mathematical models, one on open air cycle and another one on close air cycle have been derived. Furthermore, HETP for the humidifier was studied by comparing experimental data resulted from various tower's height and data from simulation of HDH system in ASPEN Plus; Estimation error of the HETP modelfor Pall Ring packed bed was less than 4%. Yield in closed air system was shown 15% better than open air system.

Convincingly, using of DE strategy, optimal parameters to obtain the choisest yield (0.3777 m^3/hr) have been provided: 0.0298 m^3/hr, 13 °C respectively for flow rate and temperature of cooling water, 0.008075 m^3/hr air flow rate (in closed air system) and 0.021497 m^3/hr for water inake.

Nomenclatures:

$D\ (m)$	diameter	$C_{pa}\ (\frac{KJ}{Kg.K})$	Specific heat coefficient of air
G(kg/s)	Gas mass flow rate	$\lambda\ (\frac{KJ}{Kg})$	Evaporation latent heat
L (kg/s)	Liquid mass flow rate	$K_a\ (\frac{W}{m.K})$	Thermal conductivity coefficient of air
$M\ (gr/min)$	Mass flow rate	$\mu\ (\frac{kg}{m.s})$	*viscosity*
Ma (kg/s)	Intake air mass flow rate at the bottom of humidification tower	$\rho\ (\frac{Kg}{m^3})$	*density*
Mcw (kg/s)	Mass flow rate of saline water Inlet to cooling tower	$\rho_a\ (\frac{Kg}{m^3})$	Air density
Md (kg/s)	Fresh water (condensed water) mass flow rate	σ	*Thickness*
Mw (kg/s)	Mass flow rate of rejected water from condenser	***Subscripts***	
$Q\ (w)$	Heat received /transferred	*a*	*air*
$R.H.\ (\%)$	Relative Humidity percentages	*ai*	*Inlet air*
$T\ (K)$	Temperature	*ao*	*Outlet air*
Tai (K)	Intake air temperature at the top of Humidifier	*cw*	*Cooling water*
Tao (K)	Outlet air temperature at the top of humidifier	*cwi*	*Inlet Cooling Water*
Tcwi (K)	Cooling water inlet temperature	*cwo*	*Outlet Cooling water*
Tcwo (K)	Water temperature of rejected water from condenser	*d*	*Distilled water*
Twi (K)	Water temperature of inlet water to humidifier	*G*	*Gas*
Two (K)	Water temperature of outlet brine water at the bottom of humidifier	*L*	*Liquid*
$V_a\ (\frac{m3}{min})$	Volumetric flow rate of air	*w*	*water*
$R\ (\frac{Pa.m^3}{Kmol.K})$	Universal Gas Constant	*wi*	*Inlet water to Humidifier*
M_{wei}	*Molecular weight*	*wo*	*Outlet water from Humidifier*

References:

[1] Malik, M., Tiweri, G.N., Kumer, A., Sodha, M.S., Solar distillation, New York, pergamon, pp. 3-4,1985

[2] Kalogirou, S.A., Solar thermal Collectors and applications, Progress in Energy and Combustion science, Vol. 30, pp. 231-295, 2004.

[3] Watson, I.C., Morin, O.J., Henthorne, L., Desalting handbook for planners, Aqua Resources International Inc., 2003.

[4] Qiblawy, H.M., Banat, F., solar thermal desalination technologies, Desalination, Vol. 220, pp. 633-644, 2008.

[5] Treybal R.E. Mass transfer operations, 3rd ed., NY: McGraw-Hill; 1980.

[6] Al-Hallaj S, Farid MM, Tamimi AR. Solar desalination with humidification–dehumidification cycle: performance of the unit. Desalination 1998;120:273–80.

[7] Farid MM, Al-Hallaj AW. Solar desalination with humidification–dehumidification cycle. Desalination 1996;106:427–9.

[8] Ben-Bacha H, Damak T, Bouzguenda M. Experimental validation of the distillation module of a desalination station using the SMCEC principle. Renewable Energy 2003;28:2335–54.

[9] Garg HP, Adhikari RS, Kumar R. Experimental design and computer simulation of multi-effect humidification (MEH)–dehumidification solar distillation. Desalination 2002;153:81–6.

[10] Nafey AS, Fath HES, El-Helaby SO, Soliman AM. Solar desalination using humidification–dehumidification processes. Part II. An experimental investigation. Energy Conversion Management 2004;45(7–8):1263–77.

[11] Nawayseh NK, Farid MM, Al-Hallaj S, Tamimi AR. Solar desalination based on humidification process. Part I. Evaluating the heat and mass transfer coefficients. Energy Conversion Management 1999;40:1423–39.

[12] Muller-Holst H. Solar thermal desalination using the multiple effect humidification (MEH) method. In: Book chapter, solar desalination for the 21st century; 2007. pp. 215–25.

[13] Al-Sahali M, Ettouney HM. Humidification dehumidification desalination process: Design and performance evaluation, Chemical Engineering Journal, 143, 257–26, 2008.

[14] Farid M.M., Parekh S., Selman J.R., Al-Hallaj S., Solar desalination with humidification–dehumidification cycle: mathematical modeling of the unit. Desalination; 151:153–64, 2002.

[15] Yamali C, Solmus I. Theoretical investigation of a humidificationdehumidification desalination system configured by a double-pass flat plate solar air heater. Desalination 205, 163–177, 2007.

[16] Zamen M, Amidpour M, Soufari SM. Cost optimization of a solar humidification–dehumidification desalination unit using mathematical programming. Desalination 239, 92–99, 2009.

[17] Dai Y.J, Wang RZ, Zhang HF. Parametric analysis to improve the performance of a solar desalination unit with humidification and dehumidification. Desalination; 142:107–18, 2002.

[18] W.F. He, L.N. Xu, D. Han. Parametric analysis of an air-heated humidification-dehumidification (HDH) desalination system with waste heat recovery. Desalination 398 (2016) 30–38.

[19] Cihan Yıldırım,Ismail Solmus. A parametric study on a humidification–dehumidification (HDH) desalination unit powered by solar air and water heaters. Energy Conversion and Management 86 (2014) 568–575.

[20] S.W. Sharshir, Guilong Peng, Nuo Yang, Mohamed A. Eltawil, Mohamed Kamal Ahmed Ali, A.E. Kabeel. A hybrid desalination system using humidification-dehumidification and solar stills integrated with evacuated solar water heater. Energy Conversion and Management 124 (2016) 287–296.

[21] Nabil A.S. Elminshawy, Farooq R. Siddiqui, Mohammad F. Addas. Experimental and analytical study on productivity augmentation of a novel solar humidification–dehumidification (HDH) system. Desalination 365 (2015) 36–45.

[22] Wang GQ, Yuan XG, Yu KT, Review of mass transfer correlation for packed columns. Ind. Eng. Chem. Res. 2005, 44, 8715-8729.

[23]Shulman, HL, Ullrich CF, Proulx AZ, Zimmerman JO, Performance of Packed Columns. 2. Wetted and Effective Interfacial Areas, Gas- and Liquid-Phase Mass Transfer Rates.AIChE J. 1955, 1, 253.

[24]Shulman, H. L.; de Gouff, J. J. Mass Transfer Coefficients and Interfacial Areas for 1-Inch Raschig Rings. Ind. Eng. Chem.1952, 44, 1915.

[25] Van Krevelen, D. W.; Hoftijzer, P. J. Studies of Gas Absorption. 1. Liquid Film Resistance to Gas Absorption in Scrubbers. Recl. Trav. Chim. Pays-Bas. 1947, 66, 49.

[26] Van Krevelen, D. W.; Hoftijzer, P. J. Kinetics of Simultaneous Absorption and Chemical Reaction. Chem. Eng. Prog. 1948, 44, 529.

[27]Mangers, R. J.; Ponter, A. B. Effect of Viscosity on Liquid Film Resistance to Mass Transfer in a Packed Column. Ind. Eng. Chem. Process Des. Dev. 1980, 19, 530.

[28]Shi, M. G.; Mersmann, A. Effective Interfacial Area in Packed Columns. Ger. Chem. Eng. 1985, 8, 87.

[29] Billet, R.; Schultes, M. Advantage in Correlating Packing Column Performance. Inst. Chem. Eng. Symp. Ser. 1992, 128, B129.

[30] Onda, K.; Takeuchi, H.; Okumoto, Y. Mass Transfer Coefficients between Gas and Liquid Phases in Packed Columns. J Chem. Eng. Jpn. 1968, 1, 56.

[31] Eckert, J. S.: Chem. Eng., 82, 70. Apr. 14, 1975.

[32] Mostafa H. Sharqawy, John H. Lienhard, Syed M. Zubair. Thermophysical properties of seawater: a review of existing correlations and data. Journal of Desalination and Water Treatment. 16 (2010) 354–380.

[33] J.D. Isdale and R. Morris, Physical properties of sea water solutions: density, Desalination, 10(4) (1972) 329–339.

[34] Francis L. Smith and Allan H. Harvey (September 2007). "Avoid Common Pitfalls When Using Henry's Law". CEP (Chemical Engineering Progress). ISSN 0360-7275.

[35] Feoktistov, V. (2006). Differential Evolution: In Search of Solutions. Springer. ISBN 978-0-387-36895-5.

[36] H. Bahmanyar, Mass Transfer. Tehran University's Jahad Daneshgahi Publication, 2003.

473748 Process to Planet Framework for Sustainable Design: Systematic Approach for Developing a Multiscale Model and for Multiobjective Optimization

Abstract:

Over the last decade, sustainable process design (SPD) research has utilized life cycle assessment(LCA) methods to incorporate an environmental dimension in process systems engineering(PSE). Conventional SPD involves a â €œbottom up approachâ€ of modeling the system of interest at a smaller scale with detailed process specific fundamental engineering models, while the upstream life cycle of the product is modeled at a larger scale using process based LCA. System boundary selection is a significant challenge for process LCA due toÂ infinite nature of the upstream life cycle network for any process [1]. Process based LCA can lead to significant truncation errors in life cycle resource consumption and emissions, leaving a wide margin for shifting of impacts across narrow system boundaries and leading to illusive optimal design solutions that may in fact be inferior due to incomplete life cycle information [2]. Hybrid LCA(HLCA) methods aim to address the boundary selection problem by combining process level life cycle inventories with environmentally extended input-output analysis(EEIO), which allows for a significant part of the product life cycle to be included in the assessment. The Process to Planet (P2P) framework [3] was developed by introducing the concept of integrated hybrid LCA [4] to SPD. Detailed engineering models of the primary process being designed and other key processes in the supply chain are integrated within a macroeconomic system represented by economic input-output (EIO) model. This ensures solution of process design problems while reducing chances of environmental impacts being neglected due to incomplete life cycle inventories. While bridging the knowledge gap between PSE and LCA, the P2P framework not only improves upon the life cycle assessment component of SPD, but also addresses a major weakness in Hybrid LCA. HLCA is limited to the impact assessment stage of life cycle assessment. Very few studies have dealt with the fourth and final stage of LCA, improvement analysis. This situation is lamentable because inventory scope and impact analysis provide little benefit if not followed by recommendations and decisions that should be adopted to reduce environmental burden. The P2P framework can be visualized as an extension of HLCA method where it enables the incorporation of design variables through process design and allows optimization of the system to improve its environmental footprint. It was successfully demonstrated by application to a corn based ethanol production plant and highly insightful results were obtained that proved the efficacy of the framework [2].

This presentation proposes extensions and modifications that address some of the inadequacies that are inherent in the original P2P framework. While strictly applied to P2P, these extensions can be extrapolated to answer challenges that exist in broad areas of literature, like HLCA. The selection of significant processes that are modeled at the process based life cycle scale and the insignificant processes that are studied at the economy scale is arbitrary. In other words, delineation of the boundary between the different scales of the P2P framework was a subjective task. Previously, this was done based on data availability, accuracy and detail requirements and the complexity of the overall problem which is proportional to the solving time [5,6]. The problem was first mentioned and some methods were proposed for addressing it in a paper by Suh [7]. Lenzen proposed Structural Path Analysis (SPA) involving the decomposition of the Leontief Inverse for addressing this problem [8]. Treloar also used this method to study energy use of individual residential buildings in Australia [9,10]. Lave et al. [11] and Lenzen [12] used SPA to quantify the truncation errors in process-based LCA. A path exchange method using SPA was also developed as an alternative to the method of combining process LCA and IO LCA within integrated hybrid LCA as proposed by Suh. We intend to address this knowledge gap by proposing a path analysis algorithm where the P2P framework superstructure model generation is guided by a hierarchical structural path decomposition method which determines the coverage of the different scales. It is one of the first applications of SPA for solving the boundary selection problem involving the distribution of different upstream processes between multiple scales.

A large number of studies have dealt with multi objective optimization in process design and energy systems involving environmental and economic impacts. A comprehensive list of such literature can be found in Diwekar et al.[13] Reviewing these articles, it becomes apparent that it is necessary to simultaneously work towards both economic profitability and environmental sustainability of the system. Previously, the P2P framework focused only on the environmental dimension of process design. In this work, we propose a multi objective optimization problem, involving the trade-off between an economic objective and an environmental objective function. The novelty in this addendum is that it makes the P2P framework more fruitful for designing any generic process while exercising control over both the economic and environmental scope.

Preliminary results have provided insightful views on the application of SPA algorithm for model generation as well as exposed existing deficiencies to work on. While SPA is successful in mapping the upstream networks of the process to be designed and calculate their contributions to the entire life cycle, the quantification of the quality of the generated model still needs to be addressed. For example, considering two scenarios where an upstream sector is modeled at the IO model scale and process LCA scale respectively, we need to develop parameters that can quantitatively compare between the two cases and rank them objectively. As of now, subjective decisions are being used to build the model, but we intend to develop an algorithm based on error analysis or uncertainty information to determine the model quality. Multiobjective optimization solutions demonstrated that, highly erroneous optimal solutions and counter intuitive results are obtained from practicing conventional SPD approach rather than P2P. In addition to this, it was observed that there is a 59% chance of choosing an inferior solution over the optimal solution when solving the design problem using conventional SPD.

References

[1] Sangwon Suh. Functions, commodities and environmental impacts in an ecological economic model. 48:451-467, 2004.

[2] Rebecca J Hanes and Bhavik R Bakshi. Sustainable process design by the process to planet framework. AIChE Journal, 61(10):3320-3331, 2015.

[3] Rebecca J Hanes and Bhavik R Bakshi. Process to planet: A multiscale modeling framework toward sustainable engineering. AIChE Journal, 61(10):3332-3352, 2015.

[4] Sangwon Suh and Shinichiro Nakamura. Five years in the area of input-output and hybrid lca. The international journal of life cycle assessment, 12(6):351-352, 2007.

[5] H Hondo and S Sakai. Preliminary life cycle inventory analysis (pre-lci) using an economic input-output table. In The Fourth International Conference on EcoBalance, Tsukuba, Japan, pages 181-4, 2000.

[6] G Norris. Selecting and evaluating boundaries for life cycle assessment using economic input-output analysis. In Proceedings from the SETAC 17th Annual Meeting, Washington, DC, pages 17-21, 1996.

[7] Sangwon Suh, Manfred Lenzen, Graham J Treloar, Hiroki Hondo, Arpad Horvath, Gjalt Huppes, Olivier Jolliet, Uwe Klann, Wolfram Krewitt, Yuichi Moriguchi, et al. System boundary selection in life-cycle inventories using hybrid approaches. Environmental Science & Technology, 38(3):657-664, 2004.

[8] Manfred Lenzen. A guide for compiling inventories in hybrid life-cycle assessments: some australian results. Journal of Cleaner Production, 10(6):545-572, 2002.

[9] Graham J Treloar. Extracting embodied energy paths from input-output tables: to-wards an input-output-based hybrid energy analysis method. Economic Systems Re-search, 9(4):375-391, 1997.

[10] GJ Treloar, PED Love, OO Faniran, and U Iyer-Raniga. A hybrid life cycle assessment method for construction. Construction Management & Economics, 18(1):5-9, 2000.

[11] Lester B Lave. Using input-output analysis to estimate economy-wide discharges. Environmental Science & Technology, 29(9):420A-426A, 1995.

[12] Manfred Lenzen. Errors in conventional and input-output based lifecycle inventories. Journal of Industrial Ecology, 4(4):127-148, 2000.

[13] Urmila Diwekar and Yogendra Shastri. Design for environment: a state-of-the-art re-view. Clean Technologies and Environmental Policy, 13(2):227-240, 2011.

462158 Application of a Shale Environmental Footprint Optimization Tool to Enhance Operational Excellence– a Case Study of Selecting Process Fuels with Lower Cost and Less Potential for Impacts to Power Chevron Operations

Abstract:

Unconventional resources represent significant business opportunities; however, their development is subject to much public and environmental scrutiny. Currently there are over 400 local bans and 2 state bans on hydraulic fracturing. Many countries besides the U.S. have placed moratoriums on hydraulic fracturing. The challenge and opportunity that the oil and gas industry is facing is: how to unlock the potential in unconventionals, achieve better economic and operational performance, while still fulfilling our commitment to communities and the environment.

The Shale Environmental Footprint Optimization Tool is a tool developed at Chevron to enable the integration of environmental considerations in process design and decision making related to shale development. It simulates shale operations at the play level, where material and energy flows of in-scope processes are aggregated together and converted to quantitative environmental indicators (i.e. energy consumption, water consumption, and air emissions). It is built with a commercial life cycle assessment (LCA) software (GaBi® Professional) that contains >4000 datasets, and adopts international standards (ISO 14040-10444). The tool also allows sensitivity and uncertainty analysis to better assess and interpret results under varying reservoir and operational conditions.

This tool has been used by business units to better characterize their operations, assess potential environmental risks associated with different decision alternatives, and understand the trade-offs. It also provides them a more holistic view across the value chain, and help them pinpoint major contributors to the total footprint. These insights help business partners prioritize R&D efforts, make informed decisions on mitigation options, and ultimately lead to reduced environmental impact. Externally, the tool can be used address stakeholder concerns and for benchmarking our performance.

In the presentation, we'll show how this tool has been used to assess real world situations (using field data and business scenarios) and generate information to support a critical decision – evaluating and identifying the preferred alternative fuel to power drilling rigs. Optimizing the selection of process fuels could result in significant environmental, social, and economic benefits. One of the scenarios we'll include is to switch from LNG to gas produced in the field for power generation, which may reduce truck traffic on the roads. Reducing trucks is beneficial to safety (MVC, avoid backing/maneuvering incident on well pad), cost (hauling and road repair), land disturbance (less roads and processing units), and emissions (air pollutants & spills). In some locations, it also means improved reliability and fewer burdens on business continuity.

GHG Life Cycle Assessment for the United Arab Emirates Electricity Sector Combining Optimization and Simulation Tools

Paper ID #: 466062

*Alberto Betancourt-Torcat, Mohammed Alkatheri, Ali Almansoori**

Department of Chemical Engineering, The Petroleum Institute, Sas Al Nakhl Campus, Abu Dhabi, P.O. Box 2533, United Arab Emirates.

*Corresponding Author's e-mail: aalmansoori@pi.ac.ae

Keywords: Greenhouse Gas, Life cycle assessment, Electricity sector, Economics, Optimization, Simulation.

Abstract

The United Arab Emirates (UAE) relies heavily on natural gas for power generation, and the demand for gas in a country with one of the highest per capita power consumption rates in the world position the electricity sector as the leading CO_2 emission source. Although the country is very rich in energy resources it is currently facing challenges to meet its domestic gas demand. Thus, the country has been forced to develop its large, but highly sour (high-sulfur content) gas reservoirs. Given the gas high-sulfur content, the sweetening process is particularly very energy and greenhouse gas (GHG) emission intensive compared with conventional gas resources. Additionally, the volume of sweet gas yield is low due to the high proportion of acid gases in the feed gas (e.g., over 35% of H_2S and CO_2 combined). This work quantifies the well-to-grid (WTG) GHG emissions per kilowatt hour (kWh) of domestic gas-based electricity generated in the UAE using mathematical models, simulation tools, and engineering principles. WTG emissions include emissions from sour gas extracting, sour gas processing, sweet gas product transportation, electricity generation, and electricity transmission in the UAE. Two types of gas resources were examined: conventional and ultra-sour gas. The results show that the WTG GHG emissions of domestic gas-based power generation in UAE ranges between 444 to 501 g CO_2 eq./kWh; placing the country in the low-mid range of values compared with the literature 413-587 g CO_2 eq./kWh.

Introduction

Over the last decade, the rapid economic and demographic growth of the United Arab Emirates (UAE) has pushed the country's electricity grid to its limits. Most of the UAE's generation capacity is natural gas-based [1]. Although natural gas is considered to be a relatively clean primary energy source, it is still a fossil fuel source that generates significant greenhouse gas (GHG) emissions. The UAE electricity demand reached a domestic record of 105 billion kWh in 2013, placing the country as one of the highest per capita power consumers in the world [2]. Nuclear and renewable power capacities are currently being added to increase the country's energy mix and reduce GHG emissions [3]. As a result of the steadily growing electricity demand, the UAE plans to increase natural gas production to meet the domestic demand. The majority of the growth will come from the country's largely sour (high-sulfur content) gas reservoirs. The UAE's gas reservoirs have a relatively high sulfur content, which makes it corrosive and difficult to process [4]. The process is particularly very energy intensive and environmentally challenging. Thus, the process is highly GHG-emission intensive, as well vast amounts of sulfur are generated as by-product [5].

There is currently a need to assess the environmental footprints of the UAE electricity sector in terms of its GHG emission life cycle. This could allow benchmarking the UAE's electricity sector against other nations, while finding suitable niches for GHG mitigation strategies. Moreover, since over 98% of the electricity generated in the UAE is gas-based, an analysis on the UAE's natural gas life cycle represents a key factor on determining the overall environmental impact of the electricity sector. The present analysis considers a well-to-grid (WTG) GHG lifecycle assessment; which includes the upstream and downstream operations. The upstream operations includes all activities before fuel is combusted at power plants. Such activities comprise: sour natural gas extraction, sour gas processing, and sweet gas transport to the power plants gate. On the other hand, the downstream operations include the electricity generation and transmission. For the present study, fundamental engineering principles were used to mathematically estimate process specific energy consumption and related GHG emissions; namely for sour gas extraction, sweet gas product transportation, and electricity transmission. Additionally, ProMax® v3.2 [6] was used to simulate the sour gas processing, while performing a parametric analysis to improve the performance of these processes. Also, an optimization mathematical model developed in the General Algebraic Modeling System (GAMS®) [3] was used to illustrate the operation of the UAE power sector, and obtain the average GHG emissions of the country's gas-based power fleet.

Methodology

In the UAE the well-to-grid life cycle starts with the extraction of natural gas from the reservoirs and ends at the electricity transmission. Natural gas is extracted using wells and sent to processing plants where: water, carbon dioxide, sulfur, and heavier hydrocarbons are removed. Then, the resulted sweet gas product is transported through pipelines to the power plants to generate electricity, and transmitted through the electricity grid. The UAE's WTG GHG emissions life cycle can be divided into two inventories: upstream and downstream. The upstream inventory includes all activities before fuel is combusted at a power plant. Such activities comprise: sour natural gas extraction, sour gas processing, and sweet gas product transportation to the power plants gate. On the other hand, the downstream inventory includes electricity generation and transmission. In this study, the GHG emissions associated to gas extraction, sweet gas product transportation, and electricity transmission were estimated based on typical average emission values found in the literature [7]. On the other hand, the GHG emissions generated by the natural gas processing and combustion in power plants were calculated using ProMax® v3.2 (simulation software) [6] and a mathematical model developed in GAMS® (modeling system) [3], respectively. There are two main types of natural gas resources in UAE: 1) ultra/highly-sour natural gas, and 2) conventional natural gas. Therefore, for the present analysis the WTG GHG emissions life cycle was estimated taking into account both natural gas resources. Typical compositions for both types of gas resources are shown in Table 1.

Table 1. Typical compositions for ultra-sour and conventional natural gas resources in UAE.

Main Components	Ultra-sour gas [4]	Conventional gas [8]
H_2S	22	5
CO_2	10	4
C1	54	76
C2	4	8
C3+	5	3
H_2O	5	4

The first stage in the WTG GHG emissions life cycle is the natural gas extraction using a gas well. This operation encompasses various processes such as: rig construction, drilling, well completion, and gas production. Energy is consumed during the drilling process generating GHG emissions. The emissions of natural gas during the extraction stage are considered episodic emissions because they are not part of the daily, steady state of the well operations. For instance, it was assumed that around 20 g CO_2 eq./kWh is emitted to the atmosphere in gas extraction (e.g., mainly point source and fugitive losses) [7]. Moreover, gas production starts after the well is completed. Gas from the wellhead is sent to surface processing, where it is sweetened, dehydrated, and compressed for pipeline transmission. The raw gas from the wellhead is first sent to a three phase gravity separator to separate the gas, light liquids (e.g., mostly sour hydrocarbon liquids C3+), and heavy liquids (mostly sour water). Since no fuel is consumed during this process, it is not considered a significant source of GHG emissions.

After the phase separator, the gas is treated in a gas sweetening unit. In this work, di-glycol amine (DGA) is used as the gas sweetening solvent for ultra-sour natural gas; while methyl-di-ethanol amine (MDEA) is used as sweetening solvent for conventional gas. The units that consume energy in the gas sweetening process are: feed pre-coolers, amine reboilers, circulation pumps, solvent coolers, and acid gas condensers. The amine re-boiler is assumed to be powered by low pressure steam (LPS) (i.e. saturated steam at 3.5 barg), the circulation pump and solvent air cooler are powered by electricity, the feed pre-cooler and acid gas condenser employ cooling water. Later, the wet sweet gas is sent to the dehydration unit; whereas the acid gases (i.e. H_2S and CO_2) are sent to the sulfur recovery unit. In the dehydration unit, water is removed from the sweet natural gas since it promotes corrosion. Tri-ethylene glycol (TEG) is used as liquid desiccant to dehydrate the sweet gas. The glycol reboiler, sweet gas pre-cooler, and pumps are the major energy-consuming units in this process. In the dehydration unit, high pressure steam (HPS) (i.e., saturated steam at 41 barg) is used to power the glycol reboiler. Moreover, electricity is used to drive the pumps, whereas cooling water is used to pre-cool the wet sweet gas. The dry sweet gas from the dehydration unit can be combusted at an onsite power plant to produce electricity. Sometimes the dry sweet gas from the dehydration unit is sent to a natural gas liquid recovery unit to recover ethane and ethylene. However, for simplicity the natural gas liquid recovery unit was not considered in this study.

The liquid hydrocarbons are treated in a liquid sweetening unit using DGA as solvent; while the sour water is treated using a sour water stripper. The sole purpose of both aforementioned treatment processes is the removal of sour components from the hydrocarbon liquids and water. The liquid hydrocarbons sweetening process is similar to the gas sweetening process; thus, involving equivalent energy consuming units. Conversely, mainly two types of units consume energy in the sour water stripper: the water stripper reboiler and pumps.

All acid gases effluent from the gas sweetening unit, liquid sweetening unit, and sour water stripper are collected and sent to the sulfur recovery units (SRU). In these units, the H_2S in the acid gas stream is converted into elemental sulfur. A variety of configurations are available for sulfur recovery. In this study, it was used the "straight-through" Claus process. This process consists of an initial combustion step at high temperature (e.g., above 980 °C), followed by a series of reactions (carried out in a Claus Bed) and condensation steps. Generally two or three such steps are used. Steam is generated in the combustion step; thereby, contributing to meet the steam requirements of the process. The units consuming energy in sulfur recovery are condensers and re-heaters. Condensers and re-heaters are powered by cooling water and HPS, respectively.

The ProMax® v3.2 simulation software was used to calculate all forms of energy required to process conventional and ultra–sour natural gas. However, fundamental engineering principles were employed to calculate the associated GHG emissions. Furthermore, the amount of fuel gas (i.e., dry sweet natural gas) required to produce process steam (e.g., high and low pressure steam needed to process natural gas) was also estimated using the simulator. It is assumed that the fuel gas first combines with air, and then is combusted in a burner. The heat generated from the combustion process is used to boil water and generate high and low pressure steam. From this analysis, it was found that the average emission per slandered of fuel gas is around 66.7 g CO_2 eq./scf. This emission rate is in agreement with values reported by the US Environmental Protection Agency (EPA) [9].

The electricity requirements for the cooling water loop can be estimated using the cooling water flow rate. This can be done calculating the power requirements in pumps and cooling tower fans. For instance, the power requirements in a cooling water pump with a volumetric flow rate (V (m^3/s)) and pressure drop (ΔP (kPa)) can be calculated as follows:

$$Power = \frac{1}{\text{efficiency}} * V * \Delta P \quad (1)$$

For the cooling water loop, it was assumed an overall efficiency of 75% [10], the pressure drop around the water loop was estimated to be 266.7 kPa [10]. Accordingly, the power requirements can be calculated as follows [11]:

$$Fan\ Power = \left(V\frac{m^3}{s}\right)\left(\frac{60s}{min}\right)\left(\frac{264.172\text{gallon}}{m^3}\right)\left(\frac{0.5ft^2}{\frac{gallon}{min}}\right)\left(\frac{0.041\text{hp}}{\text{ft}^2}\right)\left(\frac{0.7457\text{kw}}{\text{hp}}\right) \quad (2)$$

The emissions of transporting the sweet gas product from process facilities to power plants was assumed to be approximately 30 g CO_2 eq./kWh according to information reported in the literature [7]. Furthermore, a mathematical model developed in GAMS on the UAE power infrastructure [3] was used to calculate the average GHG emissions generated by the gas-based power plants fleet in the country. The simulation results show that in average 370 g CO_2 eq. are generated per kWh of power produced. This value was used to calculate the GHG emissions associated to the electricity requirements involved in the analyzed life cycle. Finally, the GHG emissions associated to electricity transmission was also taken from the literature (i.e., 3.4 g CO_2 eq./kWh) [7].

Results and Discussions

The ProMax® v3.2 simulator was used to calculate the utilities (e.g., high/low pressure steam, cooling water, and electricity) consumed in the processing of conventional and ultra–sour natural gas. Table 2 shows the utility requirements for both types of gas resources per standard cubic feet (scf) of dry sweet gas produced.

Table 2. Utility requirements for conventional and ultra–sour natural processing.

Utility Type	Units	Ultra-sour gas	Conventional gas
Electricity	(Wh/scf)	20.8	4.0
High Pressure Steam (HPS)	(g/scf)	6.1	1.2
Low Pressure Steam (LPS)	(g/scf)	41.5	11.0
Low Pressure Steam (LPS) generated from SRU	(g/scf)	23.7	3.7
Net Required Low Pressure Steam (LPS)	(g/scf)	17.8	7.3
Cooling water	(liter/scf)	1.0	0.3

As shown in the table, the processing of ultra-sour gas requires much more energy than conventional gas per unit of sweet gas yielded. For example, the energy requirements to process ultra-sour gas are roughly four times higher than that for conventional gas. This can be explained by the fact that as the amount of H_2S in the feed gas increases so does the absorption plat utilities demand. This because more acid gases must be removed by the absorber leading to higher recirculating solvent flow rates. Therefore, the power consumed to cool and pump the regenerated solvent back into the absorber, as well as the steam used to regenerate the alkanolamines increases. Both energy forms (i.e., steam and electricity) account for the majority of the utilities consumed in the absorption plant. Likewise, more sulfur must be processed by the sulfur recovery units. Moreover, the sweet gas product recovery rate is much lower for the ultra-sour gas feed. Table 3 shows the recovery percentage for ultra-sour and conventional gas resources.

Table 3. Recovery percentages for ultra-sour and conventional gas.

Feed type	Ultra-sour gas	Conventional gas
Recovery rate	60 %	90 %

As shown in the table, almost 40 % of the ultra-sour gas feed volume is loss due to the high acid gases content; which are removed to obtain the sweet gas product. Conversely, for conventional gas only 10% of the volume is loss. Consequently, more energy is required per unit of sweet gas produced from ultra-sour gas. The electricity requirements for the cooling water loop were estimated using Equations (1)-(2). Accordingly, the GHG emissions corresponding to the electricity requirements were estimated using the gas power generation emission factor. Moreover, the process steam requirements associated to sour gas processing were obtained simulating the gas sweetening plants with ProMax® v3.2 [6]. The GHG emissions values related to HPS and LPS usage, and electricity for processing the natural gas are reported in Table 4. These values are given per standard cubic feet (scf) of dry sweet gas produced.

Table 4. GHG emissions related to conventional and ultra–sour gas processing.

Type of Utility Consumed	Units	Ultra-sour gas	Conventional gas
Electricity	(g CO_2 eq./scf)	7.7	1.5
High Pressure Steam (HPS)	(g CO_2 eq./scf)	0.7	0.4
Light Pressure Steam (LPS)	(g CO_2 eq./scf)	2.7	1.1
Total GHG emissions	(g CO_2 eq./scf)	11.1	3.0

From the table it can be clearly noticed that the GHG emissions associated with ultra-sour gas processing are much higher than those for conventional gas. Additionally, Table 4 shows that more than 50% of the GHG emissions are due to electricity consumption. Using the GAMS model [3], the average amount of natural gas needed to produce one-kilowatt hour of electricity in UAE was estimated. This value was found to be approximately 6.97 scf of gas/kWh. Additionally, all GHG emissions values from natural gas processing were also expressed in terms of per unit of kWh of electricity produced. The GHG emissions involved in natural gas extraction and final product transportation were estimated using factors reported in the literature [7]. The WTG GHG emissions life cycle for UAE domestic gas-based power generation can be estimated as the summation of the GHG emissions from: gas extraction, sour gas processing, sweet gas transportation, electricity generation (i.e., combustion of natural gas in power plants), and electricity transmission. Table 5 shows the single components and total GHG emissions life cycle for electricity produced from natural gas for both type of gas resources.

Table 5. Single emissions and total GHG emissions life cycle for domestic gas-based power generation in the UAE.

Life Cycle Stage	Ultra-sour gas (g CO_2 eq./kWh)	Conventional gas (g CO_2 eq./kWh)
Gas Extraction	20	20
Gas Processing	77.6	20.6
Product Transportation	30	30
Gas Combustion (power plants)	370	370
Electricity Transmission	3.4	3.4
Total GHG	501	444

As shown in the table, ultra-sour gas involves higher GHG emissions than conventional gas; which results from greater energy requirements. It is worth pointing out that most of the GHG emissions originate from the combustion of the sweet natural gas product in the power plants. Additionally, the WTG GHG life cycle analysis shows that for ultra-sour gas the emissions are approximately 13% greater than those involved using conventional gas. Figure 1 shows the percentage share of each stage/activity in the GHG emissions life cycle.

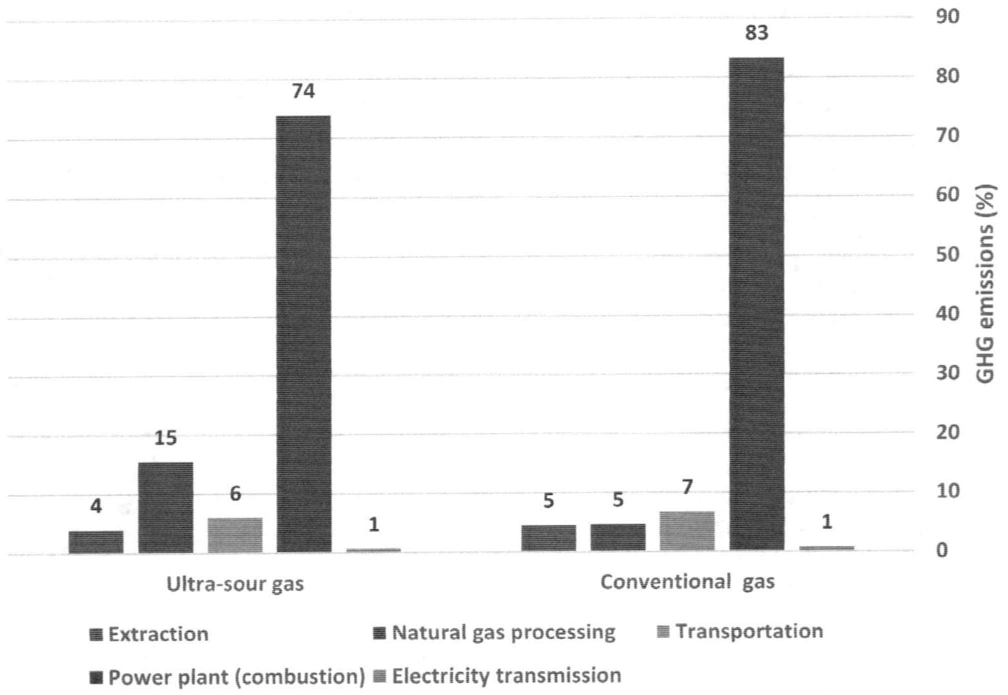

Figure 1. Percentage share per stage/activity in the GHG emissions life cycle for ultra–sour and conventional gas.

As shown in the Figure, sour gas processing and electricity generation (at the power plants) account for more than 88% of the total GHG emissions in the life cycle. Moreover, electricity generation is clearly the leading GHG emission source in the analyzed life cycle. Nevertheless, the GHG emissions generated by the electricity generation operations are rather low compared with other countries. This since the majority of the UAE's current power fleet is composed of modern gas-fired combined cycle plants; which are comparatively more efficient than other gas-based plants. Also, the total WTG GHG emissions of domestic gas-based power generation in UAE places the country in the low range (for conventional gas) and middle range (for ultra-sour gas) of values reported in the literature (e.g., 413-587 g CO_2 eq./kWh) for similar systems worldwide [7, 12, 13].

Conclusion

In this study, the well-to-grid (WTG) GHG emissions lifecycle for domestic gas-base power generation in the UAE is analyzed. The analysis results show that gas-based power generation using ultra-sour gas is more emission intensive than using conventional gas as the power plants' feed. This because much more energy is consumed in ultra-gas processing. Additionally, the yield of sweet gas product is much lower when processing ultra-sour gas compared with conventional gas. Regarding the total emissions of the life cycle, sour gas processing and electricity production account for almost 90 % of the total GHG emissions. Electricity generation is clearly the leading GHG emission source in the analyzed WTG lifecycle.

References

[1] A. Betancourt-Torcat and A. Almansoori, "Design multiperiod optimization model for the electricity sector under uncertainty – A case study of the Emirate of Abu Dhabi," *Energy Conversion and Management,* vol. 100, pp. 177-190, 8// 2015.

[2] Energy Information Administration, "United Arab Emirates " U.S. Energy Information Administration2015.

[3] A. Almansoori and A. Betancourt-Torcat, "Design optimization model for the integration of renewable and nuclear energy in the United Arab Emirates' power system," *Applied Energy,* vol. 148, pp. 234-251, 6/15/ 2015.

[4] Al Hosn Gas. (2015, 22/5/2016). *Abu Dhabi Gas Development Company Ltd. (Al Hosn Gas).* Available: : http://www.alhosngas.com/.

[5] J. Slagle, "Highly Sour Amine Sweetening Process Study for Conditions Common in the Middle East," presented at the The Abu Dhabi International Petroleum Exhibition and Conference, Abu Dhabi, UAE, 2013.

[6] Bryan Research & Engineering Inc., "ProMax," ed. Bryan, Texas, U.S.: Bryan Research & Engineering, 2013.

[7] T. Skone, J. Marriot, J. Littlefield, and B. A. Hamilton, "Life Cycle Assessment of Natural Gas Extraction, Delivery and Electricity Production," National Energy Technology Laboratory, United States of America2012.

[8] P. Pal, A. AbuKashabeh, S. Al-Asheh, and F. Banat, "Role of aqueous methyldiethanolamine (MDEA) as solvent in natural gas sweetening unit and process contaminants with probable reaction pathway," *Journal of Natural Gas Science and Engineering,* vol. 24, pp. 124-131, 5// 2015.

[9] Environmental Protection Agency, "U.S. annual non-baseload CO2 output emission rate," US Environmental Protection Agency, Washington, DC, US2015.

[10] R. Turton, R. C. Bailie, W. B. Whiting, and J. A. Shaeiwitz, *Analysis, Synthesis and Design of Chemical Process*, 3rd ed. Boston, MA, U.S.: Prentice Hall PTR International, 2008.

[11] R. H. Perry and C. H. Chilton, *Chemical Engineers' Handbook*, 5 ed. New York, US: McGraw-Hill, 1973.

[12] L. S. Abrahams, C. Samaras, W. M. Griffin, and H. S. Matthews, "Life Cycle Greenhouse Gas Emissions From U.S. Liquefied Natural Gas Exports: Implications for End Uses," *Environmental Science & Technology,* vol. 49, pp. 3237-3245, 2015/03/03 2015.

[13] World Nuclear Association, "Comparison of Lifecycle Greenhouse Gas Emissions of Various Electricity Generation Sources," World Nuclear Association London, UK2011.

468547 Data Envelopment Analysis Coupled with Thermodynamic and Life Cycle Assessment Metrics for Solvent Screening: Application to CO_2 Capture

Abstract:

The growing trend towards more sustainable products and processes calls for advanced decision-making tools for the assessment and optimisation of the economic, environmental and social impact of chemicals and chemical processes. The pressure for more sustainable processes is particularly strong in the chemical industry, in which thousands of products are developed and launched at a very fast pace, raising the question (before scaling them up) of whether they are more sustainable than the existing ones or not.

Particularly, in green chemistry applications, engineers and scientists are confronted with the task of screening chemicals (e.g. mainly solvents, additives, etc.) according to sustainability principles. In practice, it is quite hard to find a single chemical/process performing best in the whole range of sustainability metrics, as inherent trade-offs typically arise between conflictive thermodynamic, environmental and safety metrics. There has been a substantial amount of research in the field of multi-criteria assessment of chemical processes.[1-3] In contrast, the problem of screening chemicals according to several criteria has received much less attention to date. In the absence of any rigorous and systematic screening method, practitioners assess them by defining, in either an implicit or explicit manner, subjective weights that are attached to every category of interest (e.g. viscosity, density, solubility, etc.). These weights enable the calculation of aggregated indicators on the basis of which the chemicals can be ranked.[3-5]While it is straightforward to implement, this approach is highly sensitive to the weights defined and provides in turn no insight into the underlying multi-criteria problem.

In this work we propose a new methodology to screen and select chemicals according to the extent to which they adhere to sustainability principles that is based on Data Envelopment Analysis (DEA), a

technique developed in economics for efficiency assessment. Essentially, DEA is a non-parametric analytical tool for quantifying the relative efficiency of a set of decision making units (DMUs) that makes use of linear programming techniques.[6,7]DEA serves two purposes: i) it identifies, among a set of DMUs - each producing multiple outputs from multiple inputs - the efficient ones (i.e. those with the best ratio desired outputs to required inputs); and ii) it establishes efficient targets for the inefficient units that if achieved would make them efficient. DEA has recently found many applications in environmental studies, but to the best of our knowledge has never been applied to assess chemicals in green chemistry applications. In essence, we combine here DEA with a set of thermodynamic and life cycle assessment indicators so as to identify the chemicals that perform best according to technical (economic), environmental and social aspects simultaneously.

The capabilities of the DEA-based approach in the context of green chemistry are demonstrated through the screening of 125 conventional amine-based solvents for CO2 capture according to 10 performance indicators. Our approach eliminates 36% of the solvents for being inefficient, establishing quantitative targets and clear guidelines on how to improve them. The DEA models are capable of identifying the practical downsides of amines as main sources of inefficiency through the calculation of clear improvement targets for properties such as vapour pressure, acute toxicity and several life cycle impacts. The final aim of the DEA-based analysis is to facilitate the selection of more sustainable chemicals in the transition towards a more sustainable chemical industry.

References

1. C. Capello, U. Fischer and K. Hungerbühler, *Green Chem.*, 2007, **9**, 927.
2. G. Wernet, S. Papadokonstantakis, S. Hellweg and K. Hungerbühler, *Green Chem.*, 2009, **11**, 1826.
3. S. Hellweg, U. Fischer, M. Scheringer and K. Hungerbühler, *Green Chem.*, 2004, **6**, 418.
4. M. Tobiszewski, S. Tsakovski, V. Simeonov, J. Namieśnik and F. Pena-Pereira, *Green Chem.*, **17(10)**, 4773-4785.
5. A. I. Papadopoulos, S. Badr, A. Chremos, E. Forte, T. Zarogiannis, P. Seferlis, S. Papadokonstantakis, C. S. Adjiman, A. Galindo and G. Jackson, *Chem. Eng. Trans.*, 2014, **39**, 211–216.
6. W. D. Cook and L. M. Seiford, *Eur. J. Oper. Res.*, 2009, **192**, 1–17.
7. P. Zhou, B. W. Ang and K. L. Poh, *Eur. J. Oper. Res.*, 2008, **189**, 1–18.

470934 Using Multiobjective Optimization and Life Cycle Assessment for the Design of More Sustainable National and International Energy Systems

Abstract:

Using multiobjective optimization and life cycle assessment for the design of more sustainable national and international energy systems

Sabio, N[a*]., Volkart, K[b]., Densing, M[b]., Strachan, N[a].

***a* UCL Energy Institute, University College London, 14 Upper Woburn Place, WC1H 0NN**

***b* Paul Scherrer Institut, Energy Economics Group, 5232 Villigen, Switzerland**

Nowadays, an increasing energy demand driven by population and economic growth, and the need to reduce green-house gas emissions are the two major axis along which energy systems need to evolve. Not in vain, the design of sustainable energy systems has also been classified as a major challenge in the area of process systems engineering [1,2].

Energy systems models were developed after the first oil crisis using the methods of linear programming for informing policy makers concerned with the importance of energy systems planning and design at the time [3]. These models are still nowadays the main decision-support tools used for informing energy policy.

Although they were initially conceived for analysing energy security and costs, climate regulation has turned environmental factors in the major focus of energy systems analysis. For instance, after the UK committed to reduce its greenhouse gas emissions by 80% in 2050 [4], the Committee of Climate Change produced a set of carbon budgets based on energy systems analysis that outline the pathways and expected changes in the energy system needed to achieve the target [5].

Conflicting objectives lie at the heart of any design problem. Multi-objective optimization (MOO) frameworks were pervasive in PSE since the early days, the combination of multi-objective optimization frameworks combined and life-cycle analysis [6] has gained more recent attention [1,7,8]. Nevertheless, to the extent of our knowledge, this framework has not been applied to energy systems models.

One reason for this might be that multi-criteria analysis is not seen as a popular method in energy systems modelling community [9]. Although the advantages of multi-objective optimization have already been more widely recognised when applied to distributed generation problems involving a variety of stakeholders [10]. However, the challenges for the implementation of life cycle assessment in a whole energy systems model has been subject of more extensive attention in the literature [11].

In the present work we propose a framework for simultaneously optimizing several life cycle environmental impact indicators and the economic performance in energy systems models. For that purpose we use the classical ε-constraint deterministic multi-objective method that a side from being suitable for convex Pareto-sets, it allows to exploit the problem structure showing the different objective spaces without the need to articulate a priori preferences (i.e., weights), and avoiding the numerical problems arising from normalization.

Our results evaluate the effect of simultaneously optimizing several life-cycle environmental impact metrics against the more traditional total system cost optimisation with CO_2 emissions constrained for a newly developed national energy systems model of the UK (UKTM-UCL) [12] and for an international multi-regional energy systems model (GMM) [13].

***Corresponding author: n.sabio@ucl.ac.uk**

References

[1] Grossmann, I.E., Guillén-Gosálbez, G. (2010) Scope for the application of mathematical programming techniques in the synthesis and planning of sustainable processes, ***Computers and chemical engineering***, 34: 1365-1376

[2] Grossmann, I.E.(2012) Advances in mathematical programming models for enterprise-wide optimization, ***Computers and Chemical Engineering,*** 47: 2-18

[3] Stefan Pfenninger, Adam Hawkes, James Keirstead, Energy systems modeling for twenty-first century energy challenges, Renewable and Sustainable Energy Reviews, Volume 33, May 2014, Pages 74-86

[4] Acts of Parliament. ***Climate Change Act 2008***. The Stationary Office Limited, UK

[5] The Committee of Climate Change (CCC) The 5th Carbon Budget: The next step towards a low carbon economy, November 2015

[6] Guinee JB, Goree M, Heijungs R, Huppes G, Kleijn R, de Koning A, van Oers L, Wegener Sleeswijk A, Suh S, Udo de Haes HA, de Bruijn H, van Duin R, Huijbregts MAJ. Handbook on Life Cycle Assessment. Operational Guide to the ISO Standards. Dordrecht: Kluwer Academic Publishers, 2002.

[7] Azapagic A, Clift R. (1999) The application of life cycle assessment to process optimisation. ***Computers and Chemical Engineering***, 10:1509–1526.

[8] N Sabio, C Pozo, G Guillén Gosálbez, L Jiménez, R Karuppiah, Vasudevan, V., Sawaya, N., Farrell, J.T. Multiobjective optimization under uncertainty of the economic and life cycle environmental performance of industrial processes, ***AIChE Journal,*** 60 (6): 2098-2121

[9] Hall, M.H. L., Buckley, A.R. (2016) A review of energy systems models in the UK: Prevalent usage and

categorisation. ***Applied Energy,*** 169: 607-628

[10] Alarcon-Rodriguez, A., Ault, G., Galloway, S. (2010) Multi-objective planning of distributed energy resources: A review of the state-of-the-art. ***Renewable and Sustainable Energy Reviews,*** 14 (5): 1353-1366

[11] Volkart K. (2014).Development of a New Methodology for the Integration of LCA and Energy-Economic System Modelling, ***LCA XIV***, San Francisco, USA

[12]UCL Energy Institute Models Website, Available at: https://www.ucl.ac.uk/energy-models/models/uktm-ucl [10 May 2016]

[13] Turton H., Panos E., Densing M., Volkart K. (2013). Global Multi-regional MARKAL (GMM) model update: Disaggregation to 15 regions and 2010 recalibration, PSI Bericht 13-03

472120 Life Cycle Assessment and Multiobjective Optimization in a Natural Gas Based Petrochemical Complex

Abstract:

In this work, we address the optimization of a natural gas based petrochemical complex using economic and environmental criteria. The mathematical model includes linear and non-linear models for single plants, taking into account main operating variables, intermittent deliveries and inventory variable profiles. The environmental objective is measured with the global warming potential (GWP) and ReCiPe metrics according to the life cycle assessment procedures using indicators at two levels: eighteen midpoint indicators and three endpoint indicators realised in SimaPro (Pre-Consultants, 2015).The resulting mixed integer nonlinear programming (MINLP) models are implemented in GAMS. The bi-criteria MINLP model is solved with the Îµ-constraint method. Mid-point indicators are presented for current conditions. Numerical results show that the petrochemical complex can satisfy product demands while improving the environmental performance by decreasing greenhouse gases emissions in 33 %, from 1,042 to 695.33 kt CO_2-eq/year in the case of GWP minimization.

474526 Invited Presentation By 2016 Management Award Recipient--Enriching the Future of East Tennessee through Environmental Clean up and Stewardship

Abstract:

As the recipient of the 2016 AIChE Management Division Award, Ken Rueter is living testimony to the fact that chemical engineering can be a rewarding career that offers cutting edge opportunities in innovation and problem solving. His experience underscores the value of the chemical engineering profession in addressing unprecedented national technical challenges.

As President and Project Manager of URS|CH2m Oak Ridge LLC (UCOR), Rueter has led his company's success in providing positive solutions for large-scale nuclear environmental cleanup, restoring once contaminated land for future economic development purposes.

While most engineers win acclaim for the enduring projects they design and build, Ken Rueter's most recent noteworthy achievements stem from what he and his company are demolishing and cleaning up. Today, across the U.S. Department of Energy (DOE) complex, millions of gallons of liquid radioactive waste are being processed. Thousands of tons of spent nuclear fuel and special nuclear material are being disposed. Huge quantities of contaminated soil and water are being remediated. And thousands of excess facilities are being deactivated and demolished.

This massive environmental cleanup – involving some of the most dangerous materials known to humankind – results from over five decades of nuclear weapons development and production and Government-sponsored nuclear energy research. In Oak Ridge, this work addresses the legacy of the super-secret World War II Manhattan Project where uranium was enriched for use in the atomic weapons that ended the war and, later, for fuel in commercial nuclear power plants.

In 2016, UCOR completed demolition and cleanup of the gaseous diffusion facilities at the East Tennessee Technology Park in Oak Ridge, Tennessee. This was a historic moment for a community that carved its place in history during World War II and the Cold War. It marked the first time in the world that all of a site's uranium enrichment process buildings were removed anywhere, eliminating environmental hazards and paving the way for future economic development. In Oak Ridge, that task included deactivating and dismantling 4.5 million square feet of radiological and chemically contaminated facilities and shipping more than 46,000 truckloads of debris.

These facilities include the massive and historic government uranium enrichment facilities from the Manhattan Project and the Cold War eras which played a key role in ending World War II and keeping the nation safe and energy-sufficient for decades afterwards. One of those facilities – the once super-secret K-25 gaseous diffusion plant – consisted of over two million square feet and in its time was the largest building under one roof in the world.

UCOR is working closely with its multiple partners in government, labor and the private sector to accomplish this massive environmental clean up and remediation project. UCOR's partnership with the U.S. Department of Energy and its labor force, including union crafts, is unprecedented in the nuclear industry. With its collective technical competency, UCOR has accomplished a cleanup task that others will never have the opportunity to address.

This achievement brings the Department of Energy closer to its ultimate vision for the site, now known as the East Tennessee Technology Park, as a privately owned and operated industrial park and home to new job-producing industries.

UCOR has completed these first-of-a-kind demolition tasks involving some of the most hazardous materials known to man with an unprecedented safety record – amassing more than 7 ½ million hours with a Lost Day Away accident and being recognized by DOE with Voluntary Protection Program Star status, recognizing the safest worksites in the nation.

UCOR, like one of its parent companies AECOM, is built to deliver a better world. UCOR takes its inspiration from AECOM's transformative work in designing, building, financing and operating infrastructure assets for governments, businesses and organizations in more than 150 countries. As a fully integrated firm, AECOM connects knowledge and experience across its global network of experts to help clients solve their most complex challenges – from building high-performance buildings and infrastructure to creating resilient communities and environments, like the revitalized East Tennessee Technology Park in Oak Ridge.

458045 Your Career Is a Startup

Abstract:

As chemical engineers, we are equipped with highly sought after skills to tackle technological and strategic problems. But how do we apply those same skills towards the development of our own careers? What if your career was run like a startup? What would be your assets? What customers would you serve? What would make you the go-to person when certain challenges arise? How would you market yourself to your company and to your profession?

Participants will learn:

1. The difference between personal brand identity vs. brand equity
2. How to leverage those technical skills youâ€™re already great at to build your brand
3. Ways to practice executive presence, from how you hold face-to-face meetings with your manager, to presenting â€œYouâ€ in your best light to the executive levels of your company
4. How to use your skills as an engineer to analyze your current career path

475779 Chemical Engineering Careers: Community Development Gaps

Abstract:

This talk will cover three topics that I view as gaps along Chemical Engineering career paths that need volunteer support, informational resources, and collaborative address. **First:** Focus on Chemistry and Chemical Engineering. The two professions are classically separate, but in the future have the potential to innovate with far and deep reaching impact. A few examples of how practitioners of holistic design that have an impact on our world will be shared. **Second:** I will focus on being more than a chemical engineer, a chemical engineer plus some if you will (*i.e.,* ChE + S). In the past I have spoken to soft skill development but in this talk will illustrate how multidisciplinary has become a new norm with examples from my own experiences. One example for instance, is the institute has suggested that credentialing is important. While there is a push for Professional Engineering licensure, there are a myriad of accreditation processes that can aid ChEs on their career paths. **Third**: expectations for leadership and implementing change are different than they used to be. Instituting change requires standing against odds and making a difference on many scales and under adverse circumstances. Focus on this third topic will be given to the development of resources in education settings for chemical engineers.

This talk will demonstrate diversity among career paths that still require ChE community support. This talk is geared to promote collaborative discussion and action in progressive career development, please attend if you would like to support developing these topics within and for the Chemical Engineering community.

470546 Optimizing Design Parameters for Thin Film Composite Hollow Fiber Membranes and Modules for Forward Osmosis

Abstract:

Hollow fiber membranes have been considered a valuable platform for osmotic membrane processes because of their high packing density relative to flat sheet membranes. This study evaluates an approach to make thin film composite (TFC) hollow fiber membranes for forward osmosis (FO) by simply employing commercial hollow fiber ultrafiltration (UF) membranes as support material. A thin polyamide film with excellent selective properties was synthesized on the inner surface of hollow fibers via interfacial polymerization. Besides demonstrating the feasibility to be used as good TFC FO membrane supports, the commercial hollow fiber UF membranes also provide a systematic platform with consistent properties to study structure-performance relationship of FO hollow fiber membranes. A series of commercial hollow fiber membranes were used to evaluate how molecular weight cutoff (MWCO), fiber dimensions (inner and outer diameter) and module dimensions (length and width) impacted overall performance of the TFC membranes and module. With the availability of reproducible membranes and modules, we have developed a computational fluid dynamics (CFD) model with COMSOL Multiphysics in order to optimize hollow fiber module design for FO application. The comparison between the model and experimental results provided insight into both how more accurate models can be developed and how these models can be used to design better modules without costly experimental testing.

458145 Monitoring Biofouling Dynamics on Foward Osmosis (FO) Membranes Using a Clsm-Compatible Microfluidic Biofilm Flow Cell

Abstract:

In membrane biofouling studies, the current method for examining the growth of biofilms on water-purifying membranes in membrane testing flow cells has been a â€œmembrane autopsyâ€ followed by microscopic observations. However, during the preparation of membrane samples, uncontrollable and unpredictable disturbances to the biofilms on the membrane surfaces may occur, resulting in artifacts that may lead to misinterpretation of experimental results. Furthermore, the autopsy approach can only provide a snapshot of and the dynamically changing biofouling conditions over time. This limitation is mainly due to the lack of tools that allow us to monitor dynamics of biofouling without the need of dissembling the membrane testing systems. In this study, we developed a novel multichannel microfluidic membrane flow cell that enables non-destructive, real-time monitoring of biofouling dynamics using confocal laser scanning microscopy (CLSM). As a proof of concept, we examined the development of green florescent protein (GFP)-tagged bacterial biofilms on forward osmosis (FO) membranes. The temporal profiles of quantitative biofouling parameters including surface coverage, biovolume and biofilm thickness were obtained without disrupting the continuous operation of the membrane testing system. The microfluidic membrane flow cell developed in this study can be readily applied to evaluate anti-biofouling activities of FO membranes and allows direct comparison of biofouling dynamics between FO membranes with different surface modifications.

462938 Forward Osmosis Using Draw Solutions Manifesting Liquid-Liquid Phase Separation

Abstract:

Over the past decade, Forward Osmosis (FO) technology has emerged as a promising alternative to conventional Reverse Osmosis (RO) to produce fresh water through desalination of seawater. In FO, water is extracted from a saline water source (such as seawater) by a draw solution having a higher osmotic pressure than the saline water; the diluted draw solution is then split into a stream of concentrated draw solution and a stream of essentially pure water in process steps collectively referred to as regeneration. Draw agents whose regeneration is aided via liquid-liquid phase separation have gained much attention in recent years [1-4]. Specifically, aqueous solutions of these draw agents manifest either a lower critical solution temperature (LCST) phase transition or upper critical solution temperature or both. With draw solutions manifesting LCST, the FO step is typically performed at a temperature (T) below (or near) the LCST; for medium to high concentration of the draw agents, the osmotic pressure of the draw solution decreases with increasing T. Modest heating of the diluted draw solution facilitates regeneration by causing liquid-liquid phase separation yielding a water-rich phase and a second phase rich in the draw agent.

In this presentation, we describe the results from an experimental study where mixtures of two different glycol ethers (GE), tripropylene glycol methyl ether (TPM; MW=206 Da) and tripropylene glycol n-butyl ether (TPnB; MW=248 Da), were used as draw agents. These are larger in size than the GEs used in previous studies [2].

Cloud point temperatures of aqueous solutions of these GE mixtures were determined. By adjusting the relative amounts of TPM and TPnB, one can tune the cloud point temperature and the LCST. We also determined the activity of water (in the 288-323K range) at various draw solution compositions, using an Aqualab TDL water activity meter. At high GE wt% and temperature modestly below the cloud point T, the osmotic pressure of the draw solution is much higher than that of seawater, enabling their use as draw agents in FO.

The kinematic viscosities of these draw solution mixtures at different temperatures were found using a micro-Ubbelohde viscometer. Through NMR measurements, the self-diffusion coefficients of the GEs in the draw solutions at different temperatures were determined.

FO experiments were conducted using commercial CTA embedded support FO membrane (Fluid TechnologyÂ Solutions, Inc., USA) to assess water flux, reverse flux of the draw agent, and the effect of orientation of the membrane (support layer facing the draw solution â€“ the so-called FO mode; and, its inverse â€“ the so-called Pressure Retarded Osmosis (PRO) mode). These experiments were performed with fresh water on one side and draw solution on the other side of the membrane. As expected [5], the water flux was higher in the PRO mode than in the FO mode. While the initial water flux was high (~7 lit/m^{2}h), it dropped rapidly to ~ 2 lit/m^{2}h indicating appreciable membrane fouling. Even after thorough washing, the original high flux state could not be recovered.

In the PRO mode, increasing the circulation rate of the draw solution led to measurably large increase in the water flux, clearly revealing the effect of concentration polarization. FO experiments were done at three different temperatures; water flux increased when the operating temperature was increased from 295K to 305K, but decreased when the operating temperature was increased to 315K. This can be rationalized qualitatively as a consequence of the temperature dependence of the osmotic pressure, kinematic viscosity and diffusivity.

Appreciable reverse flux was recorded in the FO experiments (4.91 gr/m^{2}h in FO mode and 2.90 gr/m^{2}h in PRO mode), indicating that molecular weight in the range of 200-250 Da is not sufficiently large to suppress reverse flux. FO experiments performed with aqueous solution of polypropylene glycol (PPG 425; average MW=425 Da) showed lower reverse fluxes (2.53 gr/m^{2}h in FO mode and 1.74 gr/m^{2}h in PRO mode), suggesting that even higher molecular weights are required to suppress reverse flux of draw agent.

In summary, as a result of reverse flux and membrane fouling, using aqueous solutions of glycol ethers directly as draw solutions does not appear promising. Their indirect use, as in the process described by [6], which uses an ionic salt as a draw agent and employs an aqueous polymer solution to recover water from the diluted draw solution, appears to be more realistic.

Keywords: *forward osmosis, lower critical solution temperature, glycol ethers, liquid-liquid phase separation*

References:

1. Mok, Y., et al., *Circulatory osmotic desalination driven by a mild temperature gradient based on lower critical solution temperature (LCST) phase transition materials.* Physical Chemistry Chemical Physics, 2013. **15**(44): p. 19510-19517.

2. Nakayama, D., et al., *Lower critical solution temperature (LCST) phase separation of glycol ethers for forward osmotic control.* Physical Chemistry Chemical Physics, 2014. **16**(11): p. 5319-5325.

3. Noh, M., et al., *Novel lower critical solution temperature phase transition materials effectively control osmosis by mild temperature changes.* Chemical Communications, 2012. **48**(32): p. 3845-3847.

4. Zhao, D., et al., *Thermoresponsive copolymer-based draw solution for seawater desalination in a combined process of forward osmosis and membrane distillation.* Desalination, 2014. **348**: p. 26-32.

5. Cath, T.Y., et al., *Standard methodology for evaluating membrane performance in osmotically driven membrane processes.* Desalination, 2013. **312**: p. 31-38.

6. Rajagopalan, N. and V.A. Patel, *Solvent removal and recovery from inorganic and organic solutions*, 2014, Patent US8852436.

465747 Zwitterions Coated Hollow Fiber Membranes with Enhanced Antifouling Properties for Osmotic Power Generation from Municipal Wastewater

Abstract:

The rapid increases in energy consumption and CO_2 emissions have accelerated the exploration of renewable and green energy. Among them, osmotic energy released from the mixing of water streams with different salinities via pressure retarded osmosis (PRO) has attracted global attention owing to its abundance and renewability. Two factors primarily affect the productivity of a PRO process; namely, the PRO membrane and the feed pair. An ideal PRO membrane requires a high water permeability, high salt rejection, low concentration polarization, high mechanical strength, and minimal fouling tendency. In case of feed pairs, in addition to the conventional pair of river water and seawater, the feed pair of concentrated reverse osmosis (RO) brine and wastewater from municipal recycle plants has received great attention because it has a higher salinity gradient and can potentially produce a higher power density. In addition, this new feed pair may add extra values such as lowering the overall energy consumption for RO, diluting the seawater RO brine for ecologically-friendly disposal, and saving some seawater pretreatment costs.

However, fouling on PRO membranes leads to severe declines in water flux and power density because their porous substrates are facing the wastewater feed. Thus, inorganics, organics and microorganisms in the wastewater are prone to depositing on the substrate surface and even in its pores. In order to reduce the fouling propensity, coating the substrate surface of PRO membranes with zwitterionic materials proves to be an effective way. In our work, 2-methacryloyloxyethylphosphorylcholine (MPC), is modified and grafted onto the polydopamine (PDA) coated poly(ether sulfone) (PES) hollow fiber substrate. Both the synthesis and surface coating of MPC are simple and straightforward. Compared with the pristine PES and PES-PDA substrates, the MPC modified substrate (PES-PDA-MPC) exhibits high resistance to protein adsorption as well as bacteria adhesion. By using a state-of-the-art thin-film composite poly(ether sulfone) (TFC-PES) hollow fiber membrane as the control for power generation, the power density of the TFC-PES-PDA-MPC membrane can achieve as high as 7.7 W/m^2 while the unmodified one has only 6.0 W/m^2 after 3 hâ€™s PRO tests. In conclusion, the osmotic power generation of PRO membranes can be significantly sustained by modifying the membrane surface with zwitterions.

453481 Second Interfacial Polymerization of Thin-Film Composite Polyamide Membrane with Polyethyleneimine for Forward Osmosis Applications

Abstract:

In this work, a fast and simple surface modification-second interfacial polymerization (SIP) of thin-film composite (TFC) membranes using polyethyleneimine (PEI) of various molecular weights, was applied to improve the separation performance for forward osmosis (FO) applications. Compared to the control TFC membranes, PEI -modified membranes exhibit higher water flux and lower fouling propensity. The effects of PEI molecular weight and PEI concentration on the membrane morphology and separation performance were investigated systematically with various characterizations. Post-treatment to surface modified membranes was carried out using four aqueous solutions of different pH values. Meanwhile, the FO performance, intrinsic separation properties and antifouling properties of these post-treated membranes were also studied. In comparison with the surface-modified TFC membranes with PEI, the post-treated TFC membranes show higher water flux and improved anti-fouling properties, especially those post-treated with acidic or alkaline aqueous solutions.

455190 Integrated Seawater Reverse Osmosis and Pressure Retarded Osmosis Processes for Sustainable Water and Energy Generation

Abstract:

Pressure retarded osmosis (PRO) is a promising technology to reduce the specific energy consumption of a seawater reverse osmosis (SWRO) plant. In this study, it is projected that 25.6-40.7 million kWh/day of energy can be recovered globally, if the brines from SWRO are used as the draw solution and diluted to the seawater level in a PRO system. Detailed integrated SWRO-PRO processes are developed in this study with the option to form a closed-loop SWRO-PRO process that can substantially reduce the pretreatment cost of desalination. The governing mathematical models that describe both the transport phenomena on a module level and the energy flow on a system level are developed to evaluate the performances of the SWRO-PRO processes. The model aims to investigate the performance of the hollow fibers as dilution occurs and provides guidelines on hollow fiber module design and process operation. Determining the dilution factor and the corresponding operating pressure of PRO is the key to optimize the integrated process. The specific energy consumptions of three SWRO-involved processes; namely, (1) SWRO without a pressure exchanger, (2) SWRO with a pressure exchanger, and (3) SWRO with pressure exchangers and PRO are compared. The results show that the specific energy consumptions for the above three processes are 5.51, 1.79 and 1.08 kWh/(m^3 of desalinated water) for a 25% recovery SWRO plant; and 4.13, 2.27 and 1.14 kWh/(m^3 of desalinated water) for a 50% recovery SWRO plant, using either freshwater or wastewater as the feed solution in PRO.

460057 Bioconversion of Natural Gas to Higher Value Chemicals: Engineering Considerations for Commercialization

Abstract:

The low cost and abundant supply of natural gas have been motivators for the development of many technologies that convert this low cost carbon feedstock into compounds of greater value. Natural gas bioconversion is a developmental technology which applies synthetic biology to a naturally methane consuming bacteria, a methanotroph, to program it to produce higher value materials of interest. Intrexon has formed joint ventures to commercialize methanotrophic bioprocesses for biofuels and high value chemical intermediates. Biocatalytic processes which achieve high yields under milder processing conditions allow for reduced capital and operation expenditures while also offering greater economic viability. Currently, the production of both isobutanol and farnesene has been demonstrated in lab scale fermentation. To successfully bring this technology to commercial scale, consideration must be given not only to optimizing the metabolic pathway engineered into the methanotroph, but to key issues in the fermentor design, downstream processing and overall plant economics.

At any scale, aerobic natural gas fermentation must overcome the challenge of gas to liquid transfer of low solubility gases. Low gas to liquid mass transfer reduces the process productivity and yield, which in turn increases fermentor cost, process operating cost and may add additional equipment necessary to recycle gases. There are many fermentor designs currently available to increase the overall mass transfer coefficient of these gases; however, capital and operating expenditures of the reactor become a critical metric for evaluating commercialization feasibility. Unique fermentor configurations can greatly increase capital cost by raising the cost of fabrication and decreasing the ability to receive competitive bids. These unique fermentors also may not create a homogenous well-mixed reactor, which increases the cost of controls and instrumentation.

Downstream process selection for commercialization is a balance of capital cost, operating cost and the risk of cutting-edge over well-established techniques. Low fermentation broth concentrations of product increase the size and energy requirement of downstream equipment. Cutting-edge technologies can increase the efficiency of separation and reduce energy inputs, but these technologies can be capitally intensive and unproven at commercial scales. Bioprocesses also create metabolic byproducts which need to be evaluated for separation and purification to meet specifications for sale.

Favorable plant economics are necessary to move forward with commercialization of a technology. Scale is carefully selected based on a comparison of the production rate, fermentation productivity and yield targets to revenue, capital and operating costs. The plant location is also an important selection factor as costs are affected by utility rates in the area and/or co-location of the plant with shared infrastructure.

Traditional engineering workflows, such as a front-end loading strategy, were used to shape the commercial scale process design. The biologic, technical and economic commercial feasibilities of the end-to-end natural gas to isobutanol process were critical to Intrexonâ€™s plant design. Laboratory scale learnings were applied to select technology that was commercially feasible at demonstration scale. This demonstration scale design was scaled down to pilot scale, where the operation will yield learnings into refinement of at-scale design, equipment selection, empirical data on heating and cooling, and controls strategies.

474172 Be an Engineering Detective--Technical Assessments for Process Licensing

Abstract:

Licensing of chemical process, equipment, and catalyst technologies is an important part of the chemical manufacturing business. For each technology there can be multiple providers to select from. We will explore methods of how to evaluate them. Questions about how to find the "best" technology, or "lowest cost" technology are nontrivial to answer and much work needs to be done to get accurate and confident answers. Throughout the process we try to exercise risk reduction to potentially influence and improve the benefits of the plant and to identify any "Unproven" technology.

470115 Tablet Compression Scale-up Risk Assessment Using a Model Based Approach

Abstract:

Scale-up of processes involves uncertainties from many sources: material properties, equipment design, operation parameters, etc. The use of several models, even simple models, can constrain risks and predict failures before they occur. This talk will focus on the use of modeling in conjunction with small scale experimentation as a predictive methodology to estimate and reduce risks due to the scale-up of the tablet compaction process.

471157 Best Practices Identified While Transferring Lyophilized and Syringe Biologic Drug Products from Internal Site to a CMO

Abstract:

In this presentation, lessons learned while transferring Lyophilized and Syringe biologic drug products to a new site will be shared. The goal is to go over the critical quality attributes of the selected commercial drug product and identify best practices to maintain those attributes in the new commercial site. During the technology transfer to a second commercial site post-approval, the firm might find opportunities to improve the current manufacturing process and face challenges to implement these changes due to regulatory requirements. This presentation will cover some of those challenges as well.

472624 Minimizing Risk of Upgrading Projects on Existing Assets

Abstract:

As is the case with any upgrading projects, the addition of new units/equipment introduce uncertainties and unforeseen risk to the existing assets. In DNV GL's work processes, methodologies incorporating cutting edge technology have been developed in order to measure risk and optimize performance. One of these methodologies that will be presented in this paper will quantify and minimize the risk of the upgrading project on the existing assets as well as the risk of the existing assets on the new units in order to maximize profit.

The main objective of this work is to apply reliability, availability, and maintainability (RAM) study to optimize the performance (from production standpoint) of the integrated project. Specifically, RAM analysis will be conducted to provide a quantitative assessment of the level of performance that may be expected from the system as a whole over its design life. DNV GL will use TARO software to execute the proposed methodology.

In order to illustrate the applicability of the proposed approach, a case study addressing the major risks of an upgrading projects will be evaluated. This case study will compare the as-is performance to the future performance by evaluating the availability and utilization of all the production-critical units and their impact on production. This analysis will assist in identifying the major units contributing to production losses. In addition, the case study will quantify the impact of the increase in utility system demand (Hydrogen, steam, power...etc.) on production and identify any bottlenecks. Finally, the case study will quantify the impact of holding tanks working volumes on production and existing assets.

The results of this study will improve stakeholder confidence in the proposed design and ensure that the integrated project yields maximum return on investment. The proposed methodology will provide managers with decision support in design optimization and integration, process debottlenecking, benchmarking of performance, and mitigation of unforeseen risks.

468181 Applying Reliability Centered Maintenance in Incinerators of Industrial Solid Waste

Abstract:

The marketing growth of nations around the world leads companies to become more competitive. The service sector has a direct influence on the financial results of an industrial park in loss reduction aspect, becoming an strategic function of any business that has competition vision. Their techniques and policies should follow market developments, emerging each time different models of management and implementation. The purpose of this paper is to present the application of reliability centered maintenance on Industrial Solid Waste Incinerators â€" ISWI. Probabilistic nature of studies have been inserted in the systems selection steps and determining the frequency of activities. Establishing criteria for defining the frequency of execution of the maintenance routines. As well as the determination of their modes and causes of failure. The main result was the knowledge of the equipment failures of nature. Power systems, and combustion, and gas washing are responsible for the main failure modes that cause interruption of ISWI. Its maintenance tasks were based on time, generating high corrective maintenance rate unscheduled. The study was delimited based on historical data of failures over six years of operation. It is clear the need for improvements in equipment maintenance plan constituting these systems to reduce the likelihood of failures. Contributing in the decisions that must be made to improve operating performance.

Managing Technology Transfer Risk
Through Predictive Sciences

PSP
Predictive
Science for
Products

Tool Development

Apply different filters to measures

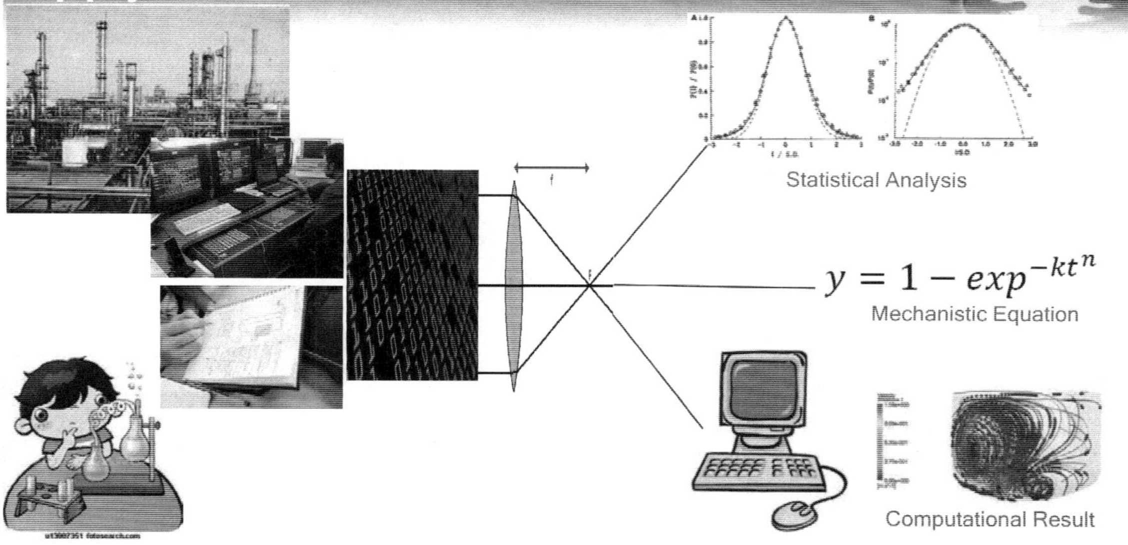

Focus on the product:

What happens to the product in processing?

What scale are we looking from?

How can we understand ,
and predict the outcome?

$F = ma$

Tool development

Build Understanding in Layers

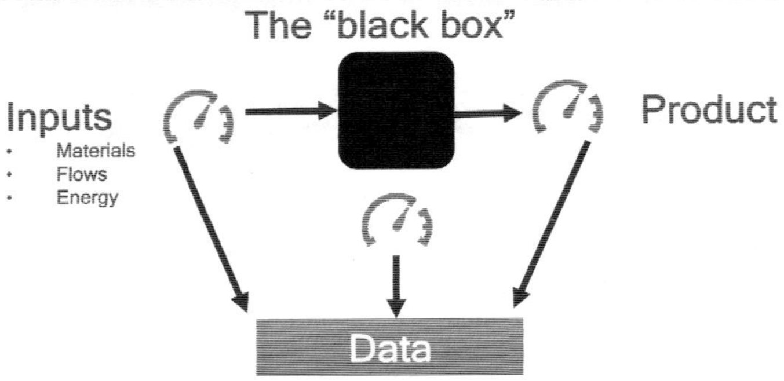

Simple Techniques:

- Ratios
- Statistics
- Trends
- Relationships

Data is the gold you mine to buy understanding.

Tool development

Deeper layer: more accuracy, complexity

Mixing Tank

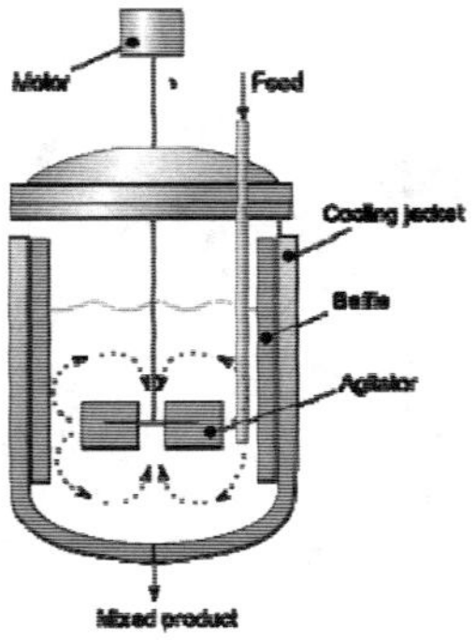

Apply Specific Knowledge:

- Geometry
- Regions
- Subprocesses

More information is available as we get smaller.

Tool development

Material, and molecular interactions

Tank entrance

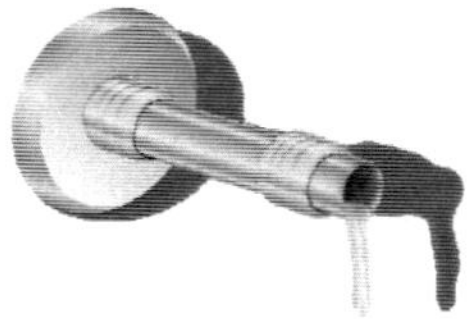

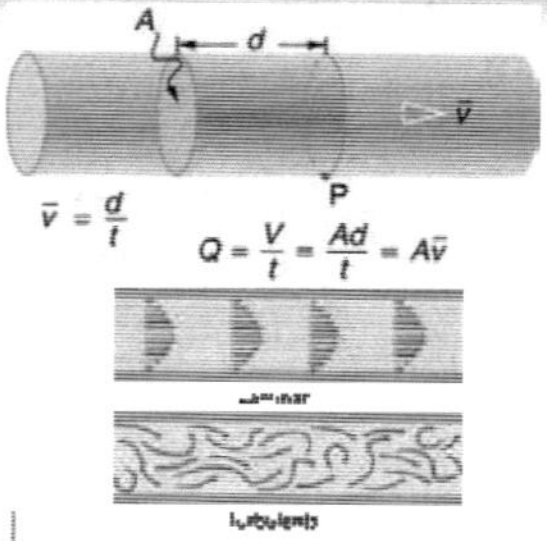

Detailed View :

- Region
- Point
- Interaction
- Variations

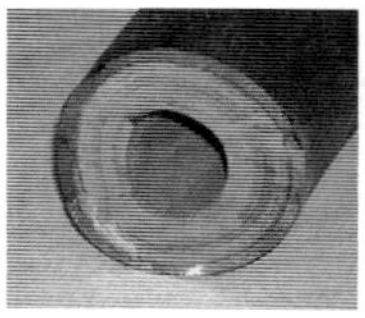

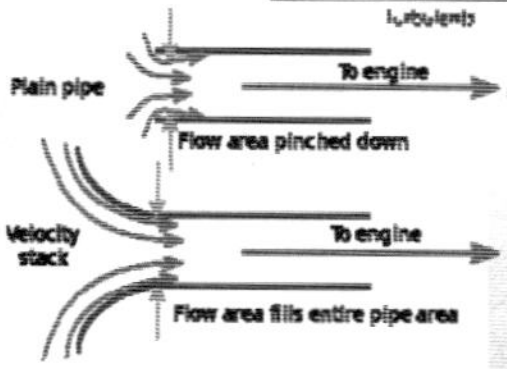

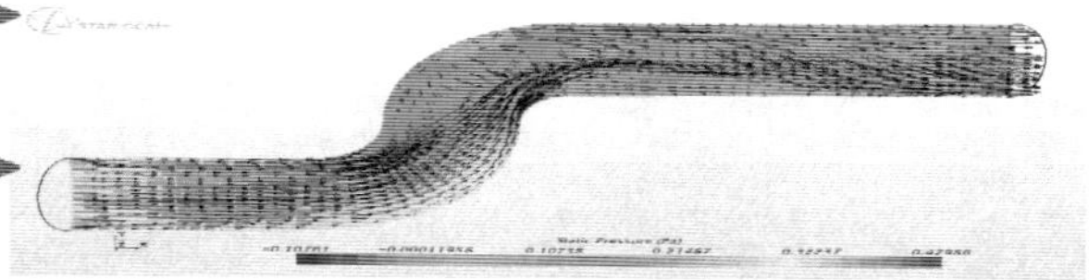

Keep asking questions until accuracy is acceptable

Tool development

How to simulate the physical process?

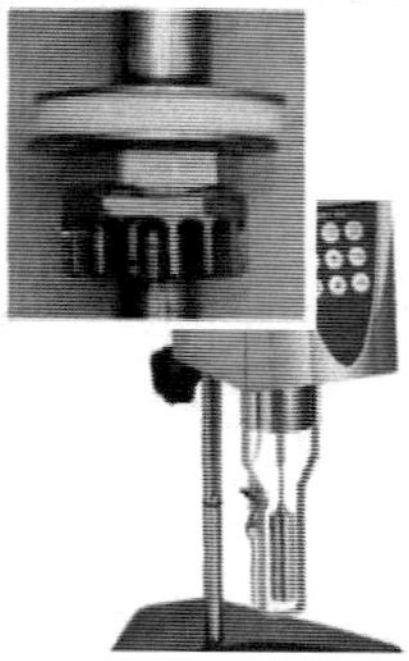

Rheometers

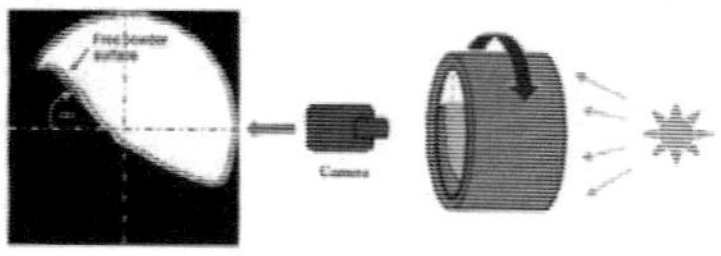

Powder flow, compaction

Tensile strength, abrasion

Shipping

Small scale tests inform processes of all sizes.

Tool development

Measures/Models can be linked

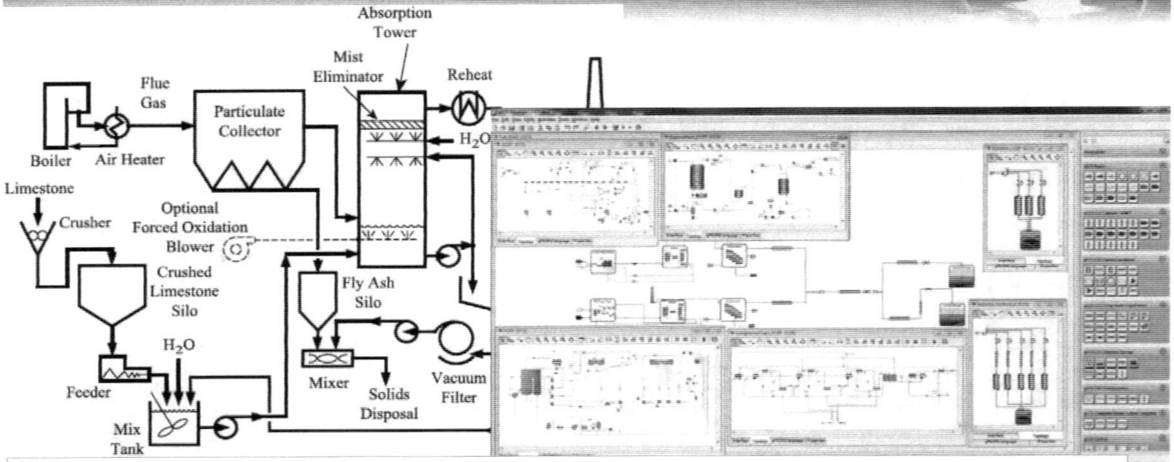

The same techniques apply to the whole plant.

Example: Fluidized Bed Modeling

Batch Type Fluid Bed Dryer

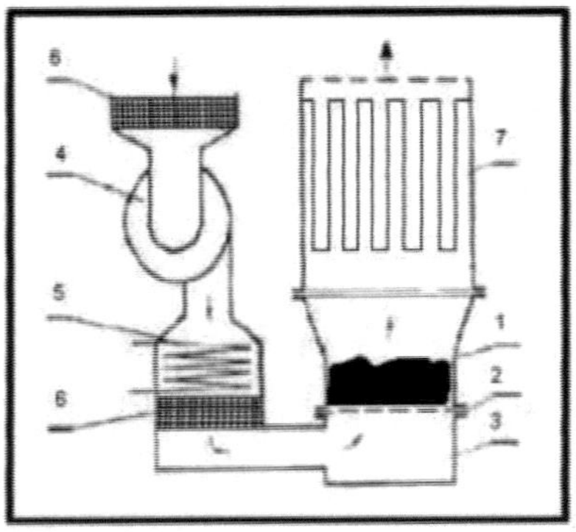

1. = vessel or chamber
2. = gas distributor
3. = plenum chamber
4. = blower
5. = heater
6. = filter
7. dust collector

Simple Techniques:

Rules:
- Pressure drop 300-500 mmH20
- Airflow ~1 m/s
- Particle size 0.1-1.0 mm
- Bulk Density 0.4-1.0 g/cc

Ratios:
- Geometric measurement
- Aiflow to screen/filter area
- Fill/Volume
- Humidity

Example: Fluidized Bed Modeling

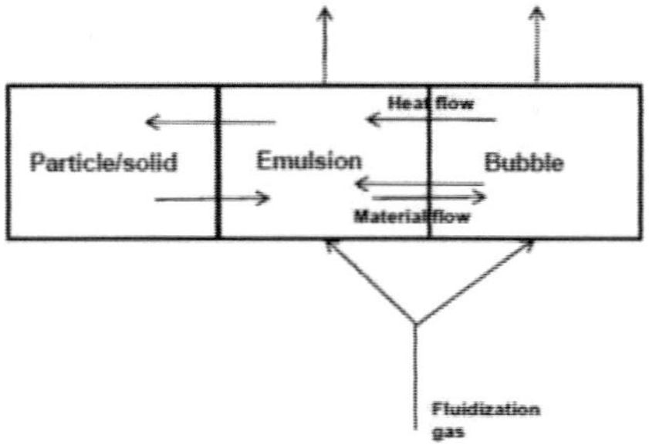

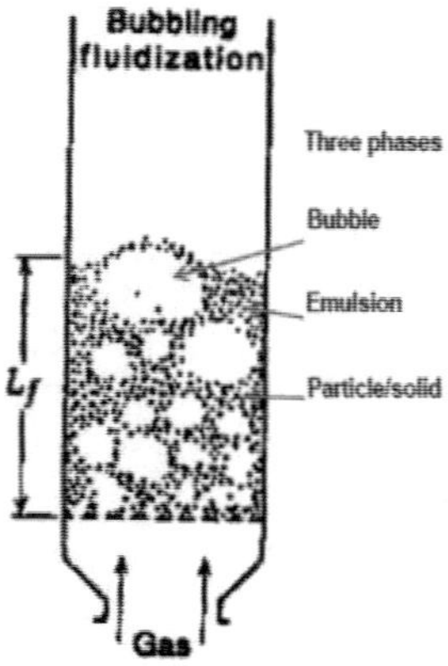

Apply Specific Knowledge:

- Geometry
- Regions
- Subprocesses

Example: Fluidized Bed Modeling

Apply Specific Knowledge:

- Build Equations

- **Correlations all originate from: Kunii and Levenspiel, "Fluidization Engineering," 2nd Ed., Butterworth-Heinemann, MA.**

Mass Transfer

$$k_p = \left(\frac{D_g}{d_p}\right)\left(2 + 1.8\left(\frac{\rho_g d_p v_{umf}}{\mu_g}\right)^{\frac{1}{2}}\left(\frac{\mu_g}{\rho_g D_g}\right)^{\frac{1}{3}}\right)$$

Mass transfer from particle to emulsion Ranz Eq. Kunni and Levenspeil Eq. 11.2

$$k_b = 4.5\left(\frac{v_{umf}}{d_b}\right) + 5.85\left(\frac{D_g^{\frac{1}{2}} g^{\frac{1}{4}}}{d_b^{\frac{5}{4}}}\right)$$

Mass transfer from emulsion to bubble Kunni and Levenspeil Eq. 10.27

Heat Transfer

$$h_p = \left(\frac{k_g}{d_p}\right)\left(2 + 1.8\left(\frac{\rho_g d_p v_{umf}}{\mu_g}\right)^{\frac{1}{2}}\left(\frac{Cp_g \mu_g}{k_g}\right)^{\frac{1}{3}}\right)$$

Heat transfer from particle to emulsion Kunni and Levenspeil Eq. 11.25

$$h_b = 4.5\left(\frac{v_{umf} \rho_g Cp_g}{d_b}\right) + 5.85\left(\frac{(k_g \rho_g Cp_g)^{\frac{1}{2}} g^{\frac{1}{4}}}{d_b^{\frac{5}{4}}}\right)$$

Heat transfer from particle to emulsion Kunni and Levenspeil Eq. 11.32

Example: Fluidized Bed Modeling

6 Ordinary differential equations and 6 unknowns - $Y_s, Y_e, Y_b, T_s, T_e, T_b$

$$r=(1-\varepsilon)\rho\left(\frac{6(1-\varepsilon_f)}{d_s}\right)k_s\left(\frac{P_{sat}(T_s)}{P_{atm}}-Y_e\right)$$

Rate of drying for constant rate period

$$r=(1-\varepsilon)(1-\varepsilon_f)\rho\,\frac{2D_{eff}}{l_{BL}}\frac{A_s}{V_p}(Y_s - [illegible])$$

Rate of drying for falling rate period

boundary layer assumed $d_s/2$

$$\rho(1-\varepsilon_f)(1-\varepsilon)\frac{dY_s}{dt}=[illegible]$$

Material balance for solid phase

$$\rho\,\varepsilon_f(1-\varepsilon)\frac{dY_e}{dt}=[illegible]+\frac{\rho\,u_{mf}(1-\varepsilon)\varepsilon_f}{\rho(1-\varepsilon_f)(1-\varepsilon)}\frac{Y_i-[illegible]}{L_{bed}}-\varepsilon\rho\,k_b(Y_e-[illegible])$$

Material balance for emulsion phase

$$\rho\,\varepsilon\frac{dY_b}{dt}=\varepsilon\rho\,k_b(Y_e-[illegible])+\varepsilon\rho\,u_b\frac{(Y_i-[illegible])}{L}$$

Material balance for bubble phase

$$\rho(1-\varepsilon_f)(1-\varepsilon)(C_{ps}+[illegible]C_{pv})\frac{dT_s}{dt}=\frac{6(1-\varepsilon_f)(1-\varepsilon)}{d_s}h_s(T_e-[illegible])-\lambda_{vap}r$$

Energy balance for solid phase

$$\rho\,\varepsilon_f(1-\varepsilon)(C_{pg}+[illegible]C_{pv})\frac{dT_e}{dt}=\varepsilon h_b(T_b-[illegible])-\frac{6(1-\varepsilon_f)(1-\varepsilon)}{d_s}h_s(T_e-[illegible])+\frac{\rho\,u_{mf}(1-\varepsilon)\varepsilon_f}{L_{bed}}\left[(C_{pg}+[illegible]C_{pv})T_i-(C_{pg}+[illegible]C_{pv})T_e\right]$$

Energy balance for emulsion phase

$$\rho\,\varepsilon(C_{pg}+[illegible]C_{pv})\frac{dT_b}{dt}=\frac{\varepsilon\rho\,u_b}{L}\left[(C_{pg}+[illegible]C_{pv})T_i-(C_{pg}+[illegible]C_{pv})T_b\right]-\varepsilon h_b(T_b-[illegible])$$

Energy balance for bubble phase

Apply Specific Knowledge:

- Refine Equations
- Program in Matlab

Example: Fluidized Bed Modeling

Axial Gas Velocity (m/s) 1700 CFM - 3 m/s

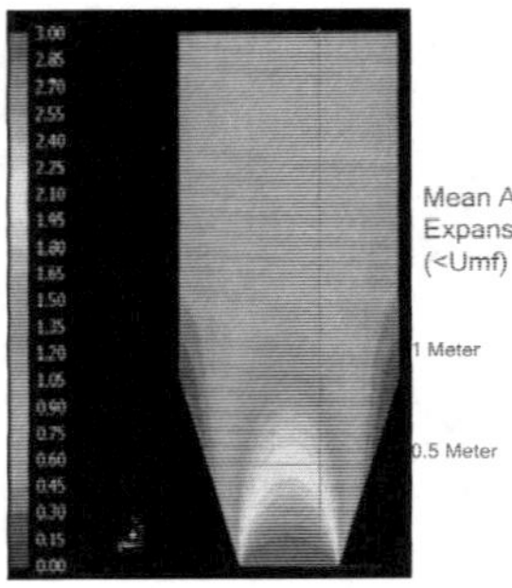

Mean Axial Velocity in Expansion Area = 0.6 m/s (<Umf)

Axial Velocity of Horizontal Planes

Axial Gas Velocity (m/s) 2450 CFM - 1.5 m/s

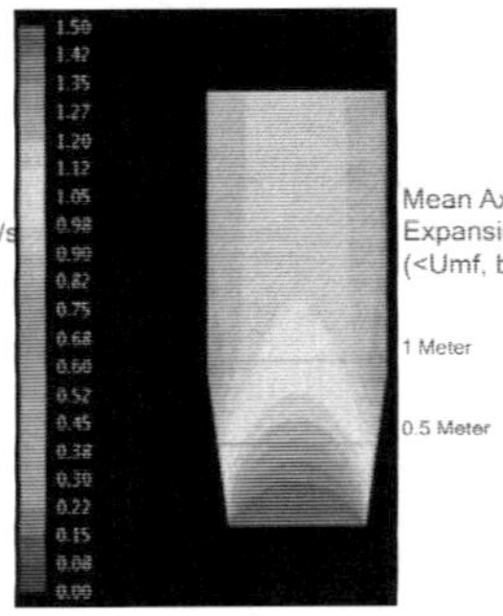

Mean Axial Velocity in Expansion Area = 0.88 m/s (<Umf, but close)

Axial Velocity of Horizontal Planes

Apply Specific Knowledge:

- Build CFD model

Example: Fluidized Bed Modeling

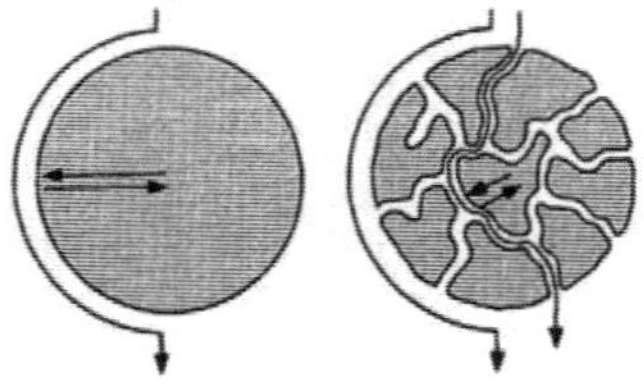

Measure diffusion of solvent
from the particle
under controlled environmental conditions

Detailed View:

- Measure property defined in the model

Example: Fluidized Bed Modeling

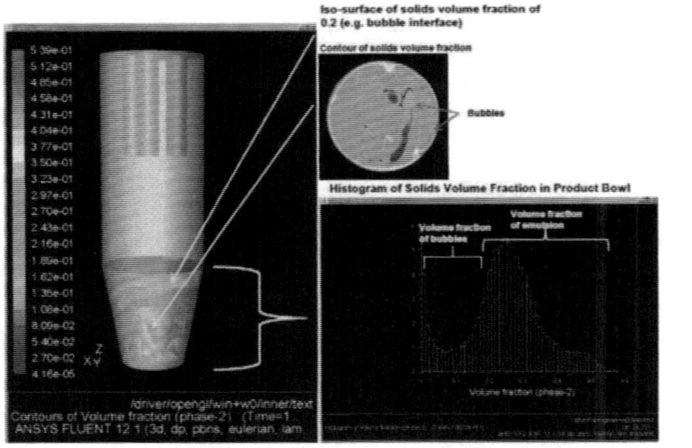

Create finer series of zones to extract bubble and emulsion volume fractions as function of vertical direction (time averaged)

Detailed View :

- Bubble size

Example: Fluidized Bed Modeling

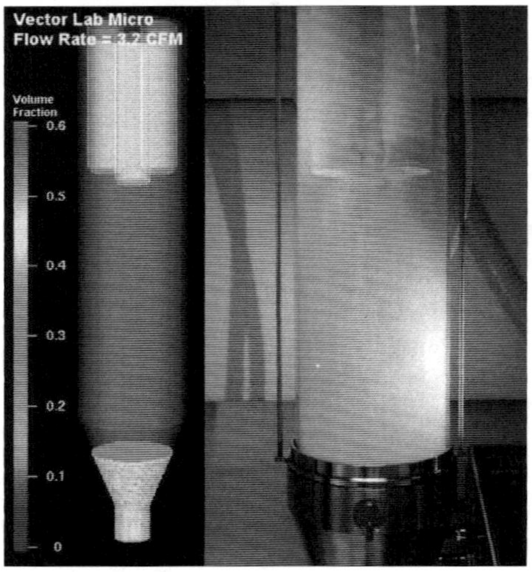

Verify:

- Compare with experiment

Fluidized Bed Drying Prediction

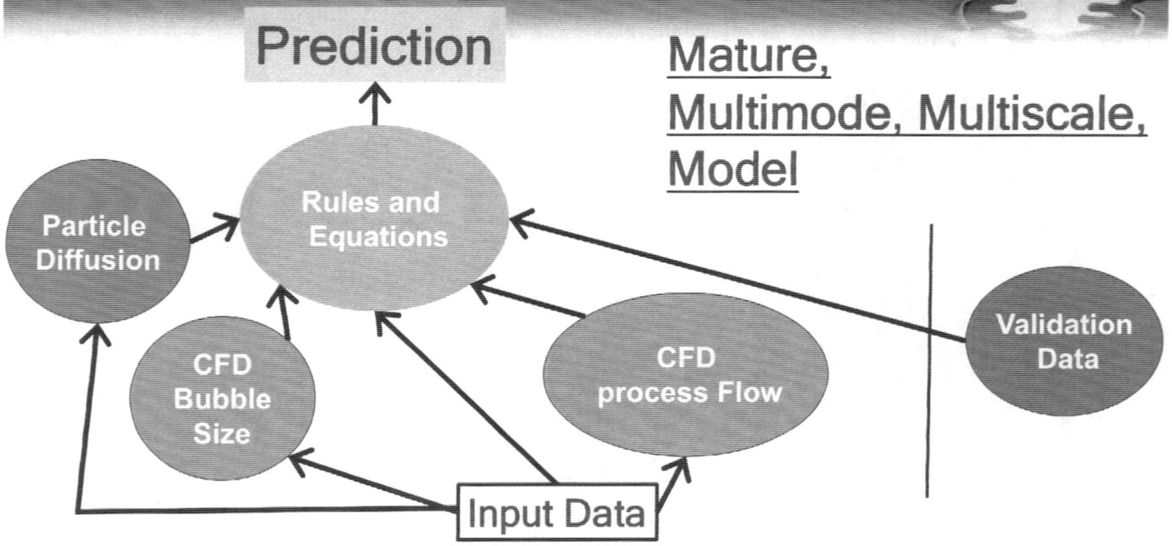

Mature, Multimode, Multiscale, Model

Fluidized Bed Drying Prediction

Stage	Maturity	Program	Impact
Simple Techniques - Initial	Experimental Design	Development on multiple programs	Save a few experiments
Apply Specific Knowledge - 1	Experimental Design, Quality	Development product with water sensitivity,	Assure quality Save Experiments Validate Model
Apply Specific Knowledge - 2	Troubleshooting	Scale up - back on track	Maintain Timing, Save 3-6 experiments, Save 450-900K$
Mature Model - 1	Equipment Selection	Scale-Up - avoid Issue, potential loss of:	Save ~6 batches at 900k$ Save equipment modification 750K$ Save project delay 6-9 months
Mature Model -2	Equipment Selection	Scale - Up	Save ~300K$ for equipment not needed

Models provide value at all stages of development

Implementation

Build Stakeholder Confidence

"all models are wrong,
but some are useful"

Box, G.E.P. and Draper, N.R.,
Empirical model-building and response surfaces , Wiley series in probability and mathematical statistics , Wiley, 1987

To have confidence stakeholders must know:

- Range of Use
- Accuracy
- Assumptions
- Risks
- Construction Details
- Training Data
- Testing
- Success

Documentation is key to acceptance.

Implementation

Gear the expectation to maturity

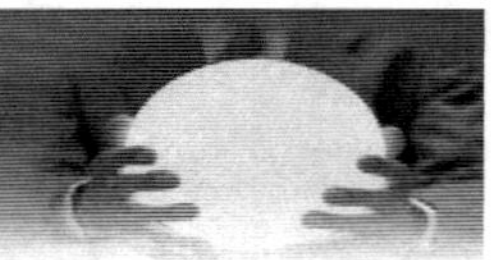

Using gardening as a model:

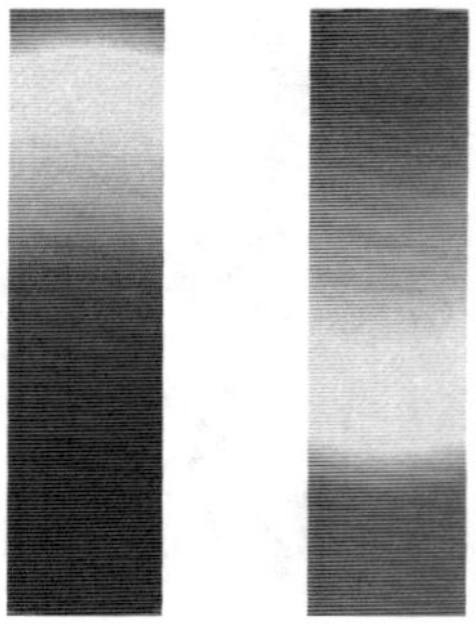

Concept-
idea that we can link measurements with outcomes using math

Seed -
limited data, simple mathematics or computation, some correlation

Sprout -
enough data to estimate error, form of the model evolved

Plant-
data validates model, multicomponent model built, high accuracy

Invasive Species-
when model and process have discrepancy-process is investigated for error

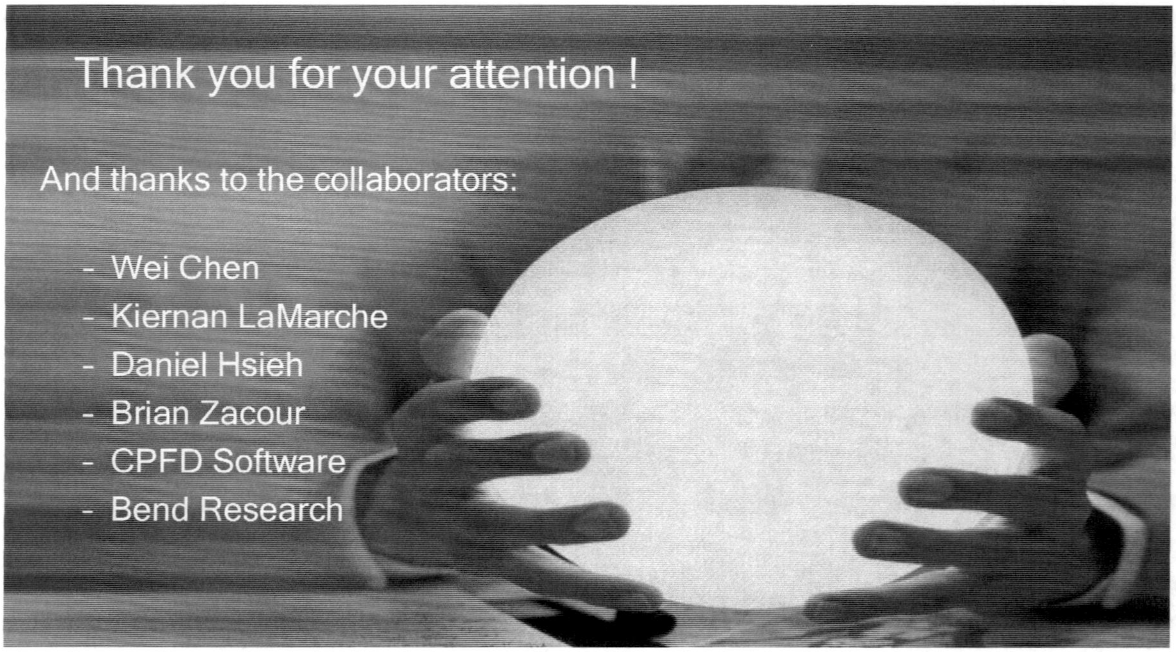

468367 Halophyte Biochar for Water Desalination Brine Concentrate Management

Abstract:

Disposal of concentrate from brackish water desalination is limited by concerns of salt leaching into and contaminating surface and ground waters. These concerns, and measures to prevent leaching, add substantially to concentrate disposal costs with little or no opportunity to recover those costs. One option being explored is the use of concentrate as irrigation water for producing halophyte biomass for forage and other applications. Continuing development of this option requires more knowledge about what happens to the salt in the concentrate once land applied as irrigation water.

The purpose of this project is to determine how much of the salt taken up by the halophyte crops can be protected from leaching by pyrolysis of the biomass, i.e. to determine the potential of sequestering salt in mineral form. Results from this research will be useful for those managing high salinity biomass streams, such as halophyte crop wastes and animal manures, and those seeking local concentrate disposal options with a potential for producing economic and/or environmental returns. The ability to sequester salt in biochar would decrease the negative impacts of land-applying high-salinity materials, including those used in desalination concentrate management systems.

455179 ^{13}c Pathway Analysis for the Role of Formate in Electricity Generation By *Shewanella Oneidensis* MR-1 Using Lactate in Microbial Fuel Cells

Abstract:

Microbial fuel cell (MFC) is a promising technology for wastewater treatment. Basically, the microorganisms sticking on the anode of the MFC utilize various organic compounds from the wastewater as the electron donors to generate electricity, while in the meantime removing the organic compounds in the wastewater. The type of electron donors fed into MFCs affects the electrical performance, and mechanistic understanding of such effects is important to optimize the MFC performance. In this study, we used a model organism in MFCs, *Shewanella oneidensis* MR-1, and ^{13}C pathway analysis to investigate the role of formate in electricity generation and the related microbial metabolism. Our results indicated a synergistic effect of formate and lactate on electricity generation, and extra formate addition on the original lactate resulted in more electrical output than using formate or lactate as a sole electron donor. Based on the ^{13}C tracer analysis, we discovered decoupled cell growth and electricity generation in *S. oneidensis* MR-1 during co-utilization of lactate and formate (i.e., while the lactate was mainly metabolized to support the cell growth, the formate was oxidized to release electrons for higher electricity generation). To our best knowledge, this is the first time that ^{13}C tracer analysis was applied to study microbial metabolism in MFCs and it was demonstrated to be a valuable tool to understand the metabolic pathways affected by electron donors in the selected electrochemically-active microorganisms.

461312 *Advanced Redox-Based Electrochemical Separations for Wastewater Treatment*

Abstract:

Water scarcity is one of the principal challenges for the 21st century, both from a geographical and an economic standpoint. Developing novel, more energetically efficient technologies for wastewater treatment, retreatment and reuse is crucial for long-term water sustainability. One of the principal challenge for water remediation, both in the U.S. and worldwide, is to target contaminants of emerging concern, such as complex organic molecules including pharmaceuticals, chemical precursors, endocrine inhibitors and pesticides.[1] These originate from chemical processes and organic synthesis, the food industry, electronic manufacturing among others. Many of these pollutants disperse in remote locations, such as agricultural locations, in which large-scale purification systems cannot be implemented. Electrochemical methods are particularly attractive for addressing these problems due to their modularity, fast kinetics and integration with renewable energy sources.[2]

We have developed an redox-mediated separation method based on organometallic-functionalized electrodes (e.g. ferrocene or cobaltocene) for the sorption and release of specific ions from the liquid-phase under oxidizing and reducing conditions.[3] Many of the organic contaminants are anionic in nature (e.g. carboxylates, sulfonates or phosphonates), thus selectivity based on chemical interactions with their functional groups is essential in chemical design. Rather than rely solely on charge or size, our immobilized redox-centers targets these compounds through the activation of a specific chemical binding. Moreover, in the charging and discharging steps, electrochemical charge can be stored by ion-selective adsorption onto the porous electrodes, thus dramatically decreasing energetic costs.

This modular system can then be easily implemented at the point-of-exit in chemical manufacturing and wastewater treatment. Here, we present a complete asymmetric redox-system in both batch and flow for (1) water remediation, through the specific separation of anionic and cationic contaminants in water, (2) tandem energy recovery due to pseudocapacitive charge storage and (3) increasing sustainability in chemical processes, through the recovery of expensive synthesis products from complex organic reactions, thus reducing organic waste and decreasing water usage during separation steps.

References

[1] a) Gilliom R.; Hamilton P. *U.S. Geological Survey*. 2014. (USGS fact sheet: http://pubs.usgs.gov/fs/2006/3028/

). b) Doughton, C.G. and Thomas, A. T. (1999). Pharmaceuticals and personal care products in environment: agents of suble change? *EHP* **107** (6), 907-938.

[2] Suss, M. E.; Porada, S.; Sun, X.; Biesheuvel, P. M.; Yoon, J.; Presser, V. *Energy & Environmental Science* **2015**. Zhao, R.; Satpradit, O.; Rijnaarts, H. H.; Biesheuvel, P. M.; van der Wal, A. *Water Res* **2013**, *47*, 1941.

[3] **Su, X.**; Kulik, H; Jamison, T.F.; Hatton, T. A. 2016. Anion-selective redox electrodes: electrochemically-mediated separation with organometallic interfaces. *Advanced Functional Materials.* Advance Article Online. DOI: 10.1002/adfm.201600079

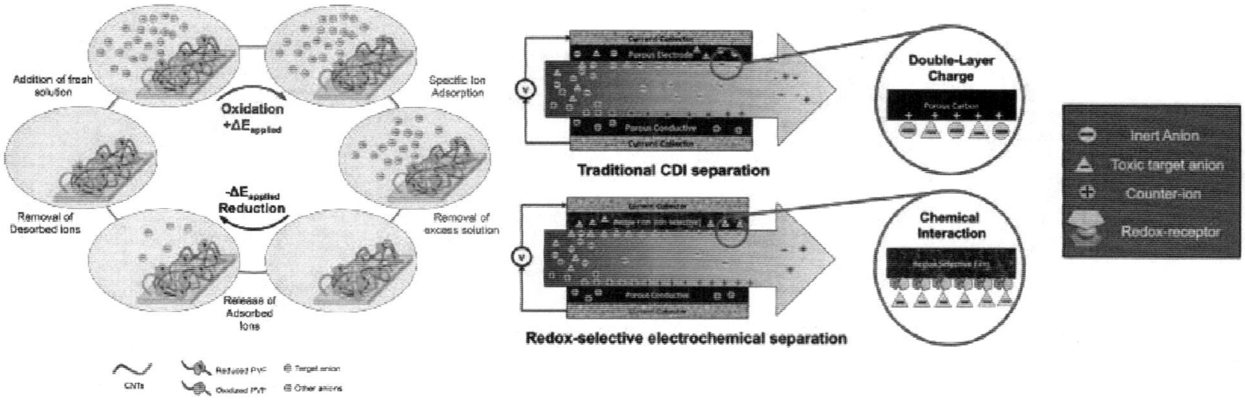

458179 Influence of Elevated Salinity and the Presence of Cytostatic Drugs on Microbial Community in FO Anaerobic Membrane Bioreactors

Abstract:

Wastewater treatment using foward osmotic anaerobic membrane bioreactors (FO-AnMBR) has a number of advantages over traditional wastewater treatment approaches. Previous studies have investigated the overall performance of lab-scale FO-AnMBR in treating synthetic wastewater. During the operation of FO-AnMBR, it is expected that salinity gradually increases, which has also been shown experimentally. In addtion, emerging contaminants such as cytostatic drugs in wastewater would also be accumulated in the FO-AnMBR. Little has been known on whether and how the increased salinity and the presence of cytostatic drugs affect the microbial community in FO-AnMBR. The objective of this study was to investigate the influence of increased salinity and the presence of cytostatic drugs on the structure and function of the microbial community in FO-AnMBR through a metagenomics approach. A lab-scale FO-AnMBR was operated at 25 ^{o}C and fed with synthetic wastewater. Total DNA was extracted from sludge samples taken at pre-determined time points and sequenced using a HiSeq 2500 platform with 250-bp paired-end sequences. Based on the analyses of the sequences, structure and potential functions of the microbial communities were obtained. We found that the increased salinity in FO-AnMBR did not reduce the microbial diversity; however, the relative abunance of methane-producing archaea decreased significantly at high salinity condition, resulting in a reduced methane yield. The general metabolism did not differ significantly under low and high salinity conditions. Intriguingly, genes responsible for cell attachment and biofilm related signalling were enriched at high salinity condition. Transition of the gene clusters for assimilatory and dissimilatory nitrate reduction was also found. The presence of cytostatic drugs only slightly affected the microbial community composition. Biofilm-forming, multidrug resistant and siderophore-producing bacteria were found more dominant after the treatment with drugs.

471970 Batch and Packed-Bed Adsorption Study for the Removal of Dye from Textile Wastewater Using in-House Prepared Silica-Based Nanosorbcats: Kinetics and Computational Modeling

Abstract:

The worldwide shortage of fresh water and the huge competing demands from variety of users stimulates an urgent need for finding alternative cost-effective and environmentally friendly wastewater treatment processes. For instance, synthetic dyes wastewater discharged from the textile and dyestuff industries poses an environmental alert since these organic pollutants are toxic to many aquatic and non-aquatic living organisms. In addition, some of these pollutants are non-biodegradable, and thus, they will exist for a long time in the environment, which may cause a real challenge to the conventional wastewater treatment processes. Accordingly, innovative and environmentally sound techniques are needed. The application of nanoparticle technology as adsorbents and catalysts (nanosorbcats), whether as a standalone or as an enabling technology, in cleaning up wastewater has recently received great attention. This is because of the unique chemical and physical properties of nanoparticles in comparison with their counterparts, which make them superior to the conventional adsorbent/catalysts.

Accordingly, the present study investigated the employment of newly in-house prepared silica-based nanosorbcats functionalized with active species for cleaning up textile wastewater. A facile co-precipitation synthesis route was used to prepare those nanosorbcats, which were characterized by different characterization techniques like XRD, BET, TPD, TPR, HRTEM and zeta potential. The prepared nanosorbcats were employed for the adsorptive removal of cationic and anionic dyes from synthetic and real textile wastewater. Batch adsorption studies were performed for kinetics and equilibrium modeling on the removal of different dyes from wastewater. The operating variables studied were contact time, initial dye concentration, and solution pH. Results showed an optimum pH value, where the maximum removal of dye was achieved. The experimental adsorption data have been modeled using Sips and BET adsorption isotherm models. In addition, a computational modeling on the selected model compounds was carried out to understand the adsorption mechanism on the prepared nanosorbcats. Breakthrough capacities were also investigated in a packed-bed column study. The most successful silica-based nanosorbcats in cleaning up the real textile wastewater in the batch experiments were prepared in extrudate form and used in up-flow packed-bed adsorption experiments. The effects of the influent concentration, flow rate, and bed height were investigated. Breakthrough time and adsorption capacity of the packed-bed increased with increasing bed height, whereas decreased with the increase in influent initial concentration and flow rate. The reliable results of this study will be elaborated and discussed further during the presentation.

463610 Optimization of Operational Parameters for Application of Zirconium/Polyvinyl Alcohol Modified Flat-Sheet Polyvinylidene Fluoride Membrane As Better Adsorptive Material in Treatment of Organic Arsenic Containing Wastewater

Abstract:

Organic arsenic in waters has been a global concern due to its toxicity. In this study, a novel polyvinyl alcohol (PVA)/zirconium modified polyvinylidene fluoride (PVDF) membrane was applied to remove organic arsenic from water. The impregnation of zirconium ions within the modified membrane was attributed to the coordination reactions among the zirconium ions, ether and hydroxyl groups. The synthesized membrane worked better at the acidic conditions and achieved the optimal uptake for both monomethylarsonic (MMA) and dimethylarsinic (DMA) at pH 2.0. The adsorption isotherm study demonstrated that the adsorption of both organic arsenic species was controlled by the mono-layer adsorption process, and the maximum adsorption capacities of 73.04 and 29.78 mg/g for MMA and DMA can be achieved at pH 2.0. The presence of humic acid had a negligible impact on the uptake of organic arsenic, whereas varying impacts on the arsenic adsorption were observed due to the presence of coexisting anions such as fluoride, phosphate, carbonate and silicate. The filtration study was also carried out to evaluate the performance of the modified membrane on the removal of organic arsenic. Based on the XPS analyses, the exchange between chloride ions on the membrane surface and organic arsenic species should be responsible for the removal of both MMA and DMA.

472779 Density Functional Theory Study of the Hydrolysis of a Sodium Borosilicate Glass

Abstract:

The long-term stability of nuclear waste glass forms is difficult to predict, and estimates rely on empirical information from natural analogs or accelerated tests. Models informed by first-principles calculations have the potential to improve the description of long-term behavior based on atomic-scale details. In this contribution, we present reaction barriers and energies for the hydrolysis of a model sodium borosilicate glass calculated from density functional theory considering acid, base, and neutral conditions. We discuss the roles of the glass composition and structure on glass stability in potential environmental conditions.

PORE SATURATION MODEL FOR CAPILLARY IMBIBITION AND DRAINAGE PRESSURES

Dr. James E. Laurinat
Savannah River National Laboratory
Aiken, South Carolina

Abstract

Leaching and transport of radionuclides from cementitious waste forms and from waste tanks is a concern at the Savannah River Site and other Department of Energy sites. Computer models are used to predict the rate and direction for migration of these through the surrounding soil. These models commonly utilize relative permeability and capillary pressure correlations to calculate migration rates in the vadose (unsaturated) zone between the surface and the water table. The most commonly used capillary pressure models utilize two parameters to relate the pressure to the relative saturation between the wetting (liquid) and nonwetting (gas) phases. The correlation typically takes the form of a power law relation (e.g., Brooks and Corey and van Genuchten)[1,2] or an exponential equation (e.g., Kosugi).[3,4]

In this study, a pore saturation model is used to derive the capillary pressure as a function of a characteristic pore pressure and the wetting phase saturation. Singularity analyses of the total energies of the wetting and nonwetting phases give residual saturations for the two phases. The model includes separate pressures for imbibition and drainage to account for capillary hysteresis.

The model successfully correlates a selected set of laboratory imbibition and drainage data for sand, from the U. S. Salinity Laboratory's UNSODA database[5,6] and other sources.

The capillary pressure model utilizes a single fitting parameter, a characteristic pore pressure, which is related to a characteristic pore diameter by the Laplace equation. Data regression shows that this pore diameter approximately equals the diameter predicted from a capillary rise analysis based on the mean particle diameter.

Nomenclature

$\bar{d}_{part}$	volume average particle diameter for porous material
d_{pore}	effective pore diameter for porous material
G_s	potential energy associated with saturation of the porous medium
ΔG_{jump}	Gibbs free energy associated with jump condition between a homogeneous porous material and a porous material with fissures
H	molar enthalpy
k	Darcy coefficient for permeability
k_n	relative permeability to the nonwetting phase
k_w	relative permeability to the wetting phase
n_i	number of moles of the ith solute in the Gibbs' adsorption theorem calculation
P_a	average pressure in the pores of the porous medium
P_c	intrinsic capillary pore pressure, equal to the gas-liquid pressure difference

$P_{c,d}$ capillary pressure measured for drainage of the wetting phase

$P_{c,d,ms}$ capillary pressure measured for metastable drainage of the wetting phase

$P_{c,i}$ capillary pressure measured for imbibition of the wetting phase

$P_{c,i,0}$ minimum capillary pressure required to imbibe the wetting phase at the residual nonwetting phase saturation

$\Delta P_{c,ov}$ maximum reversible overshoot in the capillary pressure at the minimum saturation

P_{flow} characteristic pressure for flow in capillary pores

P_g gas phase pressure in capillary pores

ΔP_{jump} pressure change associated with jump condition between a homogeneous porous material and a porous material with fissures, applied to the liquid phase

P_l liquid phase pressure in capillary pores

P_{pot} potential pressure associated with partial saturation of capillary pores

$P_{v,i}$ vapor pressure of the ith solute in the Gibbs' adsorption theorem calculation

P_o ambient pressure for liquid in contact with the porous medium

R_g universal gas law constant

s relative saturation with the wetting phase

s^* saturation for a fissured soil, scaled so that it gives a capillary pressure equal to that of a nonfissured soil

s_{cr} critical saturation where the capillary pressure for imbibition equals the capillary pressure for drainage

s_{max} maximum saturation for a fissured porous material

s_{min} minimum saturation for a fissured porous material

s_{nr} residual nonwetting phase saturation

s_{wr} residual wetting phase saturation

T absolute temperature

v_n velocity of the nonwetting phase

v_w velocity of the wetting phase

W work required for flow through the porous medium

W_g work required for flow of gas through the porous medium

W_l work required for flow of liquid through the porous medium

z distance in the direction of flow

ε porosity

θ wetting angle

μ_n dynamic viscosity of the gas or nonwetting fluid

μ_w dynamic viscosity of the liquid or wetting fluid

σ air-water interfacial tension

Introduction

Capillary pressure represents the pressure required to displace a given volume of one of the phases from a two-phase mixture in a porous medium. The capillary pressure is typically measured either by imbibition or by drainage. Imbibition pressure measurements begin with uptake of liquid by the porous medium, typically a sand or soil, which initially is saturated with gas. For drainage pressure measurements, the gas displaces the liquid. (For convenience, this paper will refer to the terms wetting phase and liquid and nonwetting phase and gas interchangeably.)

During drainage, there exists a residual wetting saturation, s_{wr}, at which no additional amount of liquid can be displaced, no matter how much pressure is applied. For imbibition, as the liquid saturation increases, the imbibition pressure decreases. The pressure drops to zero, and absorption of liquid stops, at a maximum fractional wetting phase saturation, s_{nr}, called the residual nonwetting saturation. Significant drainage does not occur until a threshold pressure referred to as the displacement pressure or the entry head is exceeded. The drainage pressure then increases as the liquid saturation decreases. At any given saturation, the drainage capillary pressure exceeds the imbibition capillary pressure. The hysteresis between the two pressures is greatest at the residual nonwetting saturation.

Historically, the capillary pressure has been correlated as a function of the liquid saturation, normalized with respect to the difference between the residual nonwetting and wetting saturations. Most capillary pressure models utilize two parameters to relate the pressure to the relative saturation. The relationship typically takes the form of a power law relation (e.g., Brooks and Corey and van Genuchten),[1,2] an exponential equation (e.g., Kosugi),[3,4] or a combination of these two forms applied over different liquid saturation ranges (e.g., Rossi and Nimmo).[7] Brooks and Corey assigned the entry head, defined as the minimum capillary pressure during drainage, as one parameter and used the other parameter to characterize the pore size distribution. Several other models have followed this convention.

Parker and Lenhard[8] addressed the hysteresis between drainage and imbibition. They developed a model that includes effects of gas entrapment during imbibition. Their model defines limiting maximum capillary heads for drainage and minimum capillary heads for imbibition. They state that the actual pressure must lie at some value between these limits that depends on the flow history. A simple method of applying this model, applied by White and Oostrom,[9] is to define separate limiting capillary pressures for imbibition and drainage.

This paper presents a departure from traditional analytic models. A probabilistic pore pressure model and global energy balances replace the semi-empirical approach of other models. The energy balances are solved for singularities in the driving forces for flow of the wetting and nonwetting phases, to obtain the residual wetting and nonwetting saturations, respectively. The pore pressure models are then used to calculate the capillary pressure as a function of saturation. Results are compared to selected capillary pressure measurements.

Derivation of Energy Balance for Capillary Pressure Model

The singularity analyses for the residual saturations are based on comparisons of the potential energy associated with the difference between the pressures of the two phases in the porous medium and the work required for flow of each phase. The analysis is loosely analogous to the Gibbs' adsorption theorem, which relates work done by surface forces to the chemical potential of a system. The model replaces the chemical potential with a mechanical potential based on the stored interfacial energy and substitutes a flow work term for the work needed to extend an interface. The model is applied to an adiabatic control volume within a larger volume of the porous medium, from which the only energy transfer occurs by flow of gas and liquid in or out. The model evaluates changes of the potential energy within the control volume and changes in the amount of flow work performed by the control volume on its surroundings.

The energy balance for the control volume takes the form

$$\Delta H = \Delta G_s + \Delta W \tag{1}$$

The energy balance is formulated in terms of the enthalpy because it is an open system in which liquid and gas may cross the volume boundaries.

Since the model assumes that the porous medium control volume is adiabatic, the enthalpy term is zero, and

$$\Delta G_s = -\Delta W \tag{2}$$

The mechanical potential and flow work terms can be expressed in terms of pressures much as the chemical potential in the Gibbs' adsorption equation. For a control volume initially comprised of one mole of gas, these expressions are

$$\Delta G_s = -R_g T \Delta \ln\left(P_{pot}\right) \tag{3}$$

and

$$\Delta W = -R_g T \Delta\left(-\ln\left(P_{flow}\right)\right) \tag{4}$$

The model employs different flow pressures for imbibition and drainage. The flow pressure for imbibition is based on liquid phase flow, while the flow pressure for drainage is based on gas phase flow.

Derivation of Pore Pressure Model

To derive the pressures for the potential energy and the flow work, a model for the pore pressure as a function of saturation must be developed. The pore pressure model assumes that each pore is filled with either a liquid wetting or a gaseous nonwetting phase. The capillary pore pressure is distributed by gas-liquid interfaces between the pores. The distribution of gas-and liquid-filled pores and the connections between pores are assumed to be random. The difference between the average pore pressure and the liquid pressure, according to this model, is equal to the product of the capillary pore pressure and the probability that a liquid-filled pore is in contact with a gas-filled pore and not with liquid in another pore. This probability, in turn, is the gas saturation. Thus, the difference between the average pore pressure and the liquid pressure is given by

$$P_a - P_l = P_c\left(1 - s\right) \tag{5}$$

Likewise, the difference between the gas pressure and the average pore pressure is equal to the product of the capillary pressure and the liquid saturation, or,

$$P_g - P_a = P_c s \qquad (6)$$

The potential energy is calculated from the difference between the average pressure in the porous medium, P_a, and the reference pressure for flow through the porous medium, P_o. For an influx of liquid into the porous medium (imbibition), the reference pressure is the liquid phase pressure, and

$$P_a - P_o = P_c(1-s) \qquad (7)$$

For flow of gas (drainage), the reference pressure is the gas phase pressure, so that

$$P_a - P_o = -P_c s \qquad (8)$$

These pressure differences apply only to the fractions of the pore volume occupied by liquid and gas, respectively. Consequently, for either liquid or gas flow, the potential energy is based on a pressure differential given by

$$P_{pot} = P_c s(1-s) \qquad (9)$$

The pressure differential is defined to be positive for both imbibition and drainage. Substitution of this pore pressure in the equation for the potential energy yields

$$\Delta G_s = -R_g T \Delta ln\left(P_c s(1-s)\right) \qquad (10)$$

Analysis of Residual Saturations

At the residual wetting phase saturation, s_{wr}, the resistance to flow of the liquid becomes infinite, and, in an overall energy balance, the resistance to flow of the gas can be ignored. It may be stated, then, that for a given change in the saturation the change in the potential energy counteracts the change in the work needed to cause the liquid to flow. In other words, at $s = s_{wr}$,

$$\frac{dG_s}{ds} = -\frac{dW_l}{ds} \qquad (11)$$

The liquid work function is derived from the Darcy equation for flow of two immiscible phases. The Darcy equation for the wetting phase is:

$$\varepsilon s v_w = -\frac{kk_w}{\mu_w}\frac{dP_l}{dz} = -\frac{kk_w}{\mu_w}P_c\frac{ds}{dz} \qquad (12)$$

or

$$v_w = -\frac{kk_w}{\mu_w \varepsilon} P_c \frac{d\,ln(s)}{dz} \qquad (13)$$

From this expression the pressure acting on a given volume of liquid is seen to be $P_c\, ln(s)$. The change in the corresponding work function is given by

$$\Delta W_l = -R_g T \Delta ln\left(-P_c\, ln(s)\right) \qquad (14)$$

As with the potential energy, this expression gives the change in the work function for small changes in the saturation.

Substitution of this work term and the change in the potential energy in the energy balance yields

$$-R_g T \frac{d\,ln\left(P_c s(1-s)\right)}{ds} = -R_g T \frac{d\,ln\left(-P_c\, ln(s)\right)}{ds} \qquad (15)$$

or

$$ln(s) = \frac{s-1}{1-2s} \qquad (16)$$

This equation is satisfied for $s_{wr} = 0.236$.

A similar line of reasoning is used to calculate the residual saturation of the gas, s_{nr}. Here, the resistance to the gas flow is infinite, and the resistance to flow of the liquid can be ignored. Thus,

$$\frac{dG_s}{ds} = -\frac{dW_g}{ds} \qquad (17)$$

The Darcy equation for the gas phase is:

$$\varepsilon(1-s)v_n = -\frac{kk_n}{\mu_n}\frac{dP_g}{dz} = -\frac{kk_n}{\mu_n} P_c \frac{ds}{dz} \qquad (18)$$

or

$$v_n = -\frac{kk_n}{\mu_n \varepsilon}\left(-P_c \frac{d\,ln(1-s)}{dz}\right) \qquad (19)$$

From this expression the pressure acting on a given volume of gas is seen to be $-P_c\, ln(1-s)$. The capillary pressure is subtracted from this pressure. The capillary pressure is supplied to the gas to convert it from a stationary state in which it is in contact with a pore wall to a mobile state in which it is surrounded by liquid. The total pressure supplied to the gas phase is, therefore, $-P_c\left(1+ln(1-s)\right)$, and the corresponding change in the work function is:

$$\Delta W_g = -R_g T \Delta ln\left(-P_c\left(1+ln(1-s)\right)\right) \tag{20}$$

Substitution of this work term and the change in the potential energy in the energy balance gives

$$-R_g T \frac{d\, ln\left(P_c s(1-s)\right)}{ds} = -R_g T \frac{d\, ln\left(-P_c\left(1+ln(1-s)\right)\right)}{ds} \tag{21}$$

or

$$ln(1-s) = \frac{3s-1}{1-2s} \tag{22}$$

This equality is satisfied for $s_{nr} = 0.884$.

Measured residual saturations for a uniform sand[10,11,12] agree almost exactly with the derived values; the measured wetting saturation, estimated by graphical interpolation, was 0.230, and the measure wetting saturation was 0.884.

Calculation of Capillary Pressures for Imbibition and Drainage

The following section describes a capillary pressure model for imbibition and drainage, based on the residual nonwetting saturation. For imbibition, the wetting phase, or liquid, supplies the motive force for displacement of the nonwetting, or gas, phase. Consequently, it may be argued that the effective capillary pressure within the porous medium is the integral of the pressure gradient for the liquid phase. During imbibition, the force to displace the gas is applied over the entire surface of the porous medium, but the medium can be considered to be aerated so that the air within the capillary pores remains at atmospheric pressure. A hydraulic advantage accompanies the force applied to the liquid in the pores, then, in inverse proportion to the liquid saturation. It follows that the measured liquid pressure is the integral of the liquid phase pressure gradient, divided by the saturation. This gives for the measured capillary pressure for imbibition:

$$\frac{P_{c,i}}{P_c} = \frac{P_{c,i,0}}{P_c} - \frac{ln(s)}{s} \tag{23}$$

where $P_{c,i,0}$ represents the minimum pressure required for imbibition of liquid into a saturated porous medium containing non-displaceable gas.

During drainage, both liquid and gas displace the liquid that leaves the porous medium. Hence it may be argued that the effective capillary pressure is the volume average of the integrals of the pressure gradients for the gas and the liquid. Again, because the air remained at atmospheric pressure during drainage, there is a hydraulic advantage applied to the measured pressure, so the effective pressure is this volume-average pressure integral, divided by the saturation.

At the maximum residual nonwetting phase saturation, the measured suction pressure drops by a step change from its value for drainage to its value for imbibition. This pressure change can be attributed to the change from a continuous gas phase at atmospheric pressure to a continuous liquid phase for which, at static equilibrium, the outside pressure equals the liquid phase pressure. Since the liquid phase pressure is less than the gas phase pressure by the characteristic capillary pore pressure, the suction pressure for imbibition must be less than the pressure for drainage by just this pressure, divided by the saturation to account for the effective hydraulic advantage. These considerations, with adjustments to account for differences between the integration constants for the liquid and volume-average pressure gradients, give, for the capillary pressure for drainage,

$$\frac{P_{c,d}}{P_c} = \frac{P_{c,i,0}}{P_c} + \frac{1}{s_{nr}} - \frac{\ln(s_{nr})}{s_{nr}} + \frac{(1-s)(1+\ln(1-s)) - s\ln(s)}{s} - \frac{(1-s_{nr})(1+\ln(1-s_{nr})) - s_{nr}\ln(s_{nr})}{s_{nr}} \tag{24}$$

The minimum suction pressure required for imbibition into a liquid-saturated porous medium, i.e., the entry head, most likely is a function of the capillary pore pressure. The most plausible explanation that fits the measured data is that an excess capillary force is required to displace air from an array of pores located on the surface in all three directions (one perpendicular to the surface and two in transverse directions). The required force is the three-dimensional vector sum of the forces required for unidirectional displacement, one component of which corresponds to the pressure. According to this interpretation, the minimum capillary pressure is the pore pressure multiplied by the square root of three:

$$\frac{P_{c,i,0}}{P_c} = \sqrt{3} \tag{25}$$

This entry head is applied at the outer surface of the porous medium, so this pressure difference is not normalized with respect to the liquid saturation.

A combination of the capillary pressure equations for imbibition and drainage yields a critical saturation where these two pressures coincide:

$$\frac{1-(1-s_{nr})(1+\ln(s_{nr})+\ln(1-s_{nr}))}{s_{nr}} = -\frac{(1-s_{cr})(1+\ln(s_{cr})+\ln(1-s_{cr}))}{s_{cr}} \tag{26}$$

Substitution of s_{nr} in this expression gives $s_{cr} = 0.301$.

Below this saturation, the suction pressure that develops during drainage is less than the suction pressure that accompanies imbibition. The logical conclusion is that the porous medium will not drain to a saturation below this critical value, unless the source of liquid is removed from contact with the porous medium or there is an external gas flow.

Comparison of Model Predictions to Capillary Pressure Data

Data from the UNSODA database maintained by the George E. Brown, Jr., Salinity Laboratory of the U. S. Department of Agriculture,[5,6] was selected for comparison with the model. The comparison is limited to data sets with paired imbibition and drainage pressures that exhibited a significant entry head $P_{c,i,0}$ and that included particle size measurements. There were three such data sources, namely Shen and Jaynes,[13] Stauffer and Dracos,[14] and Poulosavassilis.[15] An additional data source not included in the UNSODA data base, that of Smiles et al. and Vachaud and Thony,[16,17] also is analyzed. All of these data sources contained measurements made in sand. The fitting procedure entailed a calculation of the capillary pore pressure P_c, computed as the average of the measured capillary pressures divided by the capillary pressures predicted by the model at the relative saturations for each measurement. To eliminate measurements near saturation with either liquid or gas, the calculation was limited to relative liquid saturations between 0.4 and 0.75.

Figure 1 compares these data with model predictions. As this figure shows, the Shen and Jaynes data and the Stauffer and Dracos data have minimum measured saturations about half the predicted critical minimum saturation s_{cr}, while the Poulovassilis and Smiles et al. data have minimum measured saturations approximately equal to s_{cr}.

The model fits the data closely for the Poulovassilis and Smiles et al. data. The most logical explanation for the apparent discrepancy between the minimum measured saturations for the Shen and Jaynes and the Stauffer and Dracos data and the model prediction is that the porous material develops fissures that fill with gas during drainage. According to this interpretation, the fissures are larger than the pores by a sufficient magnitude that they do not offer any flow resistance but rather simply conduct fluid among the pores. The minimum saturation for a fissured material can be derived by applying a jump condition for the development of fissures. Following the logic used in the analysis of residual wetting saturations, the pressure applied to the liquid phase across such a jump, ΔP_{jump}, is

$$\Delta P_{jump} = P_c\left(\left(1 - s_{min}\right) - \left(1 - s_{cr}\right)\right) = P_c\left(s_{cr} - s_{min}\right) \tag{27}$$

The minimum saturation is the saturation that will minimize the free energy associated with this jump, ΔG_{jump}. This free energy, evaluated after the appearance of the fissures, is

$$\Delta G_{jump} = -R_g T \ln\left(P_c s_{min}\left(s_{cr} - s_{min}\right)\right) \tag{28}$$

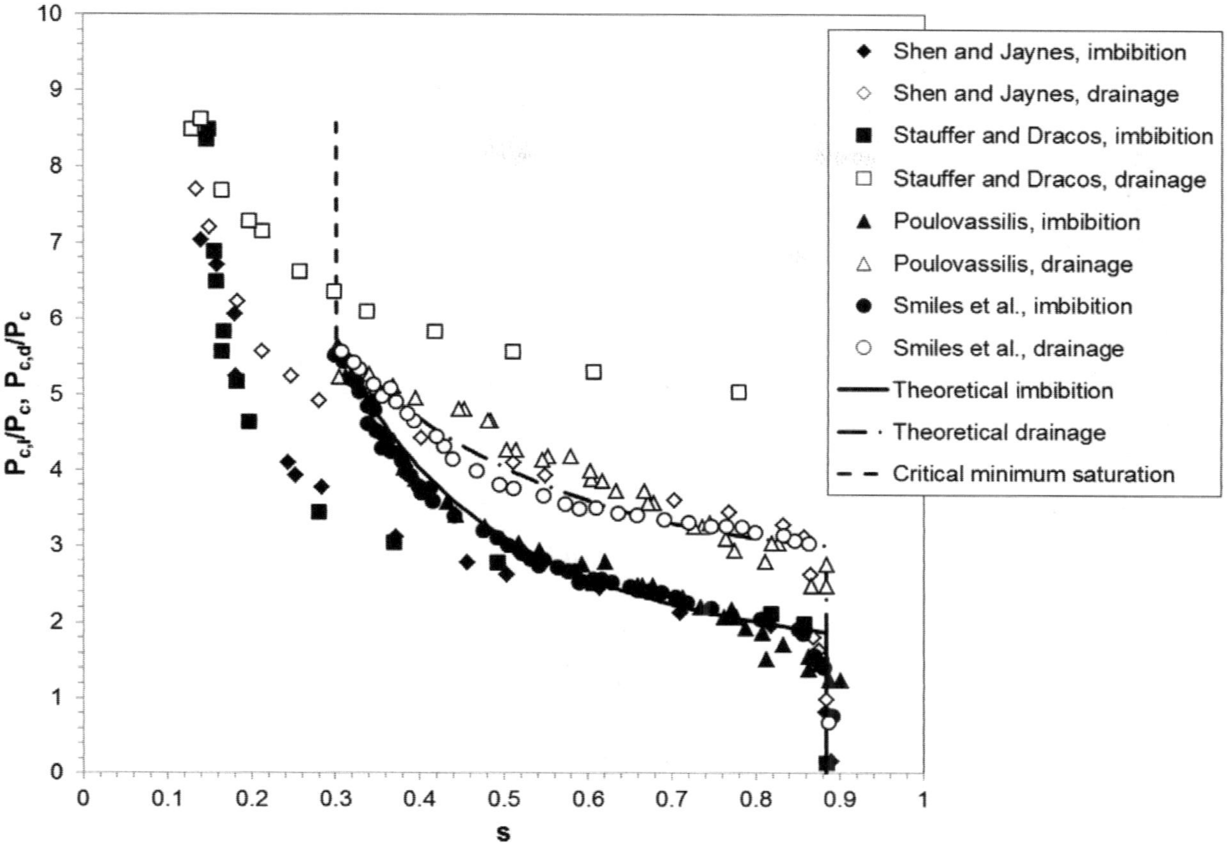

Figure 1. Variation of Imbibition and Drainage Pressures with Relative Saturation

This condition is satisfied by

$$s_{min} = 0.5 s_{cr} \tag{29}$$

A similar analysis, combined with the observation that the maximum saturation for a homogeneous porous material is the residual nonwetting saturation, gives for the maximum saturation for a fissured porous medium s_{max},

$$s_{max} = 0.5 + 0.5 s_{rn} \tag{30}$$

The capillary pressure model describes the saturation for just the pores of a fissured material. Because the fissure volume merely conducts pressure among the pores, the overall saturation for the total volume of porous material is a linear function of the pore saturation. If the maximum saturation is defined to be equal to the residual nonwetting saturation to conform with the model, then a linear interpolation between this saturation and the minimum saturation yields the following expression for the overall saturation of a fissured material, s^*, as a function of the model saturation for a homogeneous porous material, s, the residual nonwetting saturation, s_{nr}, and the critical saturation, s_{cr}:

$$s^* = \frac{0.5 s_{cr} s_{nr} - s_{cr} s + s_{nr} s}{s_{nr} - 0.5 s_{cr}} \tag{31}$$

Figure 2 compares the capillary pressure model with the Shen and Jaynes data and the Stauffer and Dracos data defined by s^*. The model fits the adjusted imbibition data closely, but the Stauffer and Dracos drainage data deviate significantly from the model. The difference between the measured and predicted drainage pressures remains approximately constant as the saturation increases. This suggests the presence of a metastable equilibrium that originates from the upstream (dry) side of the capillary pressure gradient.

The most plausible explanation for the overshoot and apparent metastability phenomena is that at the minimum liquid saturation, the interstitial gas in the fissures begins to exert pressure on the residual liquid in the pores. As explained in the capillary pressure model derivation, this pressure would be the product of the capillary pore pressure P_c and the fraction of the volume occupied by the interstitial gas, $0.5(1 - 0.5 s_{cr})$, divided by the residual pore saturation, $0.5 s_{cr}$. Thus, the overshoot pressure $\Delta P_{c,ov}$ is given by

$$\frac{\Delta P_{c,ov}}{P_c} = \frac{1 - 0.5 s_{cr}}{s_{cr}} \tag{32}$$

The apparent metastability, according to this interpretation, results when the overshoot in the capillary pressure at minimum saturation propagates downstream in the nonwetting (gas) phase. The fraction of the overshoot pressure that propagates would be equal to the fraction of the pore volume occupied by gas, $1 - s_{cr}$. If this overshoot pressure propagates, then the metastable drainage pressure $P_{c,d,ms}$ would be given by

$$\frac{P_{c,d,ms}}{P_c} = \frac{P_{c,d}}{P_c} + \frac{(1 - s_{cr})(1 - 0.5 s_{cr})}{s_{cr}} \tag{33}$$

As shown by Figure 2, the predicted overshoot pressure approximately equals maximum measured pressures, and the Stauffer and Dracos data closely follow the predicted metastable drainage pressure curve.

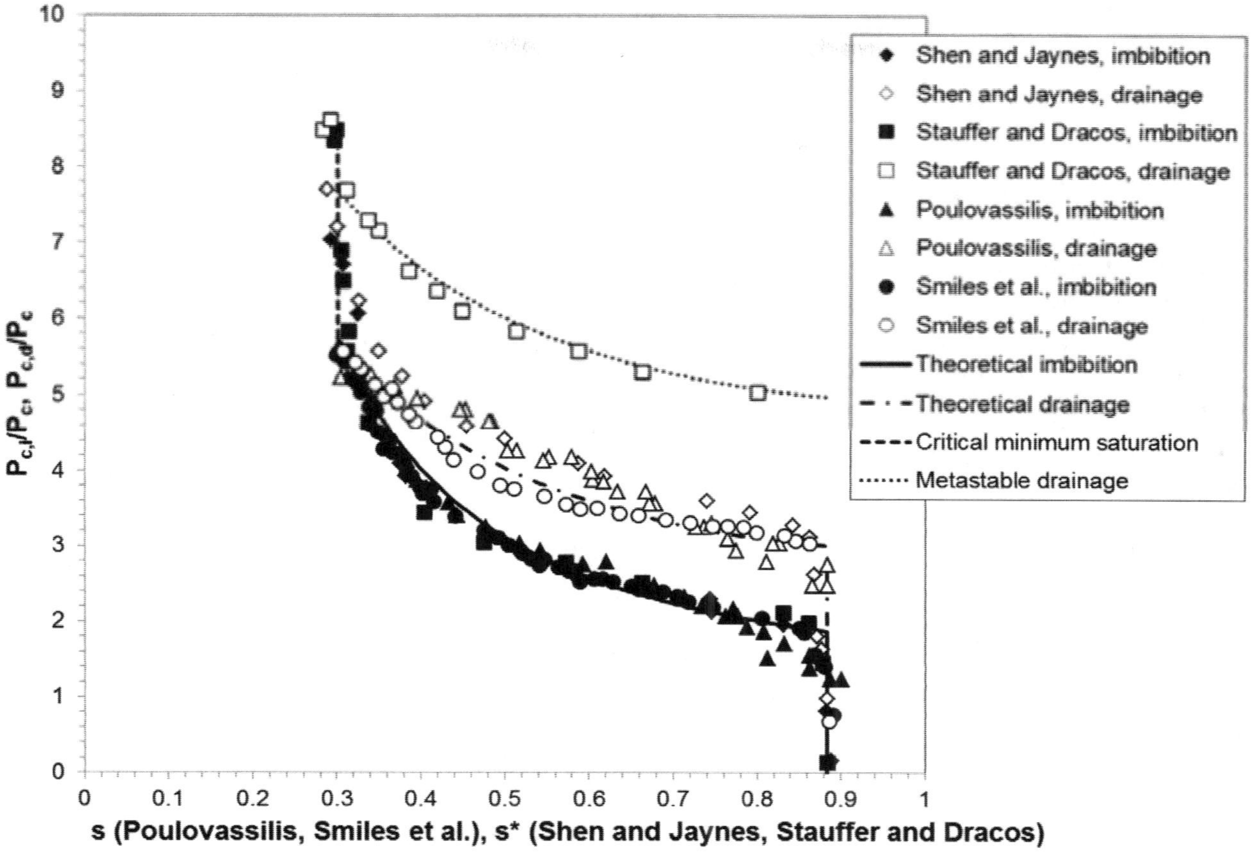

Figure 2. Variation of Imbibition and Drainage Pressures with Relative Saturation, Corrected for Fissure Effect

Evaluation of Characteristic Pore Pressures

The final step in development of the capillary pressure model is to relate the characteristic pore pressure P_c derived from the data fit to the particle size of the porous material. (One may recall that, for each data set that was regressed, the pressure multiplier is $\frac{1}{P_c}$.) From an analysis of the capillary rise in a powder, based on the specific surface area of the powder, Dunstan and White derived the following relationship between the average particle diameter and the pore diameter:[18,19]

$$d_{pore} = \frac{2\varepsilon}{3(1-\varepsilon)}\bar{d}_{part} \tag{34}$$

Average particle diameters were calculated from sieve tray data using an interval-censored method[20] based on an assumed log normal size distribution. For each data set, the void fraction was either given, calculated from the bulk density of the sand, or estimated by fitting the residual nonwetting saturation to the capillary pressure data.

The pore diameter also is related to the interfacial tension, σ, the liquid wetting angle, θ, and the capillary rise pressure, $P_{c,i,0}$, by the Laplace relation:

$$d_{pore} = \frac{4\sigma \cos(\theta)}{P_{c,i,0}} \qquad (35)$$

The surface tension of water at room temperature is 72.2 dyne/cm^2.[21] The measured wetting angle is highly variable and depends on the measurement method. For air-dried sand, a dynamic measurement method angle yields a wetting angle of 45°.[22]

Table 1 compares pore diameters calculated from the average particle diameter (Equation 34) and the capillary rise pressure to the characteristic pore diameter for the capillary pressure model presented in this study (Equation 35). As may be seen, these pore diameters are in relatively close agreement; the pore diameter calculated from the capillary pressure model differs from the pore diameter calculated from the average particle diameter by an average of 7%, with a standard deviation of ±25%.

Table 1. Comparison of Pore Diameters from Particle Diameter and Capillary Pressure Measurements

Data Set	$\bar{d}_{part}$, μm	d_{pore} from $\bar{d}_{part}$, μm	d_{pore} from $P_{c,i,0}$, μm	Deviation, %
Shen and Jaynes	354	200	160	-20
Stauffer and Dracos	312	167	135	-19
Poulovassilas	537	185	219	18
Poulovassilas	551	154	205	33
Smiles et al.	264	100	122	23

Conclusions

A model has been developed that uses a characteristic pore pressure to predict residual wetting and nonwetting saturations and capillary imbibition and drainage pressures. The model is based on an energy balance that equates changes in potential surface energy to changes in pressure-volume work. A simple probabilistic distribution expresses capillary forces as a function of the relative saturation and the characteristic pore pressure. The model implicitly assumes that the porous medium is homogeneous and isotropic.

The model predicts a residual wetting saturation, s_{wr}, of 0.236 and a residual nonwetting saturation, s_{nr}, of 0.884. These predicted residual saturations are in almost exact agreement with measurements. In addition, the model predicts that there is a critical saturation, s_{cr}, of 0.301, below which the imbibition pressure exceeds the drainage pressure. A porous medium will not drain to any lower saturation, provided that the porous medium remains in contact with liquid and there is no external gas flow.

The capillary pressure model successfully correlates limiting imbibition and drainage pressures for selected laboratory tests that used uniformly packed, sieved sand, although for certain tests an adjusted saturation is required to account for fissures. For the data that was correlated, the characteristic pore diameter from the capillary pressure model approximately equals the pore diameter calculated from the average particle diameter using a capillary rise model. This agreement suggests that, for uniform packed sands with a relatively narrow particle size distribution, a capillary pressure curve can be estimated from a measured particle size distribution. The applicability of the model to soils and rocks with widely varying pore size distributions has not been demonstrated.

Acknowledgment

This work was funded by the U. S. Department of Energy Office of Environmental Management under contract number DE-AC09-08SR22470.

References

1. Brooks, R.H. and A.T. Corey (1964), "Hydraulic properties of porous media," Hydrol. Pap. 3, Civ. Eng. Dept., Colorado State University, Fort Collins, Colo.
2. van Genuchten, M.T. (1980), "A closed-form equation for predicting the hydraulic conductivity of unsaturated soils," Soil Sci. Soc. Am. J. 44, 892-898.
3. Kosugi, K. (1994), "Three-parameter lognormal distribution model for soil water retention," Water Resour. Res. 30(4), 891-901.
4. Kosugi, K. (1996), "Lognormal distribution model for unsaturated soil hydraulic properties," Water Resour. Res. 32(9), 2697-2703.
5. Nemes, A., M.G. Schaap, and F.J. Leij (1999), "The UNSODA unsaturated soil hydraulic database, version 2.0," U. S. Salinity Laboratory, U. S. Department of Agriculture, Agricultural Research Service, Riverside, California.
6. Nemes, A., M.G. Schaap, F.J. Leij, and J.H.M. Wosten (2001), "Description of the unsaturated soil hydraulic database UNSODA version 2.0," J. Hydrol. 251(3-4), 151-162.
7. Rossi, C. and J.R. Nimmo (1994), "Modeling of soil water retention from saturation to oven dryness," Water Resour. Res. 30(3), 701-708.
8. Parker, J.C. and R.J. Lenhard (1987), "A model for hysteric constitutive relations governing multiphase flow: 1. Saturation-pressure relations," Water Resour. Res. 23(12), 2187-2196.
9. White, M.D. and M. Oostrom (1996), "STOMP, subsurface transport over multiple phases: theory guide," U.S. DOE Pacific Northwest National Laboratory Report PNNL-11217, UC-2010.
10. Demond, A.H. and P.V. Roberts (1991), "Effect of interfacial forces on two-phase capillary pressure-saturation relationships," Water Resour. Res. 27(3), 423-437.
11. Morel-Seytoux, H.J. and J. Khanji (1975), "Prediction of imbibition in a horizontal column," Soil Sci. Soc. Am. Proc. 39, 613-617.
12. Morel-Seytoux, H.J., J. Khanji, and G. Vachaud (1973), "Prediction errors due to uncertainties in the measurement and extrapolation of the diffusivity function," Civil Engineering Report, Colorado State University, Fort Collins, Colorado 80523, CEP72-73, HJM48.
13. Shen, R. and D.B. Jaynes, (1988), "Effect of soil water hysteresis on simulation, infiltration, and distribution," Shuili Xuebao: J. Hydr. Eng. (China) 10, 11-20.
14. Stauffer, F. and T. Dracos (1986), "Experimental and numerical study of water and solute infiltration in layered porous media," J. Hydrol. 84(1-2), 9-34.

15. Poulovassilis, A. (1970), "Hysteresis of pore water in granular porous bodies," Soil Sci. 109(1), 5-12.
16. Smiles, D.E., G. Vachaud, and M. Vauclin (1971), "A test of the uniqueness of the soil moisture characteristic during transient, nonhysteric, flow of water in a rigid soil," Soil Sci. Soc. Am. Proc. 35, 534-539.
17. Vachaud, G. and J.-L. Thony (1971), "Hysteresis during infiltration and redistribution in a soil column at different initial water contents," Water Resour. Res. 7(1), 111-127.
18. White, L.R. (1982), "Capillary rise in powders," J. Colloid Interf. Sci. 90(2), 536-538.
19. Dunstan, D., and L.R. White (1986), "A capillary pressure method for measurement of contact angles in powders and porous media," J. Colloid Interf. Sci. 111(1), 60-64.
20. So, Y., G. Johnston, and S.H. Kim, "Analyzing Interval-Censored Data with SAS® Software," Paper 257-2010, Proceedings, SAS Global Forum 2010, Seattle, Washington, April 11-14, 2010.
21. Lide, D.R., ed. (1994), CRC Handbook of Chemistry and Physics, 75th ed., CRC Press, Boca Raton, Florida.
22. Weisbrod, N., T. McGinnis, M.L. Rockhold, M.R. Niemet, and J.S. Selker (2009), "Effective Darcy-scale contact angles in porous media imbibing solutions of various surface tensions," Water Resour. Res. 45(4), W00D39.

469001 Technetium Leaching Cementitious Materials

Abstract:

Technetium (Tc), a nuclear fission product with no stable isotopes, represents much of the long-term radioactivity associated with nuclear waste. Under oxidizing conditions, Tc persist in the environment as an anion in the +7 oxidation state, i.e., pertechnetate (TcO_4^-). In soil partitioning studies, the oxidized species displays limited retention generally associated with amphoteric soil oxides. However, Tc(VII) is subject to chemical reduction to the +4 oxidation state, which is less soluble and mobile in the environment. One disposition strategy for immobilizing ^{99}Tc containing waste is the production of cementitious materials that contain reducing agents, such as blast furnace slag (BFS), to enhance the retention of redox sensitive contaminants. At the Saltstone Disposal Facility (SDF) on the Department of Energyâ€™s Savannah River Site (SRS), low-level radioactive saltwaste is mixed with BFS, fly ash and portland cement to produce a solidified cementitious material known as Saltstone, which is deposited in a series of vaults for long-term disposal.

In the current study, contaminant mass transfer rates for ^{99}Tc and other contaminants from saltstone simulants spiked with ^{99}Tc and rhenium (Re), often used as surrogate for ^{99}Tc, and actual SDF saltstone samples were assessed using EPA Method 1315, a method for evaluating contaminant leaching from intact monolithic materials. Since redox state plays a key role in controlling ^{99}Tc solubility, the monoliths were exposed to one of three different test environments during leaching: aerobic (ambient air), anoxic (N_2), and reducing (H_2/N_2). Rhenium leaching rates (and other poorly sorbing contaminants like NO_3^- and iodine) were much higher than ^{99}Tc, which was attributed to poor initial reduction or more rapid oxidation, indicating that Re is not a suitable chemical analog for evaluating Tc partitioning under reducing conditions.

456482 Remediation of Uranium-Contaminated Groundwater By Functionalized Magnetic Mesoporous Silica Nanoparticles

Abstract:

U(VI) species display limited adsorption onto sediment minerals and synthetic sorbents in pH < 4 or pH > 8 groundwater. In this work, magnetic mesoporous silica nanoparticles (MMSNs) with magnetite nanoparticle cores were functionalized with various organic molecules using post-synthetic methods. The functionalized MMSNs were characterized using N_2 adsorption-desorption isotherms, thermogravimetric analysis (TGA), transmission electron microscopy (TEM), ^{13}C cross polarization and magic angle spinning (CPMAS) nuclear magnetic resonance (NMR) spectroscopy, and powder X-ray diffraction (XRD). The functionalized MMSNs were effective for U removal from pH 3.5 and 9.6 artificial groundwater (AGW). Functionalized MMSNs removed U from the pH 3.5 AGW by as much as 6 orders of magnitude more than unfunctionalized nanoparticles or silica and had adsorption capacities as high as 38 mg/g. They removed U from the pH 9.6 AGW as much as 4 orders of magnitude greater than silica and 2 orders of magnitude greater than the unfunctionalized nanoparticles with adsorption capacities as high as 133 mg/g. These results provide an applied solution for treating U contamination that occurs at extreme pH environments and a scientific foundation for solving critical industrial issues related to environmental stewardship and nuclear power production.

Further, after U adsorption, U L_3-edge XANES and EXAFS spectra of all functionalized MMSNs were collected using the Materials Research Collaborative Access Team (MRCAT) Sector 10-ID beamline at the Advanced Photon Source (APS) in order to understand molecular mechanisms of different U species binding with functional organic ligands. Such molecular mechanistic studies can further guide the future selection of organic ligands to develop more robust nano sorbent materials that are more efficient and cost-effective for U extraction from environmental aqueous systems.

472311 Kinetics of Simultaneous Transformation of Co-Existing Mixed Contaminant Media of Cr(VI) and TCE Using Nanoscale Zero Valent Iron (nZVI)

Abstract:

Industrialization has helped advance technology but at the cost of contaminating the natural environment. Several products and by-products of industrial processes are very toxic and hazardous to the health of terrestrial and aquatic life forms. Wastewater and other industrial effluents leach through the earth crust, adsorb on to the surface of soil particles and contaminate subsurface aquifers resulting in the pollution of ground water and many drinking water sources. Among the numerous pollutants co-existing in groundwater and subsurface aquifers, Hexavalent Chromium, Cr(VI), and Trichloroethylene, TCE, have been classified by the USEPA as known carcinogens. Thus, remediation efforts to remove/convert them to non-toxic forms are a priority. A survey of more recent research literature, shows many successful attempts using nanoscale Zero Valent Iron (nZVI), to individually reduce Cr(VI) to its benign form Cr(III) and de-halogenate TCE, but none of these studies have attempted to investigate simultaneous transformation of co-existing Cr(VI)â€“TCE mixed contaminant media and the associated synergistic or antagonistic effects. Using nZVI to remediate Cr(VI)-TCE mixtures has significant advantages including: very low to zero toxicity to the environment, abundance and presence of ZVI in the subsurface, and the ability to degrade TCE and Cr(VI) to less toxic forms.

Results of kinetic analysis using nZVI as a single transforming agent to degrade a mixed contaminant media of Cr(VI)-TCE in aqueous and soil systems will be presented for various pH and temperature values. Application to subsurface hazardous waste remediation contaminated with multiple contaminants will be discussed.

471994 Numerical Simulation and Optimization of Environmental Treatment Design Using Parallel Computational Methods: Theory and Application

Abstract:

ABSTRACT

Environmental remediation has been ongoing in the Unites States for several decades and has produced some successes. However, the more difficult sites remain challenging and resistant to efficient and effective remediation. These are precisely the environmental problems that remain as persistent fiscal and technological challenges as the success of the heuristic trial-and-error approach, which assisted in remediating the simpler sites, has reached it a limit of applicability on these cases. These tools herein enable formal, physically consistent and robust analysis of optimal remedial designs for environmental contamination problems of any level of complexity.

This paper discusses and demonstrates fast computational simulation and optimization technologies used for optimally managing environmental remediation project and programs by reducing remediation design analysis timeframes and costs. Those responsible for managing or remediating chemical or radionuclide impacted sediments, soils, surface or groundwater including designing remedial design and operation projects will benefit from tools that will help them find acceptable, efficient and effective solutions to complex and multi-faceted planning challenges at efficient cost points. A new universal optimization system is developed which has successfully been deployed on environmental and radionuclide impacted sites across a variety of computational environments including desktops and the high performance elastic Cloud-based computing platforms.

This computational advancement removes the impediment of burdensome and unpredictable runtime requirements for optimizing the design of these types of projects, which have typically consumed weeks or months of calendar time using heuristic trial-and-error or even automated serial computation approaches. These new optimization methods return results in minutes to hours for simulations and optimal solutions in hours to at most a few days for medium and larger size projects which here-to-fore required months to solve.

Computational Approach

PBMO™ is a universal optimization tool. Capabilities include optimizing linear or non-linear elliptic, parabolic and hyperbolic models and objective functions that are continuous, non-continuous and mixed integer. It is a tool system that helps users perform complex high-level tasks with amazing simplicity. Modes of operation include standard and expert modes, and it runs on computers with single and multiple CPUs, including local area grids and the Cloud. Because it is programmed as a set of modules using standard FORTRAN, it is platform independent. It has the capability for including applications with mixed operating systems and can deploy with multiple, independent or blended process models including subject matter expert (SME), data-driven (DD), Physics-based (PB) or integrated model (IM). Analysis can begin with the best result from a trial-and-error approach, or from a random starting point so efforts engaged to date by the design teams can be fully used. The optimal solution search relies on evaluations of candidate solutions, which can be entirely physics-based or response function or minorant approximated solution search (Deschaine, 2014).

This is accomplished by developing both a parallel optimal search strategy and the numerical matrix solver in the groundwater flow and transport modeling code. The parallel distributed optimization method will execute on one to several hundred (or more as needed) CPUs while the parallel numerical solver that has been added to the subsurface flow and transport simulator MODFLOW-SURFACT™ has demonstrated speedups as high as a factor of 9.8. Hence, the integrated parallel simulation and optimization method uses the breadth and depth search approach; iteratively producing sets of tens or 100s of candidate investigation points that are evaluated concurrently instead of the previous and widely used sequential approach (i.e. one model evaluation at a time). This enables 80-100% globally optimal solutions are regularly found expending only 5-10% of the common computational effort using this new parallel optimization method (reducing computation effort by 10 to 20 times and calendar time by a factor of 100 or more).

Additionally, practitioners conducting conceptual site model development, model calibration by auto-calibration methods and remedial design by heuristics can enjoy solution speedups by using the parallel numerical matrix solver in the physics-based simulator as well as automated uncertainty analysis using the Latin Hypercube, Monte-Carlo or geostatistical realization techniques. Combining investigation and design analysis with breadth and depth parallel optimization methods, projects at the fringe of practical optimization, along with projects that here-to-fore were beyond the reach of formal optimization methods are now wholly feasible and tractable optimization problems.

Remedial Design Approach

PBMO™ project design involves selecting the appropriate combination of its fully comprehensive solution options, including the full capability of mixing and matching various approaches.

- Select proper descriptive model or model set that adequately describes the system to be optimized
- Select optimization algorithm or set of proper optimization algorithms to optimize the system best.
- Select computer operation system or a mixture of operating systems, to enable the integrated system to run as-is.

It is important to note that "optimization" refers not just to cost minimization but also to the effective and efficient balance of cost, performance, risk, management, and societal priorities along with uncertainty analysis. To wit, the PBMO™ process formally integrates the stakeholder input into the analysis and provides not only the "optimal solution" but also "what-if" capability that includes management override control on remedial design analysis. This approach integrates all of these elements into a single decision framework that is tractable, traceable, and defensible. The approach is modular and scalable. It can be applied either as individual components or in total. This solution methodology represents a significant improvement over the non-optimal "trial and error" approach to developing environmental response(s).

METHODS

Understanding optimization rationale and theory, methodological strength, applicability, and limitations, is critical for understanding why optimization is valuable and how to successfully and deploy computational optimization analysis on projects. Project implementation experience has shown that several effective optimization methods exist that are routinely successful when used on projects and are included in this system. Exhibition of case histories demonstrates their value for developing optimal solutions not obtainable by traditional subjective design methods (a.k.a. manual trial and error).

PBMO™ uses a breadth and depth search strategy. The breadth search identifies good solution quickly, whereas the local search drills down to find the optimal value. The optimization methods use include:

- Derivative-based global methods
    - Linear programming (LP)
    - Sequential linear approximation (SLA)
    - Sequential Quadratic Programming (SQP)
    - Generalized Reduced Gradient (GRG)
- Derivative-free global methods
    - Adaptive Design of Experiment (ADOE)
    - Adaptive Global Random Search (AGRS)
    - Branch and Bound (B&B)
    - Global Adaptive Random Search (GARS)

Table 1 depicts the applicability of the PBMO™ system (Deschaine, 2014).

Table 1. PBMO™ Solution Scope and Analysis Results

Scope	Description	Affects	Analysis Results
Optimization of environmental *response program*	Optimal FS/CMS planning-level design considering all reasonably viable alternative remediation responses including waste isolation and monitoring.	Total response cost, multi-period fiscal budget planning and allocation strategies; calculations and contingencies	Optimal and balanced solution that best meets needs of site and stakeholders
Optimization of environmental *remedial*	Optimal detailed design of FS/CMS selected	Total response cost, specific capital	Optimal design basis for site including

design	response program	expenditures, annual costs, plant operations	constructability assessment and predicted performance
Optimization of *long-term monitoring*	Optimal design of monitoring well network and sampling program	Total response cost, annual costs, stakeholder confidence that the remedial design and installation are operating as intended.	Proactive management of stakeholder expectations, early warning detection if as-built and operated systems' performance begins to deviate from design intent.
Optimal *contingency plans*	Optimal and pro-active considerations for addressing remedial design *and observed* environmental system response	Total response cost, annual costs, stakeholder confidence that the remedial program and design are robust and amenable to refinement, if warranted.	Optimal and robust remedial system designs, installations and monitoring that incorporate feedback from the operation. Efficiently allow for modifications or inclusion of additional components or processes.

The solution from the PBMO™ methodology has always exceeded the subjective engineering solution when available for comparison (Deschaine, 2014).

CONCLUSIONS

The basic PBMO™ methodology has been previously tested, verified, validated and documented using a broad range of industrial projects. Those results indicate that it meets or exceeds the best available solution either in use (i.e. as an RIP), proposed by the subject matter experts or developed by other optimization methods. These two new parallelization methods now allow for the optimization of remedial designs for the more complicated problems. These are precisely the environmental problems that remain in the environmental contamination site portfolio as persistent challenges. The earlier successes of the heuristic trial-and-error approach - which assisted in remediating the simpler sites - has reached it limit of applicability in these cases. By simultaneously combining project level optimization with optimal programmatic design, the optimal design and implementation of environmental challenge response are now computationally tractable, practical and viable.

REFERENCES

Deschaine, L. M. (2014). Decision support for complex planning challenges. Ph.D. Dissertation, Chalmers University of Technology, Göteborg, Sweden. 233p.

Deschaine, L. M., Lillys, T. P., & Pintér, J. D. (2013). Groundwater remediation design using physics-based flow, transport, and optimization technologies. Environmental Systems Research, 2(1), 1-21.

http://www.environmentalsystemsresearch.com/content/pdf/2193-2697-2-6.pdf

468615 Investigation of Ultrafiltration, Nanofiltration,Ozonation and Biofiltration for the Removal of Microcystin-LR from Water

Abstract:

Investigation of Ultrafiltration, Nanofiltration,Ozonation and Biofiltration for the Removal of Microcystin-LR from Water

On August 2, 2014, the greater Toledo area woke up to a Do Not Drink or Boil Water Advisory. The advisory was due to the presence of a cyanotoxin (algal toxin) produced by cyanobacteria in Lake Erie called microcystin-LR in the drinking water supply that has a WHO provisional guideline of 1 mg/L. Upon entering the City of Toledo Collins WTP crib at the Lake Erie intake, potassium permanganate is added to the water for mussel control. While potassium permanganate is needed to control mussels, it lyses cyanobacteria cells, releasing algal toxins to the water. The water is then pumped nearly three miles to the Low Service Station, where powdered activated carbon (PAC) is added to the water for taste and odor control, and the water is transported approximately six miles to the WTP (High Service Station). PAC is generally effective for removal of algal toxins through adsorption onto its surface. At High Service, alum, lime and soda ash are added to the water for coagulation-flocculation, softening, and removal of metals. The water is then sand filtered, carbonated and chlorinated before being sent to the distribution system. However, traditional physicochemical water treatment processes, such as coagulation-sedimentation-filtration, have been shown to only be partially effective for the removal of whole algal cells and not effective for the removal of algal toxins. Furthermore, chlorination is effective for oxidizing algal toxins at relatively high free chlorine concentration as long as the pH is below 8; however, for corrosion control, Toledo water is kept at a pH above 9. Therefore, the treatment process was not enough of a barrier to prevent microcystin-LR from entering the drinking water supply. By August 4, the water treatment plant increased its PAC dosage by nearly four times, and while the toxin was removed, a significant amount of PAC sludge was produced and the cost of PAC addition was unsustainable. The objective of this project is to study alternative water treatment processes, namely membrane separations, biofiltration and ozonation, for the effective removal of algal toxins.

462400 On the Operational Flexibility of Process Systems for Membrane-Based Water Treatment and Desalination

Abstract:

Membrane-based processes are continuing to be applied for water treatment and desalination of a growing portfolio of underutilized, geographically distributed water sources. With diversified source waters, the conventional paradigm of centralized water treatment via reverse osmosis (RO) membranes using large-scale process systems is increasingly being challenged. The alternative is the use of smaller distributed but more robust RO water treatment/desalination systems. The flexibility of a single RO process system to operate over a wide range of feed water conditions is therefore becoming increasingly important. High operational flexibility that maintains high energy efficiency and compact system footprint, however, remains challenging due to the inherent limitations of conventional membrane system designs. Membrane process systems are composed of integrated process components that are highly interdependent. Optimal operation of each system component is often limited to a narrow flow range, which in turn restricts the operational range of the integrated process system. Pumps (and associated motors) and energy recovery devices, for example, typically have narrow flow ranges at which energy efficiencies are optimal (i.e., best efficiency points). Furthermore, membrane modules are inherently limited with respect to the flow ranges for optimal operation close to the thermodynamic limit, while maintaining conditions that protect physical integrity, maximize membrane area utilization, minimize concentration polarization, and avoid excessive membrane fouling and/or mineral scaling. Such module-level operational limits can significantly restrict water recovery and permeate productivity ranges of the integrated system, which is highly dependent on the specific membrane array design. Existing methods for improving operational flexibility rely on RO concentrate recycling (to RO feed); this provides an additional degree of freedom for steady-state process regulation and also for enabling unsteady-state cyclic operation. At present, however, the impact of operational flexibility on systems energy and membrane footprint has not beem fully explored and remains unclear in the published literature. Accordingly, the present work presents a fundamental analysis framework for quantifying operational flexibility and its realizable improvements. Using the framework, the interdependence between system

operational flexibility with energy consumption were explored. The resulting analysis indicated that concentrate recycling, whether applied in a steady-state or cyclic mode of operation, can significantly enhance system operational flexibility. However, increased operational flexibility leads to either higher energy consumption or larger system footprint when compared to a less-flexible system with no concentrate recycling. This finding is consistent with the higher expected axial average osmotic pressure in the membrane array of systems with concentrate recycling. Analysis also shows that an energy recovery device can impose a restriction on operational flexibility; such restriction can be overcome through process decomposition. It is concluded that high operational flexibility is well suited for small- to medium-scale applications or distributed deployments, especially where effective handling of feed water variability to maintain system optimality is critical. In such systems, it is imperative to balance system operational flexibility with the expected increased energy consumption and/or membrane footprint. Less-flexible membrane systems with high efficiencies with respect to energy consumption and membrane area utilization may better serve large-scale, centralized applications with low feed water variability or where feed water equalization tank/basin can be cost effective.

471085 Polyamide Thin Film Nanocomposite (TFN) Membranes for Simultaneous Boron Removal and Desalination of Seawater

Abstract:

Production of potable water has become a worldwide concern. For many countries the available water resources are not adequate due to population growth and industrialization. Desalination of brackish or seawater has emerged as a solution to the limited water supplies. Reverse osmosis (RO) today is the leading technology that is responsible for producing more than half of the worldâ€™s desalination capacity. Polymeric membranes have dominated the RO market since they offer high salt rejection, durability, small footprint and simple operation. An emerging problem is removing boron from seawater in order to meet the World Health Organization (WHO) recommendation of 2.4 ppm for potable water and 0.5 ppm for water used for irrigating crops such as citrus fruits and nuts. Since boron is present in seawater as boric acid whose effective diameter is nearly the same as that of a hydrogen-bonded cluster of water molecules, conventional RO membranes cannot effective achieve the WHO recommendations for boron removal. Recently, a new concept of mixed-matrix composite RO membranes was advanced whereby nanoparticles are embedded into the selective layer of thin film composite polyamide RO membranes fabricated on porous supports via interfacial polymerization. These mixed-matrix RO membranes offer considerably promise for r achieving desalination while concurrently reducing the boron concentration to the WHO recommendations. Incorporating materials such as zeolites and functionalized carbon nanotubes into the selective thin layer of polyamide RO membranes offer the potential for improving the boron rejection as well as the, flux and anti-fouling properties. However, this requires optimizing the conditions for loading, size, type, and functional groups of the nanomaterials as well as the interfacial polymerization and post-treatment protocols. This study reports the first assessment of the performance of thin film nanocomposite (TFNC) desalination membranes for more effectively removing boron from seawater.

463696 Hollow Fiber Lumen Modification Via Environmentally Friendly Poly(zwitterion) Grafting

Abstract:

Membrane fouling is a major obstacle to the practical applications of membrane-based techniques such as ultrafiltration, nanofiltration, reverse osmosis, forward osmosis and pressure-retarded osmosis. To prevent organic fouling, surface modification with hydrophilic polymers has been identified as a potential strategy because many organic foulants are hydrophobic in nature. Efficiency of fouling prevention depends on surface materials and modification techniques. Although grafting poly(ethylene) glycol improves fouling propensity, the layer is susceptible to cleaning solutions, which becomes critical in long-term operations. Zwitterionic polymers are demonstrated as a promising alternative non-fouling material because of their high capacity to generate a stable hydration layer which resists to non-specific protein adsorption. On the other hand, an environmentally friendly grafting technique was developed with a capability of modifying the lumen of hollow fiber membranes, where common techniques such as UV or plasma treatments are unlikely to be feasible and other techniques requiring thermal control are also complicated. This work is an important contribution to reduce fouling and eliminate constrains, which are currently seen as hurdles for the use of hollow fibers in specific membrane-based applications.

469224 Interfacial Transport in Thin Film Nanocomposite Membranes for Water Purification

Abstract:

Current commercial reverse osmosis membranes are limited by the properties of the polymeric semipermeable membranes used. In the past, a variety of nanoparticles have been added in order to improve the membrane flux and selectivity. Some of these thin film nanocomposite membranes have exhibited enhanced membrane properties, allowing for improved reverse osmosis process performance. Previous work in our laboratory has demonstrated that the addition of zwitterion-functionalized carbon nanotubes (CNTs) to the selective layer of thin film composite membranes increases the overall flux of the membranes while maintaining high salt rejection.[1] The CNTs also have also shown improvements in anti-fouling behavior. There is also evidence that the improvement in flux is not necessarily due to transport inside the CNTs, but could be due to transport at the polymer matrix/nanotube interface. Due to the high cost of CNTs, other nanoparticle additives are currently being studied. Cellulose nanocrystals (CNCs) are of interest due to their lower cost and availability as well as their potential for functionalization. In the current work, polyamide/CNC based thin-film nanocomposite membranes have been fabricated on polysulfone supports via vacuum filtration and interfacial polymerization and their water flux and salt rejection have been studied for RO desalination applications. Results have shown that TEMPO-oxidized CNCs may increase the flux of the membranes by up to 40% while maintaining high salt rejection. The flux and salt rejection can be tuned via careful control of the CNC surface functionality and functional group density, and via control of polymer-surface interactions.

[1] Chan, Wai-Fong, Eva Marand, and Stephen M. Martin; *J. Membrane Science* **509** (2016): pp. 125-137.

465759 Fabrication of Polyphenylenesulfone (PPSU) and Sulfonated Polyphenylenesulfone (sPPSU) Membranes for Water Treatment and Reuse

Abstract:

Polymeric membranes fabricated via non-solvent induced phase inversion method have been employed for various water treatment applications including ultrafiltration (UF), nanofiltration (NF), forward osmosis (FO) and reverse osmosis (RO). Polyphenylenesulfone (PPSU) is one of the most widely used materials due to its good chemical resistance, hydrolysis stability and great mechanical strength. Sulfonation of PPSU could further improve the hydrophilicity of the fabricated membranes and therefore enhance their water permeability and anti-fouling performance. In this study, membranes were fabricated with PPSU and sulfonated PPSU (sPPSU) polymers with various degrees of sulfonation. The rheological properties and phase separation behavior of PPSU and sPPSU were investigated. These characteristics of dope solutions greatly influence the morphology and separation performance of the fabricated membranes. sPPSU polymers exhibit enhanced rheological properties compared to PPSU. In addition, the sPPSU polymers show greater non-solvent (i.e., water) tolerance and delayed demixing during phase separation. For the flat-sheet membranes cast by the non-solvent induced phase separation method, sPPSU membranes possess higher water permeability and hydrophilicity than PPSU membranes. This fundamental study may contribute to the design strategies of preparing hydrophilic membranes from sulfonated polymers for water treatment applications.

450977 Removal of Organic Compounds from Aqueous Systems Using Carbon Nanomaterial-Based Aerogels

Abstract:

Continued efforts are underway to develop new and improved techniques for removing contaminants from aqueous environments. Contaminants vary and include organic, metal, and biological compounds. The use of organic compounds (e.g. fertilizers) in agricultural settings continued to cause concern due to the runoff and buildup in downstream environments. Buildup of high concentrations of these compounds can potentially lead to adverse health effects in humans and animals. Adsorbents, such as activated carbon, are well known given their ability to adsorb various contaminants. However, activated carbon is limited in its effectiveness when introduced in environments where pH and temperature swings reduced activated carbon's uptake capacity. Therefore, alternatives to current state-of-the-art adsorbents are desired. In this work, the adsorption capability of a lightweight, carbon-based aerogel is presented. The aerogel is composed of 0 to 2 wt% graphene nanoplatelets and single-walled carbon nanotubes (SWCNT). Two organic compounds, 2,4-dichlorophenoxyacetic acid and 1-pyrenebutyric acid, served as the contaminants. A design of experiment approach was taken to optimize the aerogel for maximum performance. Results show that the aerogel containing 0.2 wt% GnP-SWCNT achieved an uptake capacity of 0.22 mg 2,4-D/mg aerogel and 0.083 mg PBA/mg aerogel. Compared to activated carbon, the aerogels produced in this work demonstrated 39 times higher adsorption of 2,4-D and 5 times higher adsorption of PBA making them a viable candidate as a next generation adsorbent.

471101 VMD Desalination Using Nanocomposite Membranes Prepared By Hydrophilic Nano-Additives

Abstract:

Desalination has attracted many efforts over the past few decades. Lack of fresh water all over the world from south west of the United States to the Middle East and extended to Australia has drawn much attention toward the desalination as an alternative to compensate the water shortage. Among the processes that are used for this purpose, a larger portion of the market is occupied by the membrane-based technologies, i.e. 63.7%, and Reverse Osmosis (RO) has placed on top by managing approximately 60% of the market. In spite of its lower water production cost compared to the other desalination techniques, RO is considered to be a quite energy intensive process, which approximately 40% of the total water cost is spent for electrical energy consumption within the process. However, Membrane Distillation (MD) has demonstrated outstanding characteristics, which is potentially considered an appropriate alternative for RO in global market, although the process has not yet been commercialized, mainly due to the huge thermal energy consumption in MD besides lack of a champion in manufacturing the high performance durable MD membranes at industrial scale. To improve the performance of the MD membranes by enhancing their permeability, hydrophilic nanoparticles have shown a great potential via optimizing the membrane structure. Hydrophilic nanomaterials, such as $CaCO_3$, CuO, and amine modified SiO_2, were able to increase the pure water flux in a Vacuum Membrane Distillation (VMD) process by 102.3%, 153.4%, and 2456%, respectively, at an appropriate concentration, i.e. 2.0 wt.% for $CaCO_3$ and CuO and 7.0 wt.% for silica nanoparticles. Surface hydrophobicity of the MD membranes was not compromised as a result of the addition of the hydrophilic nanoparticles into the polymeric PVDF matrix and nanocomposite membranes possessed adequate Liquid Entry Pressure of water (LEP_w) to be used in seawater VMD desalination. Nanomaterials incorporated membranes demonstrated high selectivity by rejecting more than 99.9% of the sodium chloride when an artificial seawater with a NaCl concentration of 35 g/L was fed to the system.

472319 Invited Presentation: Carbonaceous Nanoparticles and Their Interactions with Biological Cells

Abstract:

Particles originating from human activities have existed for millennia, e.g., smoke from combustion, but the recent development of industry and combustion-based engine transportation has increased anthropogenic particles pollution. At the same time, technological advancement has also changed the character of these particles, increasing the proportion of nanometer-sized particles --"nanoparticles"-- and expanding the variety of chemical compositions. Indeed, the manipulation of matter at the scale of atoms, "nanotechnology," is creating many new materials with characteristics not always easily predicted from current knowledge.

In this work we report on the interactions of carbon-based nanomaterials (both from combustion and synthetic sources) with biological cells, using atomistic simulations. As engineered nanomaterials, we will focus on carbonaceous quantum dots, which have recently emerged and ignited tremendous research interest. Their favorable characteristics include size- and wavelength-dependent luminescence, resistance to photobleaching, bio-conjugation, and functionalization to produce chiral nanostructures. Carbon-based quantum dots show promise in areas such optoelectronics, catalysis, bioanalysis and drug delivery.

463630 Nanomaterials: Occupational Exposure and Toxicity in Context

Abstract:

At the forefront of particulate health concerns are nanomaterials. This talk will propose some considerations for evaluating nanomaterials and their health effects.

1. Is the material as prepared truly â€œnanoâ€? For example, even though SWNTs satisfy the accepted definition of being nanoscale in 2-dimensions, in practice and certainly as synthesized, they are highly coalesced into aggregates of micron in size.

2. What distinguishing characteristics apply to the particular nanomaterial? High surface area is a universal hallmark of a nanomaterial. Yet fumed silica with a surface area of > 200 m^2/g is rated biologically safe and even used a food additive.

3. What are the synthesis methods? Even if synthesis is conducted within a closed system, subsequent processing steps of harvesting and processing can create far greater exposure risk.

3. Do the current markets and/or applications warrant concern? For example quantum dots contain heavy metals, yet they are a commercial product used in limited quantities for biological imaging.

4. What are the exposure routes and levels likely to be encountered in the manufacture, product development and consumer use of nanomaterial based products? Examples in which the nanomaterial, manufacturing process and product are generally considered safe include nanoclays used in various commercial polymers, TiO_2 in sunscreen and carbon black in car tires.

5. Is there any relevant information on related materials and/or nanoparticles and if so, is it applicable?

Examples of nanomaterials and synthesis conditions will be used to address these questions.

The talk will balance these considerations with those nanomaterials derived accidentally such from combustion-based stationary power generation or transportation. For the general population these sources overwhelm their potential exposures to nanomaterials, even after decades of study and emission controls. In fact recent studies have linked soot with asthma, heart arrhythmias and even death.

465578 Dose-Dependent Intracellular Reactive Oxygen and Nitrogen Species Production from Particulate Matter Exposure: Results from Ambient Samples and Chamber Experiments

Abstract:

Epidemiological studies have found associations between elevated particulate matter (PM) concentrations and increased incidences of cardiovascular and respiratory diseases. PM-induced oxidative stress has been suggests as a possible mechanism leading to these health effects. Here we present chemical and cellular oxidant measurements from ambient PM samples and controlled chamber experiments. We investigated reactive oxygen and nitrogen species (ROS/RNS) production from ambient PM samples and laboratory-generated secondary organic aerosols (SOA). Ambient samples (n = 104) were collected during multiple seasons from rural and urban sites around the greater Atlanta area as part of the Southeastern Center for Air Pollution and Epidemiology (SCAPE). SOA were generated from the photooxidation of six commonly emitted volatile organic compounds (VOCs). These experiments were conducted in the Georgia Tech Environmental (GTEC) facility under both dry and humid conditions in the presence of ammonium sulfate seed using H_2O_2 as an OH radical precursor. Murine alveolar macrophages were used to measure intracellular ROS/RNS production, while dithiothreitol (DTT) was used to quantify the concentration of redox-active species in each sample. For each sample, intracellular ROS/RNS was characterized using response parameters obtained by fitting dose-response data using the Hill equation. These response parameters (maximum response, EC_{50}, Hill slope, threshold, and area under the dose-response curve) were compared with DTT activity to determine whether chemical assays represent cellular ROS/RNS responses and with water-soluble PM components to elucidate species associated with PM-induced ROS/RNS production. From both ambient and chamber filters, we found that ROS/RNS production was highly dose-dependent, non-linear, and could not be represented by a single concentration measurement. We demonstrate that there is no simple correlation between DTT activity/PM composition and cellular ROS/RNS responses. For summer ambient samples, we found that DTT activity and organic species were significantly correlated

with cellular ROS/RNS production, which motivates further studies involving PM-induced ROS/RNS production from summertime photochemically driven SOA. Finally, from chamber filters, we found that SOA generated from different precursor VOCs contained different concentrations of redox-active species and induced different levels of cellular ROS/RNS.

462217 Biofuel and Reference Diesel Particles: Differences in Inflammatory and Oxidative Effects

Abstract:

Particulate matter (PM) from engine exhaust has become an increasing concern due to its links to numerous adverse effects on human health and the environment. Recent efforts to reduce PM emissions and the associated climate change impacts from combustion include new technologies, such as the installation of diesel particulate filters and the use of biofuels. However, studies comparing the biological effects of PM from biodiesel and petroleum-based diesel provide conflicting results, and interpretation of these results is difficult because of different methods for particle generation and for biological assays.

This study aims to systematically evaluate the effect of particles generated from two biofuel blends, one biodiesel (BD) and one alcohol diesel (AD) and one reference standard diesel particle sample (NIST SRM 2975) on selected biological outcomes associated with responses to diesel particles in lung cells and in animals. These include: (a) CYP1A1 and 1B1 enzymes that can both detoxify and bioactivate PAHs in human airway epithelial cells, human bronchial epithelial cells and human macrophage cells lines, and (b) IL-8 and TNF-a that play critical roles in pulmonary host-defense and inflammatory responses. We also examined the role of activation of the transient receptor potential ankyrin-1 (TRPA1) cation channel, which is a target for diesel exhaust and similar combustion-derived particles and mediator of pro-inflammatory and irritant responses. Because PM deposits in small airways and the alveolar space, cell lines derived from small airway epithelial cells (H441) and modeling alveolar macrophages (THP-1) were selected for this study. PM was generated from simple surrogate fuels and characterized under well-controlled combustion conditions.

The results show that particle extracts caused similar responses compared to whole particles in both upper airway

epithelial and macrophage cells. This suggests that the soluble organic components adsorbed on the particle surface are highly associated with the cellular responses we examined. When comparing the particles from the different fuel sources, BD particles are capable of activating key enzymes involved in PAH metabolism to very high levels compared to RD and AD. BD particles also had the greatest surface area, soluble organic fraction and oxidative potential. In contrast, RD was more pro-inflammatory than BD or AD, an effect that involved TRPA1 activation. This study adds to the growing body of evidence that biodiesel PM results in adverse lung health effects, potentially via different pathways than conventional diesel PM.

470397 Real Life PM Emissions from Traffic and Human Exposure Implications

Abstract:

The contribution of vehicle emissions to urban air pollution has been extensively investigated. Such emissions have been found to account for almost 90% of the total amount of substances released into urban ambient air. The ability of solid microparticles to penetrate deep into the human respiratory tract results in increased morbidity and mortality. There is also evidence that car exhaust gases are directly involved in the pathogenesis of several health endpoints, including respiratory, cardiovascular, neurological, and allergic diseases, as well as asthma. With regard to smaller particles that penetrate deeply across human respiratory tract, we need to highlight that particle translocation into systemic circulation is one of the major mechanisms related to air pollution attributed cardiovascular disease. Taking into account all the above, regulatory frameworks have been developed worldwide. Obviously, during the operation of a car, the emission control technologies efficiency declines with mileage, resulting in increased exhaust emissions; however, emissions of conditionally new cars that have a mileage of up to 1,000 km are not studied and unfortunately, not regulated. Considering all the above, the aim of this study is to characterize particulate (segregated by size and active surface) released by cars with low mileage and compare the results with those obtained from in-use vehicles with a mileage of 60,000-280,000 km, as well as to investigate the potential implications to human exposure.

The cars with high mileage and the ones with low mileage were kindly provided by the author colleagues. Conditional codes were introduced for the cars so as to avoid specifying the car manufacturers and models, covering various individual vehicle characteristics including manufacturing year, engine capacity and fuel type. For the aims of the study, the exhaust gas suspension (EGS) method was selected. The measurement methods were patented and used according to the following plan:

1. The test car was started and ran for 1-3 minutes before all extraneous dust and soot particles that had

settled there during the downtime and brought from the outside in the tailpipe were removed.

2. Then, the engine was shut off and the hose was connected to the tailpipe of the test car. After that, the hose was dipped into the plastic container, which was filled with 10 liters of deionized water.

3. After that, the car was started and worked at the neutral speed for 20 minutes. This time interval was chosen considering that the engine had been warmed up for approximately ten minutes (according to a coolant temperature sensor). Then, the warmed up engine idled for the remaining ten minutes.

Once the measurements had been completed, the distilled water container through which the exhaust gases had passed, was sealed with a lid and then sent to the laboratory. A significant advantage of this method is the lack of transformation of aerosols and solid particles. To measure particle size distribution, a laser particle size analyzer supplied with Fritch MaS software was employed. PM size distribution of the wet or dry dispersion units can be measured separately or simultaneously, with automated switching features. Optimal dispersion was accomplished in the NanoTec integral wet dispersion unit by using a combination of a robust, variable speed centrifugal pump with powerful ultrasonification.

To translate the different emission profiles identified in the study into actual exposure and uptake, taking also into account the deposition across HRT, a comprehensive methodological scheme has been developed, taking into account:

a. The amount of PM emitted per km travelled by the vehicles investigated, using the COPERT 4 methodology. PM emission factors (in mass of PM per km traveled) for the different vehicles is necessary so as to account of the overall mass of PM emitted by each individual vehicle type.

b. The PM size distribution of the emitted PM, as identified by the laser particle size analyzer.

c. The intake fraction (IF) of the emitted PM. Intake fraction is a metric of the emission-to-inhalation relationship, facilitating comparisons among sources (in our case the individual vehicles) in terms of their exposure potential. For a given emission source and pollutant, intake fraction is the cumulative mass inhaled by the exposed population divided by the cumulative emissions. Considering that IF depends on several parameters affecting the emission-to-intake process, (e.g. prevalent wind, emissions strength, population density), it is expected that it would vary with location and time. In a recent global intraurban intake fraction study, it was found that population-weighted mean intra-urban intake fraction for cities is equal to 32.

d. The deposition of PM across the HRT. To better understand the exposure implications of size-segregated particles distribution emitted by the investigated vehicles, HRT particle deposition modeling was applied to reckon the PM fraction deposited (DF) to the three parts of HRT. Major mechanisms of PM deposition across HRT include diffusion, sedimentation and impaction. Secondary mechanisms involve interception and electrostatic deposition. Different HRT regions involve different deposition mechanism as a function of PM size as follows:

- Naso-pharyngeal region (or upper respiratory tract – URT): impaction, sedimentation, electrostatic (particles > 1 μm)

- Tracheo-bronchial (TB) region: impaction, sedimentation, diffusion (particles < 1 μm)

- Pulmonary (P) region: sedimentation, diffusion (particles < 0.1 μm)

Several parameters that affect HRT deposition have been taken into account, including PM properties (concentration and size distribution), air flow parameters (lung capacity and breathing frequency) and HRT physiology (structure and morphology). HRT deposition was carried out using the Multiple Path Particle Deposition (MPPD) v. 2.1 model. Software inputs include morphological parameters of the pulmonary system – functional residual volume (FRC), tidal volume (TV), upper respiratory tract (URT) volume, as well as breathing frequency (BF) for each age group.

The results of the study indicated that the mean diameter of particulate emitted from high mileage cars ranges in hundreds of microns. Smaller particles were found in cars that were manufactured over 10 years ago and have considerable mileage. Thus, particles were detected in five of the 17 cars. The same percentage of cars (30%, 5 out of 17) is the source of PM50 class. Particles with a mean diameter of more than 100 μm were found in most samples of the cars without mileage (58%, 10 out of 17), but 42% of the cars without mileage are a source of particles of up to 20 μm. It appears that these aerosols correspond to different agglomeration stages. Data on the surface area of the exhaust gas particulate showed that entirely new cars emit extremely small particles with a large surface area (up to ca. 90,000 cm^2/cm^3).

The qualitative analysis of the exhaust gas particulates showed that there are limited differences in their composition between cars with high and low mileage respectively.

In terms of implications for population exposure and public health, age-dependent physiological differences are also of special importance. Taking into account the vehicle emission profile (mass of PM emitted per km travelled and PM size distribution), the respective intake fraction and the HRT deposition based on the size distribution, the amount of PM retained by all HRT regions was estimated. Among the newest vehicles, the highest amount of PM retained is associated to the emissions of the diesel vehicle VT 2012; the very low mean diameter of the emitted PM implies that its vast majority is within the respirable size spectrum. In general, among the newer vehicles, 0.0096 μg were deposited per km travelled for the diesel vehicles, while the respective amount for the gasoline cars was 0.0074 μg. Regarding older cars, gasoline cars contribution was the same to the newer ones (0.0074 μg per km traveled), while for diesel vehicles the amount (0.031 μg per km traveled) was the highest among all investigated vehicle categories. This observation is attributed to the higher amount of overall PM emitted (per km travelled) by older vehicles. Considering that newer diesel vehicles emit PM of lower aerodynamic diameter compared to older diesel vehicles, a higher fraction of the PM taken up is expected to deposit to the lower HRT.

Among the investigated vehicles, the diesel vehicle encoded VT 2012 combines all characteristics that may be potentially harmful for public health; the lowest average PM diameter (3.84 μm) that results in significant PM deposition across HRT (and especially the lower part), the highest active surface (89,871 cm^2/cm^3), and the highest adsorption potential of organic (e.g. PAHs) and metals.

A key finding of the study is that car mileage has virtually no effect on the size of the solid particles released into the environment, and that even cars with very low mileage (up to 1,000 km) are not "clean" in terms of emissions. On the contrary, special attention has to be paid to the lower aerodynamic diameter related to newer diesel vehicles, their higher specific surface (as well as the higher capacity of toxic substances adsorption) and how this is translated into actual increased human exposure and, consequently, health risk.

467482 Disinfection of Water Using Silver and Copper Nanoparticle Impregnated Activated Carbon

Abstract:

Access to safe drinking water is one of the biggest challenges across the globe. In this context, activated carbon is widely used as a water purification media for producing drinking water, which although removing most of the contaminants (responsible for taste, colour, odour, organics, heavy metals etc.), however is incapable of killing pathogens. To mitigate this, antibacterial properties of silver and copper nanoparticles (Ag-NPs and Cu-NPs) have been utilized. Therefore, the objective of this research is to combine both purification and disinfection effects of activated carbon and nanoparticles together and thereby produce drinking water in a continuous, flow-column set-up.

To this end, first we synthesized Ag-NPs by UV reduction of aqueous silver nitrate solution, in presence of tri-sodium citrate. The latter acts both as a reducing and a stabilizing agent. Following this, Ag-NPs were selectively impregnated on the external surface of plasma treated AC granules. This hybrid is referred to as Ag-AC hybrid, having 0.8 wt% of Ag-NPs of 28 nm mean size. Subsequently, microbial cell-killing experiments were performed in two set-ups: (i) shake flask (called batch experiments) and flow-column (called continuous-mode). We have used *E. coli* K12 (MTCC 1302) as a model organism (as it is typical of contaminated drinking water), always at a concentration of 10^4 CFU cells/ml in the inlet water. Batch results using 8 mg Ag-AC hybrid/ml of cell suspension showed that, all cells were killed within 25 minutes of contact time. Maintaining same *E. coli* concentration and Ag-NP loading percentage in Ag-AC (as in the batch experiment), a glass column (8 cm diameter and 25 cm in height) packed with Ag-AC was used for monitoring cell-killing performance over a long duration. Again, inlet water having 10^4 CFU of *E. coli*/ml could be completely disinfected to produce treated, outlet water having zero cell count, with a contact time of 23 minutes, very close to the designed value based on the previous batch cell-killing kinetics. Therefore, a continuous throughput of treating 1.62 L of contaminated water per hour could be maintained for at least up to 16 days, beyond which experiments were not conducted presently. Moreover, the Ag concentration in the outlet water reached only up to a maximum of 29.8 µg/L at steady state, which is well within the recommended maximum Ag concentration limit of 100 µg/L for drinking water (EPA). Thus these results demonstrate the viability of such a column-design to produce decontaminated water at a reasonable rate per hour, over a long period and with minimal loss of Ag in the treated water.

Cu-NPs being reported for its antibacterial activity too, are actually cheaper to make than Ag-NPs. For some of the Ag-resistant microorganisms, Cu-NP also shows better activity compared to Ag-NP. Thus, Cu-NPs have been synthesized using the thermal reduction method where L-ascorbic acid reacts with cupric chloride dihydrate to form Cu-NPs. In this green route, L-ascorbic acid works as a reducing agent. Cu-NPs of mean diameter 5.17 nm were synthesized and subsequently impregnated on plasma treated AC by wet impregnation. The Cu-AC showed a copper loading of 2.1 wt. %. Disinfection studies were performed using these Cu-AC, again in both batch and continuous modes, using the same wild type model microorganism *E. coli* K12 (MTCC 1302), starting with the same initial cell concentration of 10^4 CFU/ml. Similar to Ag-AC, on using 8 mg/ml of Cu-AC in the batch experiment, zero cell count was achieved in only 10 minutes. Based on this batch mode cell-killing data, the same glass column (but packed with only Cu-AC now) was used for continuous, flow experiments. When 10^4 CFU/ml of *E. coli*cells were introduced in the inlet water to this column, zero cell count in the outlet water was achieved in a contact time of 9 minutes, giving a maximum treated water flow rate of 3.77 L/h. Therefore, compared to 1.62 L/h of water in case of Ag-AC, a much higher amount of 3.77 L/h of water could be produced in the same column with Cu-AC. Simultaneously, in steady state continuous flow-column experiments, the maximum copper concentration in the outlet water was around 83.4 µg/L, which is far below the upper recommended limit of 1000 µg/L (EPA). Thus, Cu-AC shows better cell-killing performance compared to Ag-AC. This is primarily because the Cu loading (2.1 wt%) is higher in Cu-AC, compared to Ag loading of 0.8 wt.% in Ag-AC. Moreover, the steady state release concentration of Cu was much higher (83.4 µg/L) compared to Ag release concentration (29.8 µg/L) from Ag-AC. Therefore, Cu-AC shows higher antibacterial activity compared to Ag-AC and reaches zero cell count within a shorter contact time of 9 minutes in the continuous flow-column, in contrast to 23 minutes required for Ag-AC column to achieve zero cell count.

As Cu-AC shows better antibacterial activity compared to Ag-AC, further experiments are being pursued now with a mixture of Ag-AC and Cu-AC. Both these samples are mixed at 1:1 wt. ratio, that is, Ag-AC having 0.8 wt% Ag loading and Cu-AC having 2.1 wt% Cu loading. These have also been further used in both batch and continuous, flow-column experiments for cell-killing. Starting with the same 10^4 CFU/ml *E. coli*, zero cell count has been achieved in 8 minutes, using 8 mg/ml of the hybrid mixture. This implies that Ag-AC and Cu-AC combination potentially gives a synergistic effect in terms of *E. coli*cell-killing and is hence superior to the usage of only Ag-AC or only Cu-AC. Currently, further continuous column experiments are in progress, using the combination of Ag-AC and Cu-AC to optimize its performance towards the fastest possible cell-killing kinetics, which will maximize the production rate of drinking water.

Hence, water disinfection for potable quality (drinking) water (zero *E. coli* count with < 100 µg/L Ag and < 1000 µg/L Cu) could be achieved in a continuous manner at a large flow rate of nearly 4 L/h over a long duration, by using mixed packing of Ag-AC and Cu-AC.

457443 Metal Organic Framework Composites for Chemical Protection

Abstract:

Several MOFs have shown remarkable capabilities of degrading chemical warfare agents (CWAs) and their simulants. However, the exploration of MOFs in engineered materials such as textiles and clothing for chemical protection has not been prominently explored. It is envisioned that the next generation of protective clothing can shield environmental contaminants by integrating reactive materials, such as MOFs, into the fabric, where in addition to providing a physical barrier, the protective clothing can also undergo self-decontamination. Toward that end, we have engineered a nonwoven fiber composite containing reactive Zr-based MOFs that show the ability to degrade G- and V-type nerve agents. Our approach allows us to load a high weight content of MOFs on the fabric and offers access to reactive sites on the MOFs. We demonstrate through simulant and chemical agent reactions that MOFs in the fabric matrix remain highly reactive and mimic reactions of native MOF powders. These newly developed textiles are further evaluated for breathability, moisture vapor transport, and toxic gas permeation rates across the composite. We use DRIFTS to quantify the interactions between the substrate and reactive sites on the MOF to gain understanding of their reactivity. Our approach will serve as a platform for engineering MOFs into usable textiles without compromising MOF reactivity and performance.

464667
Flare Minimization for Simultaneous Turnaround Operations of Two Olefin Plants

Abstract:

Flare minimization for olefin plant turnaround operations such as start-up and shut-down is a double-win practice, which brings not only industrial economic benefits, but also local environmental credits. By far, almost all of flare minimization studies focus on single units or single plant normal or abnormal operations. No systematic studies have been reported on flare minimization coupling abnormal operations of two chemical plants. Conceivably, if material and energy exchange between two chemical plants could be effectively integrated and scheduled, flare emissions from one plant might be adopted by another plant, so that flare minimization could be achieved to its maximum extent. In this paper, a systematic method has been developed to study the flare minimization for simultaneous turnaround operations of two olefin plants: one plant is experiencing start-up while another plant is conducting shut-down operations.

When one olefin plant undergoes its shut-down operation, its flare minimization strategy is conducted as: (i) keep the minium number of furnace running by shuting down extra furnaces; meanwhile, reduce liquid inventories in tanks, drums, and columns; (ii) shut down the remaining furnaces and keep CGC (crack gas compressor) running with vapor recycles; meanwhile, vaporize liquid remains in tanks, drums, and columns; (iii) shut down CGC after recovering liquids and vapor inventory as much as possible; (iv) perform decommissioning for each facilities including left liquid discharge, vapor discharge, and nitrogen purge. For another plant performing its start-up operation, the general operating procedure includes: (i) perform nitrogen replacement with hydrocarbons; (ii) input liquids to columns to build up lthe vapor - liquid balance in various vessels; meanwhile columns are engaged in the full-reflux mode; (iii) start CGC with vapor recycles; meanwhile, precooling the chilling train and warm up columns; (iv) start up furances gradually; (v) gear up each column operation to their normal status.

In this paper, flare minimization opportunities coupling shutdown operating steps (ii) through (iv) together with start-up operating steps (ii) and (iii) are exploried. To study the feasibility of this coupling turnaround operation and calculate flare emission sources, plant-wide dynamic simulations for these two plants have been conducted. The developed duo-plant flare minimization strategy has been virtually tested and demonstrated. The obtained flare minimization strategies and simulation results provide an indepth understanding on synergies of two plant abnormal operations for flare minimization.

469916 Consequence Assessment for Event Tree Analysis in Transitions Using Dynamic Simulation

Abstract:

Incidents that occur in industrial facilities can lead to the release of hazardous materials, uncontrolled fire and explosions, economic losses and environmental contamination. Consequence assessment is an examination of the undesirable impacts of an accident. Fuzzy logic combined with Event Tree Analysis (ETA) is used for quantitative risk assessment (QRA).

In this work dynamic simulation was used to assess the consequences of event tree branches for two examples of equipment switching transitions in chemical plants. The first example was switching between two solid desiccant dryers in an ethylene plant. The other example was the switching of a reflux pump with its spare in a reformate stabilizer column. Both examples have standard operating procedures (SOP) commonly accomplished manually by operators. Numerous operator induced fault cases were considered. These SOPs and fault cases were simulated in Aspen HYSYS V8.0 using the event scheduler feature. The consequences were identified by monitoring variables during dynamic simulation of each case. Thus, the fuzzy ETA algorithm was used to evaluate the risk of equipment switching transition faults while dynamic simulation was used to assess the consequence.

AIChE Annual Conference 2016

Division: Process Development

Session: Process Research and Development for Industrial Sustainability

Title: Effect of chemical price uncertainty on project viability: an economic evaluation

Authors: Gaurav Arora, Mohammad Shafiei, Mike Huckman

Abstract: A case study is presented for evaluating economic viability of a chemical process. The discussion includes process design choices, cost estimates, and economic evaluation of the process. The economics are found to be very sensitive to the feed and product prices. Three specific scenarios for process are analyzed in detail – at current chemical price, low feed-high product price and high feed-low product price. While these scenarios are helpful in understanding the cost, revenue breakdown, and economics of the project, a better insight is obtained by analyzing the effect of chemical prices over the last 15 years on the economic viability of the project. Margin required to meet an IRR of 15% is calculated followed by a market price analysis to calculate the probability of this occurring in the last 15 years.

Keywords: Economic Analysis

463989 Improving the Energy Efficiency of Carbon Capture Processes – Combining Enzyme Accelerated Solvent Systems and Improved Contacting Equipment

Abstract:

In order to achieve the ambitious targets specified by the United Nations Framework Convention on Climate Change in 2011 and reduce GHG emission in the EU by 2020 at least 20% below those in 1990 carbon dioxide (CO_2) emissions have to be reduced significantly. Fossil-fired power plants are identified as major targets for carbon capture processes, since they are responsible for 40 % of the total global CO_2 emissions (Mondal et al., 2012). Efficient post-combustion capture of $CO_2$is one of the most promising options to face this crucial challenge and to achieve at least the specified short-term targets (Rochelle, 2009).

As a result, a wide range of different fluid separation processes, including (reactive) absorption, adsorption, cryogenic distillation and gas permeation, have been intensively studied in recent years (Mathias et al., 2013). Nevertheless, the current state-of-the-art process for CO_2 capture is the reactive absorption in packed columns using amines. Especially primary amines, like Monoethanolamine (MEA), are often used because of the covalent bonding of the CO_2, resulting in high absorption rates (Kenig & Gorak, 2005). However, the recovery of the solvent, which is required for a continuous operation, becomes more challenging the stronger the $CO_2$is bound (Yildirim et al., 2015).

Therefore, the best absorption process will represent an optimal trade-off between fast absorption rates and energy efficient solvent recovery. The latter causes up to 80 % of the total capture costs for the MEA process and results in a significant loss in overall power plant efficiency (Notz et al., 2011). Consequently, solvent recovery represents the bottleneck of the reactive absorption/desorption process, which is practically not considered for large scale industrial application (Mondal et al., 2012). In order to overcome this restriction a lot of research effort is carried out with a focus on more energy efficient solvents. While tertiary amines or potassium carbonate require less energy in the regeneration step, application of these solvents is restricted by their relatively low absorption rate (Kunze et al., 2015).

Process intensification (PI) can help to overcome this restriction in different ways, which are investigated in this study. In order to compensate for the low absorption rates of tertiary amines enzymes are integrated as efficient biocatalysts to the solvent system. Here, especially the enzyme Carbonic Anhydrase (CA) is considered as the fastest catalyst for the hydration of CO_2, which is the prevailing CO_2consuming reaction in tertiary amine systems (Gundersen et al., 2014). A significant enhancement of the absorption performance was observed in previous studies with a wetted wall column set-up, validating the direct influence on the reaction rate (Kunze et al., 2015). However, besides the possible improvement, the application of carbonic anhydrase as a biocatalyst also introduces additional restrictions to the process due to sensitivity towards temperature and pH (Gundersen et al., 2014).

In addition to an intensification of the solvent system, combining high capacity solvents with the efficient biocatalyst, the mass transfer can be further improved by means of intensified contacting devices (ICD), other than making use of conventional columns. The use of a membrane contactor provides a well-defined interfacial area, with packing densities being a magnitude higher as in conventional equipment (Zhao et al., 2016). Due to the fully separated liquid and vapour flow, membrane contactors also allow operational modes different from conventional equipment, which are limited by flooding or loading. The modular nature and operational flexibility make membrane contactors a promising alternative to conventional equipment. However, mass transfer resistance might be increased due to the membrane and the operational window of the membrane contactor is limited as well, such that especially transmembrane pressure differences have to be controlled in order to avoid breakthrough of one of the two phases, which will lead to undesired phase dispersion.

Another ICD that can improve mass transfer in ab- and desorption is the rotating packed bed (RPB) technology. In RPB processes centrifugal forces are exploited to increase acceleration of the liquid. In a counter-currently operated RPB, gas is pushed through the torus-shaped rotating packing from the outside to the middle of the rotor. Liquid is sprayed into the eye of the rotor and flows counter-currently to the gas outwards, driven by the centrifugal forces. RPBs offer mass transfer improvements due to high turbulences and extend the operating window, having the rotational speed as an additional degree of freedom. These advantages result in high capacity combined with compactness (Sudhoff et al., 2015). Yet, application of RPBs is still restricted due to lack of reliable data on the performance efficiency.

While both, ICD and the use of an improved solvent system with CA as catalyst, have the potential to significantly intensify the CO_2capture process, literature studies and research on the combination of both is scarce. In order to investigate if the advantages of the different PI techniques can be exploited and if a combination is feasible, first investigations are performed in order to characterize the operating windows of such intensified processes and to evaluate the potential benefits for an actual implementation. In preliminary work (Kunze et al., 2015), the tertiary amine Methyldiethanolamine (MDEA) has shown promising results to be more energy-efficient compared to MEA and is therefore chosen as solvent (30 wt.-%) for this work. Absorption experiments in a column and the different ICD are performed to determine the operating windows and compare separation efficiency. Subsequently CA is added to the solvent system in order to investigate the potential synergetic intensification of ICD and CA versus the acceleration by CA in a conventional packed column set-up. The results enable the discrimination between the different means for process intensification depending on the determined operating windows and separation performances. Finally, the obtained results are evaluated and benchmarked against the current state-of-the-art process on a theoretical basis. Future work will also investigate the application of the different means for process intensification to the desorption step in order to provide a complete picture for a potential industrial application.

REFERENCES

Gundersen, M.T., Von Solms, N., Woodley, J.M., 2014, Energy Procedia, 63, pp. 624-632.

Kenig, E.Y., Górak, A., 2005, Integrated Chemical Processes: Synthesis, Operation, Analysis, and Control, pp. 265-311.

Kunze, A.-K., Dojchinov, G., Haritos, V.S., Lutze, P., 2015, Applied Energy, 156, pp. 676-685.

Mathias, P.M., Reddy, S., Smith, A., Afshar, K., 2013, Energy Procedia, 37, pp. 1863-1870.

Mondal, M.K., Balsora, H.K., Varshney, P., 2012, Energy, 46 (1), pp. 431-441.

R. Notz, I.Tönnies, H.P. Mangalapally, S. Hoch, H. Hasse, 2011, Int. J. Greenh. Gas Control, 5, 3, 413-421

Rochelle, G.T., 2009, Science, 325 (5948), pp. 1652-1654.

Sudhoff, D., Leimbrink, M., Schleinitz, M., Górak, A., Lutze, P., 2015, Chemical Engineering Research and Design, 94, pp. 72-89.

Zhao, S., Feron, P.H.M., Deng, L., Favre, E., Chabanon, E., Yan, S., Hou, J., Chen, V., Qi, H, 2016, Journal of Membrane Science, 511, pp. 180-206.

Yildirim Ö., Kiss A. A., Hüser N., Leßmann K., Kenig, E.Y., 2012, Chemical Engineering Journal, 213, pp. 371–391.

ACKNOWLEDGEMENT

The research leading to these results has received funding from the European Union Seventh Framework Programme FP7/2007-2013 under grant agreement n° 608535.

464321 Boosting the Multienzymatic Conversion of CO2 through Spatially Separated Immobilization Strategy

Abstract:

Multienzymatic catalysis is considered to be the next generation of biocatalysis. As a central part, rational design and construction of multeinzyme systems have attracted great attention and become a hot research topic nowadays. When designing an efficient multienzyme system, three crucial issues should be taken into consideration: 1) synergistic catalysis among multienzymes; 2) efficient mass transfer of substrates and products; and 3) easy recyclability of multienyzmes. Inspired by the existence form and synergistic catalytic behavior of multienzymes in organisms, we put forward to the strategy of constructing spatially isolated multienzyme systems. In our recent study, a series of spatially isolated multienzyme systems (containing formate dehydrogenase and formaldehyde dehydrogenase) enabled by nanospheres/microcapsules are constructed by combining biomimetic mineralization with other biomimetic platform techniques (*e.g.*, biomimetic adhesion, Pickering emulsion, surface segregation and layer-by-layer (LbL) self-assembly). When utilized for the conversion of CO_2, these systems exhibit desirable catalytic performance and easy recyclability.

Firstly, a spatially isolated multienzyme system enabled by active nanospheres was successfully constructed through the synergy between biomimetic mineralization and biomimetic adhesion, achieving the efficient process intensification. The first enzyme (formate dehydrogenase) was entrapped accompanying the formation of nanospheres through biomimetic titanification. After *in situ* surface functionalization of nanospheres with oligodopa, the second enzyme (formaldehyde dehydrogenase) was immobilized on the surface of active nanospheres through amine-catechol adduct reaction. When utilized for the conversion of CO_2, the as-constructed multienzyme system exhibited a formaldehyde yield of higher than 60.0%. Besides, after evaluating the performance of different-sized multienzyme systems, it was found that smaller sized multienzyme system possessed higher enzymatic activity but lower recyclability.

Secondly, a spatially isolated multienzyme system enabled by nanosphere-stabilized capsules (NSSCs) was constructed through the synergy of biomimetic mineralization and Pickering emulsion, achieving the efficient process intensification. The first enzyme-containing nanospheres were monodispersely spread on the surface of oil droplets through Pickeirng emulsion, maintaining their structural characteristics and integrity. Then, the gel titania, that was produced through sol-gel process of tetrabutyl titanate, was in charge of crosslinking the nanospheres, forming a stable and continuous capsule wall. The second enzyme was immobilized on the surface of capsule wall through amine-catechol adduct reaction between oligodopa and enzyme. During the construction process, the released butanol as a result of the sol-gel process of butyl titanate could be adsorbed on the surface of nanospheres, further inducing the self-assembly of the nanospheres at the oil/water interface. This process played key role on the formation of Pickering emulsions. When utilized for the conversion of CO_2, the as-constructed multienzyme system acquired a formaldehyde yield of higher than 50.0%. Meanwhile, the lower density of the oil core than that of water endowed the multienzyme system with excellent recyclability, especially the formaldehyde yield kept almost unaltered after 10 times recycling.

Thirdly, a spatially isolated multienzyme system enabled by mesoporous hybrid microcapsules was constructed through the synergy of biomimetic mineralization and surface segregation, achieving the efficient process intensification. The first enzyme was entrapped accompanying the formation of the PAH-segregated $CaCO_3$ template. Then, the PAH polymeric network was formed through crosslinking between GA and PAH. Afterwards, this polymeric network was utilized for implementing biomimetic mineralization with *in situ* entrapping the second enzyme. After removing the template, a spatially isolated multienzyme system was acquired. During the construction process, the $CaCO_3$ microsphere served as the dual templates for formation of both capsule lumen and mesopores on the capsule wall, whereas PAH played the following two crucial roles: (1) forming the bulk polymer network to render confined space for biomimetic mineralization and (2) triggering the mineralization of Ti-BALDH to produce titania nanoparticles. When utilized for the conversion of CO_2, the as-constructed multienzyme system exhibited a formaldehyde yield of higher than 70.0%. Meanwhile, the reaction rate was nearly two times higher than the average level. And no catalytic activity was lost after 10 times recycling. Besides, the mesoporous hybrid microcapsules can be utilized in various other applications and exhibited desired performance.

Fourthly, a spatially isolated multienzyme system enabled by mitochondria-like microcapsules was constructed through the synergy of biomimetic mineralization and LbL self-assembly, achieving the efficient process intensification. Inspired by the double membrane structure of mitochondria, we proposed to utilize process and structure biomimicism strategy instead of just process biomimicism strategy for constructing multienzyme systems. The first enzyme was entrapped accompanying the formation of the PSS-doped $CaCO_3$ template. The organic inner membrane was acquired *via* LbL self-assembly of oppositely charged polymers/proteins on the template. The silica template layer was then formed onto the inner membrane through biomimetic silicification with *in situ* entrapping the second enzyme, while the organic-inorganic hybrid outer membrane was acquired *via* biomimetic titanification subsequently. After removing the $CaCO_3$ template and the silica template, a spatially isolated multienzyme system was established. When utilized for the conversion of CO_2, the as-constructed multienzyme system exhibited a formaldehyde yield of higher than 82.0%, with a formaldehyde selectivity of 90.0%. Meanwhile, 70.0% of the initial activity was retained after 10 times recycling.

By using of a series of cases as the evidences, it can be conjectured the spatially separated immobilization strategy can boost the multienzymatic conversion of CO2. This study will shed light on the new strategies for the design and preparation of multicomponent catalysts.

466064 Evaluation of Consumption Energy in CO_2 Absorption Processes with Two-Liquid Phase Separation Type

Abstract:

The global issue of climate change is receiving ever-increasing attention. One of the main focuses of this attention is the reduction of CO_2 emission from activities such as power generation, iron and steels making process. The technology that has the potential to make a significant impact on emission is CO2 capture and storage.

The aqueous of amine CO2 capture proses can remove CO2 from exhaust gas in some industrial process. The major aqueous amine absorbent which is applied in industrial process are primary amines, the secondary amines and the tertiary amines. The aqueous of amine CO2 capture proses are consumed thermal energy for CO2 desorption. A previous study has revealed that a lot of energy for CO2 desorption are CO2 desorption heat, sensible heat of the aqueous for desorption temperature and the heat of vaporization the aqueous.

We suggest new CO_2 absorption processes with two-liquid phase separation type for reducing consumption energy. Absorbing liquid in new process consist of H2O, amines and ether or alcohol additive. In CO2 absorption section, the CO2 react to amine in absorber column. The CO2-rich solution liquid is transferred to the decanter. In the decanter, the CO2-rich solution liquid was separated hydrophilic liquid and hydrophobic liquid. Carbamate and bicarbonate ion exist in hydrophilic liquid. Hydrophobic liquid is only additive. The hydrophilic liquid is transferred to the stripper where comparatively high temperature is used to desorb the CO2. The hydrophobic liquid is transferred to mixer. It is mixed with hydrophobic liquid returned to stripper. The new CO_2 absorption processes can be reduced energy consumption for sensible heat of the aqueous for desorption temperature and the heat of vaporization the aqueous. In this work, we select 2-(Ethyl amino) ethanol, water and diethylene glycol diethyl ether to Absorbing liquid. We examine the CO2 absorption characteristics. We study the effect of the new Absorbing liquid on energy consumption

for process. We determine the comparison of absorbing liquid. (H2O, EAE, diethylene glycol diethyl ether = 10, 30, 60 wt%) We measure the solubility of CO2 on this absorbing liquid. In experiment, we obtained amount of CO2 are measured to be 0.3 (mol-CO2/mol-amine) at absorber temperature 40 degree C. The new absorbing liquid has the enough loading amount of CO2. We find out that the enough performance is obtained as absorbing solution.

Consumption energy of the new process is calculated by process simulator. The consumption energy is calculated to be 2.5 (GJ/t-CO2). This value is lower than that obtained for MEA absorption process. Thus, CO_2 absorption processes with two-liquid phase separation type can be valuable in CO2 capture and separation technology.

466110 The Influence of DissociationÂ Constants on the Kinetics of Carbon Dioxide Absorption in Aqueous Tertiary Amines Solutions Containing Carbonic Anhydrase

Abstract:

American Institute of Chemical Engineers (AIChE) Annual Meeting

The influence of dissociation constants on the kinetics of carbon dioxide absorption in aqueous tertiary amines solutions containing carbonic anhydrase

Bin Liu[a], Xiao Luo[a*], Zhiwu Liang[a,b*], Paitoon Tontiwachwuthikul[a,b*]

[a]Joint International Center for CO_2 Capture and Storage (iCCS), Hunan Provincial Key Laboratory for Cost-effective Utilization of Fossil Fuel Aimed at Reducing CO_2 Emissions, College of Chemistry and Chemical Engineering, Hunan University, Changsha 410082, PR China

[b]Clean Energy Technologies Research Institute (CETRI), Faculty of Engineering and Applied Science, University of Regina, Regina, Saskatchewan, S4S 0A2, Canada

Keywords: Tertiary amines; CO_2 absorption; carbonic anhydrase

* Author for correspondence: Dr. Zhiwu Liang E-mail: zwliang@hnu.edu.cn

Tel: +86-13618481627 and +86-73188573033

ABSTRACT

The chemical absorption of CO_2 into primary amine and secondary amine, such as monoethanolamine (MEA), diethanolamine (DEA), is currently the most widely accepted commercial approach to carbon capture.[1] This solvent is extremely effective in forming stable carbamates when reacted with CO_2, leading to a very efficient removal of CO_2 with a 2:1 reaction stoichiometry.[2,3] However, regeneration of the carbamate species requires a large parasitic supply of energy to release the CO_2 during stripping.[1] This high energy requirement accounts for up to 80% of the operating costs for CO_2 capture and recovery.[4] It is also a highly corrosive substance that readily undergoes degradation in the presence of oxygen, emitting harmful volatile organic compounds.[5]

Considering the high absorption heat and high energy consumption during CO_2 capture process, there are a number of tertiary amines such as methyldiethanolamine (MDEA), dimethylaminoethanol (DMEA), in place of primary and secondary amine for carbon capture. While many of tertiary amines solvents are able to reduce the stripping energy requirement, the reaction rate for CO_2 sorption in the absorber is often slow. This means that it is not possible to recover high purity carbon dioxide within a reasonable column height. Carbonic anhydrase (CA), a naturally occurring enzyme and very efficient catalyst that enhances the reversible reaction of CO_2 to HCO_3^-, was first identified in 1933 in red blood cells and known to catalyze the conversion of CO_2 into bicarbonate (HCO_3^-) at extremely high turnover rates, [6-7] which may be useful for promoting the absorption rates of CO_2 from gas streams when these alternative solvents are used. There is great potential for using CA to hydrate CO_2, especially when it is known that the hydration step is the rate determining step in the CO_2 absorption process. Carbonic anhydrase is not just a single enzyme form, but a broad group of zinc metallo-proteins (enzymes) that exists in three genetically unrelated families of isoforms (¦Á, ¦Â and ¦Ã). Its catalyzed mechanism of CO_2 hydration has been introduced by Lindskog et al. [6-7] and simplified as shown in Figure 1. Penders et al. [8] has been working on CO_2 absorption into amine solutions catalyzed by human carbonic anhydrase in string cell reactor since 2012. This work focuses on the study of effect of dissociation constants (pKa) on the kinetics of carbon dioxide absorption in aqueous tertiary amines solutions containing carbonic anhydrase at various temperature using stopped-flow techniques. The examined alkanolamines were (Figure 2): Triethanolamine (TEA), Methyldiethanolamine (MDEA), Dimethylethanolamine (DMEA), 2-(Diethylamino)ethanol (DEEA), 1-Diethylamino-2-propanol (1DEA-2P), 3-Diethylamino-1-propanol (3DEA-1P) and 2-(Dimethylamino)isobutanol (2DEA-2M-1P). The values of pKa of different amine was determined using conventional pH method as shown in Figure 3 and its final results was presented a function of temperature in Figure 4.

It can be seen from the results (Figure 5) that DMEA with a higher pKa value than MDEA and TEA hence showed a higher reactivity with carbon dioxide in the absence of enzyme. It can also been found that the reaction rate constant (k_0) in amine-CO_2-H_2O system was significantly enhanced in the present of BCA, in which the enzyme activity is higher in TEA solutions. This is because Enzymes often display a pH-dependent changes in activity, in characterizing the pH dependence of the enzyme activity, the experimenter often observes a bell-shaped curve in plot of activity versus pH or s-shaped curves in plots of activity versus pH (either falling from optimal activity or rising to optimal activity) with an inflection point at some pH value. This result would provide a useful information to understand the reaction activity of CO_2 absorption into various amines system.

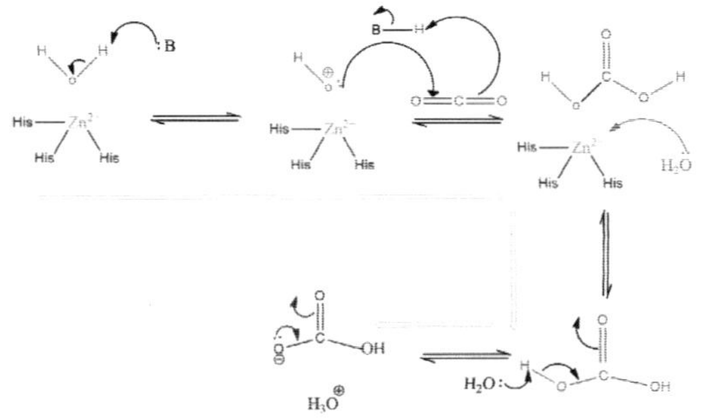

Figure 1. The catalytic Mechanism of enzyme in aqueous solutions

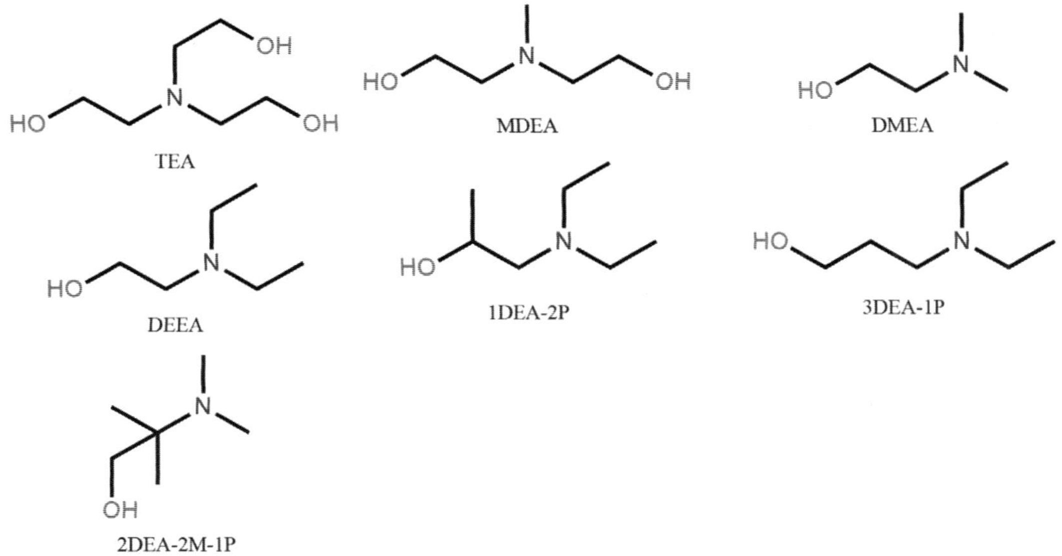

Figure 2. The chemical structure of amines studied in this work

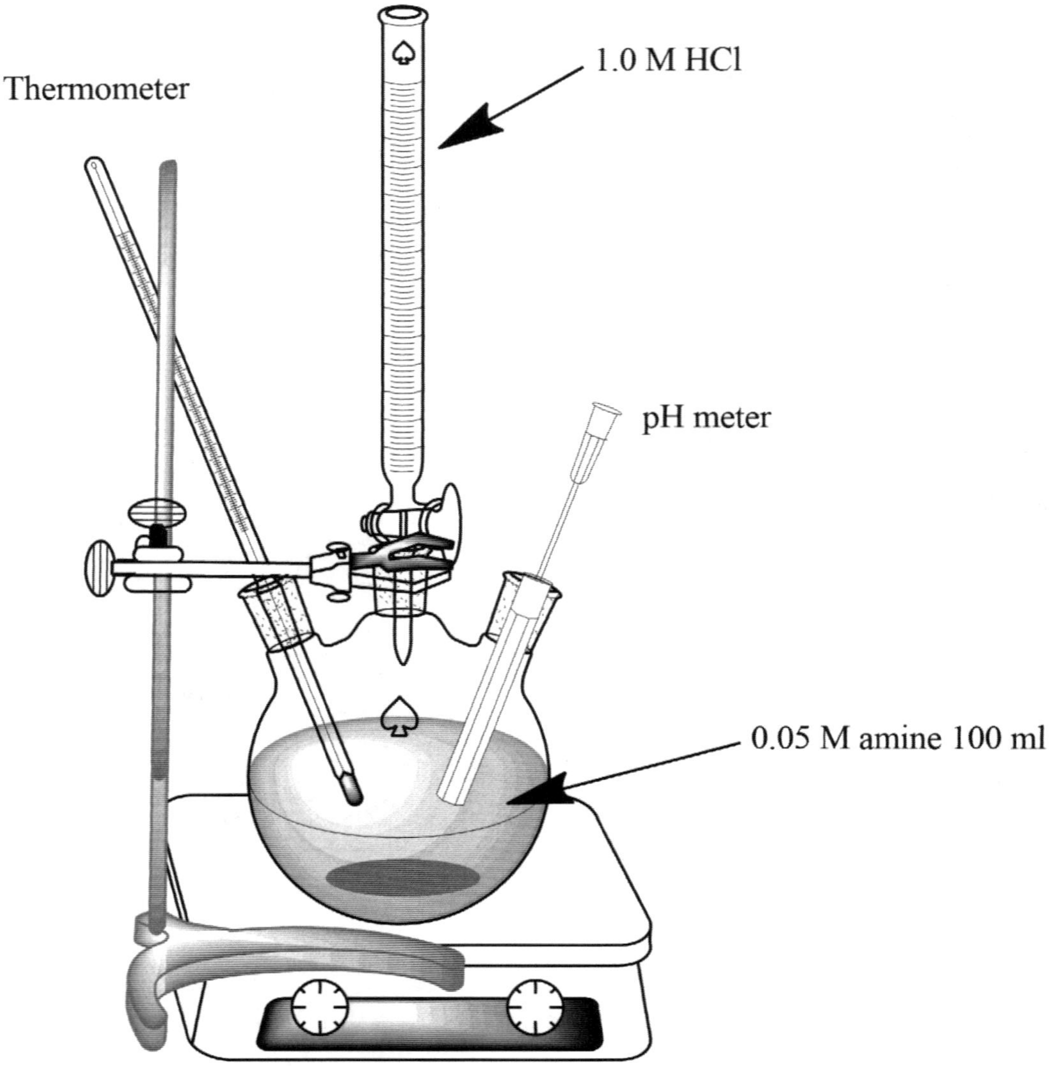

Figure 3. String cell for determination of amine pKa

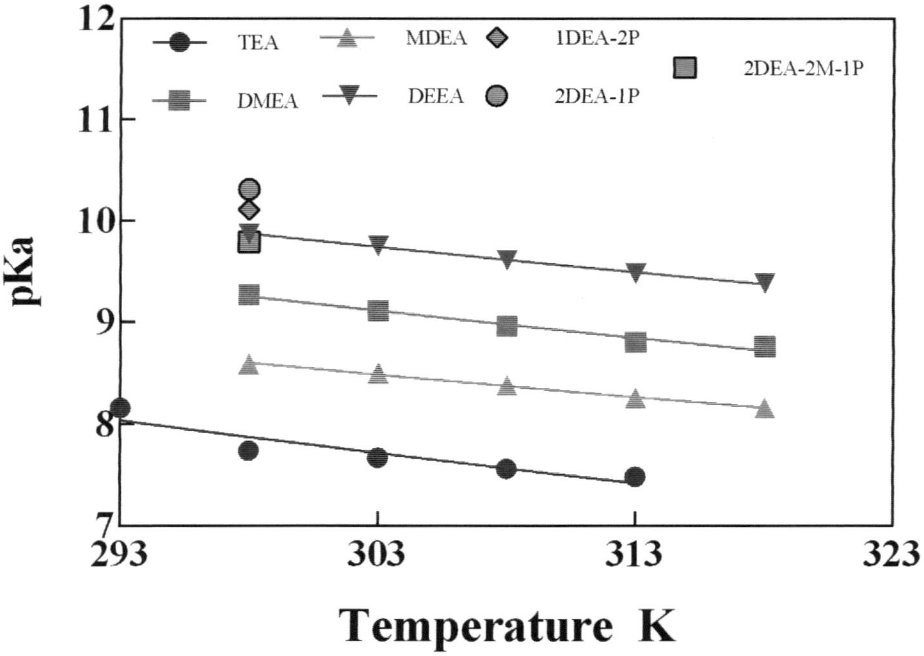

Figure 4. The dissociation constants show as a function of temperature

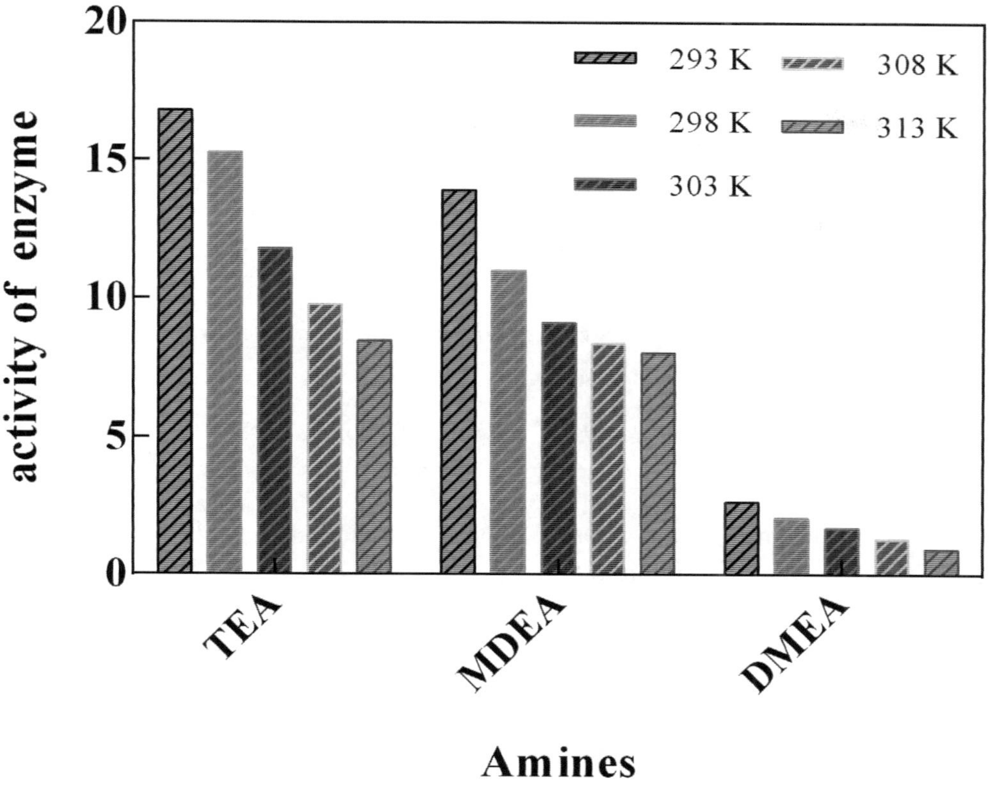

Figure 5. Comparison of enzymes activity in different aqueous amine solutions

Acknowledgment: The financial supports from the National Natural Science Foundation of China (Nos. 21476064 and 21406057), National Key Technology R&D Program (Nos. 2012BAC26B01 and2014BAC18B04), Innovative Research Team Development Plan-Ministry of Education of China (No. IRT1238), and China¡¯s State ¡°Project 985¡± in Hunan University-Novel Technology Research and Development for CO_2 Capture as well as Hunan University to the Joint International Center for CO_2 Capture and Storage (ICCS) is gratefully acknowledged.

References

1. Tontiwachwuthikul P, Idem R, Gelowitz D, Liang ZH, Supap T, Chan CW, et al. Recent progress and new development of post-combustion carbon-capture technology using reactive solvents. Carbon Management. 2011; 2: 261-3.

2. McCann N, Phan D, Wang X, Conway W, Burns R, Attalla M, et al. Kinetics and mechanism of carbamate formation from CO2 (aq), carbonate species, and monoethanolamine in aqueous solution. The Journal of Physical Chemistry A. 2009; 113: 5022-9.

3. Yu C-H, Huang C-H, Tan C-S. A Review of CO2 Capture by Absorption and Adsorption. Aerosol and Air Quality Research. 2012; 12: 745-69.

4. Liang ZH, Sanpasertparnich T, Tontiwachwuthikul P, Gelowitz D, Idem R. Part 1: Design, modeling and simulation of post-combustion CO2 capture systems using reactive solvents. Carbon Management. 2011; 2:265-88.

5. Reynolds AJ, Verheyen TV, Adeloju SB, Meuleman E, Feron P. Towards commercial scale postcombustion capture of CO2 with monoethanolamine solvent: key considerations for solvent management and

6. Lindskog S. Structure and mechanism of carbonic anhydrase. Pharmacology & therapeutics. 1997; 74: 1-20.

7. Lindskog S, Silverman DN. The catalytic mechanism of mammalian carbonic anhydrases. The Carbonic Anhydrases: Springer; 2000. p. 175-95.

8. Penders-van Elk NJMC, Derks PWJ, Fradette S, Versteeg GF. Kinetics of absorption of carbon dioxide in aqueous MDEA solutions with carbonic anhydrase at 298 K. International Journal of Greenhouse Gas Control. 2012; 9: 385-92.

A study of porous support amine-RTILs binary system for CO_2 capture

Min Xiao [a], Helei Liu [a,b], Raphael Idem [a,b], Paitoon Tontiwachwuthikul [a,b], and Zhiwu Liang [a,b]**

[a] *Joint International Center for CO_2 Capture and Storage (iCCS), Provincial Hunan Key Laboratory for Cost-effective Utilization of Fossil Fuel Aimed at Reducing CO_2 Emissions, College of Chemistry and Chemical Engineering, Hunan University, Changsha 410082, PR China*

[b] *Clean Energy Technologies Research Institute (CETRI), Faculty of Engineering and Applied Science*

University of Regina, Regina, Saskatchewan, S4S 0A2, Canada

Abstract

In this work, room temperature ionic liquids (RTILs) 1-butyl-3-methylimidazolium tetrafluoroborate ([BMIM]BF_4) and 1-butyl-3-ethylimidazolium tetrafluoroborate ([BEIM]BF_4) are synthesized. The cation structure of ionic liquids is confirmed by NMR spectroscopy. The thermal decomposition temperature of synthesized ionic liquids is measured and both ionic liquids show good thermal stability. The viscosity of ionic liquids is determined over the temperature range of 298 K to 353 K. A mixed system made by blending amine and ionic liquid is proposed for CO_2 capture. Investigation of the performance of this binary system on a porous support medium for CO_2 absorption shows that amine is able to greatly intensify the ionic liquids' ability to absorb CO_2. Furthermore, when blending tertiary amine and ionic liquids to absorb CO_2, it was found that only physical absorption takes place and that the viscosity of the absorbent increases slightly after CO_2 absorption. Based on the results of this work, the binary system of tertiary amine and RTILs is proven to be an alternative absorbent for CO_2 capture.

Keywords: *amine-RTILs; CO_2 capture; low viscosity; physical absorption*

* Author for correspondence:
Email: zwliang@hnu.edu.cn (Z. Liang) & lhl0925@hotmail.com (H. Liu)

1. Introduction

Global warming, which brings extreme weather, has become a serious issue and is more and more drawing people's attention. As one of many greenhouse gases, carbon dioxide is mainly blamed for the rise in global temperatures. Large amounts of CO_2 are directly released to the atmosphere through the burning of fossil fuels to meet the global demand for energy. There is an urgent need to reduce CO_2 emissions to protect the environment. Meanwhile, CO_2 is also a potential resource which can be applied to Enhanced Oil Recovery (EOR) or reformed with methane to produce synthesis gas etc. Thus, much research has been done on separating CO_2 from combustion exhaust gases. Although absorption of CO_2 with alcohol amine has been widely studied for years and proven to be able to capture CO_2 efficiently, its wide application is limited by a number of remaining problems such as corrosivity, volatility, degradation, and especially high energy consumption. The amount of energy required for regenerating rich amine solvents accounts for as much as 70-80% of the total operating cost of the system. It is thus of primary importance to reduce this energy consumption in order to bring down the cost of CO_2 capture.

A new kind of compound named room temperature ionic liquids (RTILs) has aroused people's interest because of their many unique properties. RTILs usually exist as liquid at room temperature with low vapor pressure, something that commends them as environmental friendly. Besides, RTILs show good thermal/chemical stability and tunable physicochemical characteristics, properties that have allowed them to be successfully used as green solvents in reactions and as catalysts for organic reactions [1]. There is also some work focusing on using RTILs for CO_2 capture. However, using pure RTILs for CO_2 capture as an answer to the world's CO_2 problem still has a very long way to go. When physical absorption dominates CO_2 capture with RTILs, the procedure is usually performed under extreme conditions, such as a very high pressure (90 bar or higher), and with a very long time being required to reach equilibrium (up to 24 h) [2]. To intensify CO_2 capture performance, a kind of ionic liquids named task-specific ionic liquids was developed. From the work of Bates et al. [3], the molar uptake of task-specific ionic liquids, which was modified with amino group, approached 0.5 mol CO_2•mol amine^{-1} during 3 h. However, high viscosity appears to be a new problem making those particular ionic liquids challenging for application in CO_2 capture [4]. The viscosity of ionic liquids is commonly high and will dramatically increase after CO_2 absorption, especially for task-specific ionic liquids [2].

To address these problems, a kind of low viscosity amine-IL mixed system is proposed for CO_2 capture in this work. Imidazole based ILs 1-butyl-3-methylimidazolium tetrafluoroborate ([BMIM]BF_4) and 1-butyl-3-ethylimidazolium tetrafluoroborate ([BEIM]BF_4) were chosen for their relatively low viscosity compared with other ILs. Conventional amines, Monoethanolamine (MEA) and Methyldiethanolamine (MDEA) were selected as the amines for blending with the ILs. The structures and abbreviation of amines and cation of ILs are shown in **Figure 1**.

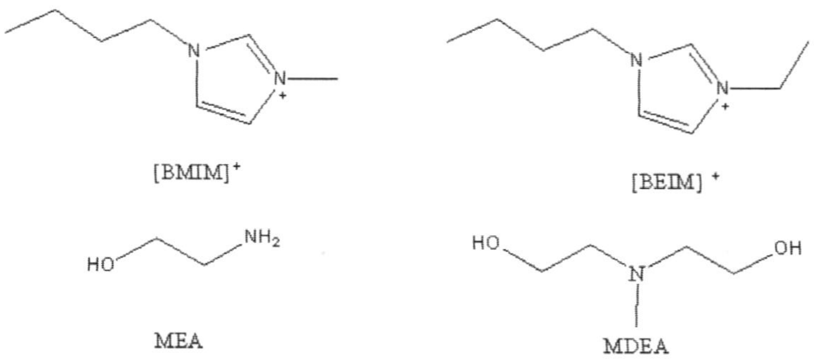

Figure 1 The structure and abbreviation of conventional amines, [BMIM]$^+$ and [BEIM]$^+$

2. Materials and experiment

2.1. Materials

CO_2 with purity of 99.9% was obtained from Changsha RiZhen Gas Co., Ltd. MDEA with purity of 98% was from Aladdin Industrial Corporation. MEA with purity of 99% and n-Butyl bromide (≥ 98%) was obtained from Sinopharm Chemical Reagent Co., Ltd. 1-Methylimidazole with purity of 99% and 1-Ethylimidazole with purity of 98% were supplied by Shanghai Macklin Biochemical Co. Ltd.

2.2. Experiment

$[BMIM]BF_4$ and $[BEIM]BF_4$ were synthesized on basis of imidazole. The general preparation was similar to the work of Harris et al [5]. Complete anion exchange was confirmed when no bromide could be detected with silver nitrate. After evaporating any unreacted components and water, the RTILs were stored in a vacuum drier. A ^{13}C NMR Spectrometer (Bruker 400MHZ Advance) was employed to characterize the cation structure of the RTILs and the absorbents after CO_2 absorption. Thermogravimetric Analysis (Shimadzu DTG-60) was introduced to evaluate the water content and thermal stability of the RTILs. Then the dynamic viscosities (Brookfield Viscometer DV2T) of pure RTILs and the amine-RTIL binary systems were measured.

The CO_2 absorption experiment process is briefly described below. Approximate 0.8 mL absorbent was stabilized into a porous filter candle which was then placed into a gas-washing bottle. The gas-washing bottle was immersed into a water bath to control the temperature. The pure CO_2 flow rate was fixed at 50 $mL•min^{-1}$ by a mass flow controller, and fed into a conical flask with concentrated sulfuric acid to get rid of all trace amounts of water. Then, the CO_2 was fed to the gas washing bottle continuously. The amount of absorbed CO_2 was determined by detecting weight change using an electronic analytical balance with accuracy of 1 mg. When the weight of the absorbent stopped increasing, equilibrium was considered to be achieved.

3. Results and discussion

3.1. Thermal stability of RTILs

To evaluate the thermal stability of $[BMIM]BF_4$ and $[BEIM]BF_4$, Thermogravimetric Analysis (TGA) was introduced. The results for pure $[BMIM]BF_4$ and $[BEIM]BF_4$ are shown in **Figure 2**. Water content of the RTILs was confirmed to be less than 1% and thermal decomposition temperatures (10% mass loss) of $[BMIM]BF_4$ and $[BEIM]BF_4$ were found to be 635.65 K and 646.96 K, respectively. Results showed that both RTILs have good thermal stability. Because the side chain of $[BEIM]^+$ was one carbon longer than $[BMIM]^+$, it showed better thermal stability.

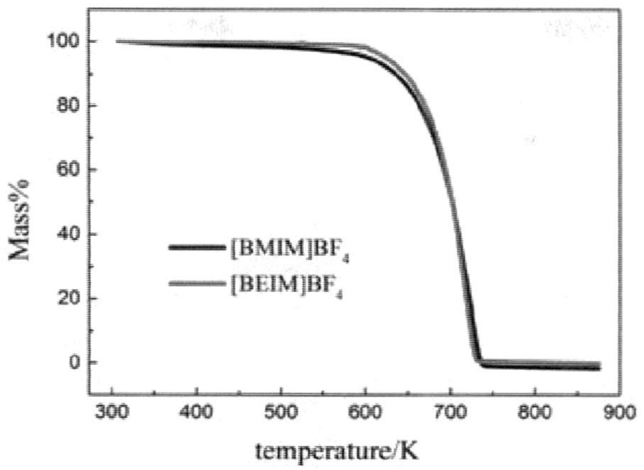

Figure 2 Thermogravimetric Analysis on $[BMIM]BF_4$ and $[BEIM]BF_4$

3.2. *Viscosity of RTILs*

The viscosities of $[BMIM]BF_4$ and $[BEIM]BF_4$ were measured over temperature ranges from 298 K to 353 K. It was found that $[BMIM]BF_4$ shows viscosity of 101.2 cP at 293 K, which agreed well with Harris's work with a deviation of 2.6% [5]. The viscosity of $[BEIM]BF_4$ is slightly higher than $[BMIM]BF_4$ due to its longer carbon chain. An extended Arrhenius equation (1) is applied to correct viscosity with temperature:

$$\ln\eta = \frac{E_a}{RT} + \ln\eta_\infty \cdots\cdots\cdots\cdots(1)$$

The R-square of viscosity fitting for $[BMIM]BF_4$ and $[BEIM]BF_4$ is 0.997 and 0.998, respectively. The equation is able to correct the experiment data well. The summaries of E_a and η_0 are listed in **Table 1**.

Table 1 Summaries of E_a and η_0 for pure $[BMIM]BF_4$ and $[BEIM]BF_4$

	characteristic parameter	
	$E_a/(kJ\cdot mol^{-1})$	$\eta_0/(\cdot 10^2 cP)$
$[BMIM]BF_4$, Harris's work [5]	33.53	0.14
$[BMIM]BF_4$, this work	34.15	0.0079
$[BEIM]BF_4$, this work	32.62	0.015

3.3. *CO_2 absorption*

In this work, the CO_2 absorption by RTILs and some binary systems was studied at the temperature of 298 K under atmospheric pressure. The absorption curves of the pure RTILs are shown in **Figure 3**. CO_2 equilibrium solubility of $[BMIM]BF_4$ and $[BEIM]BF_4$ at 298 K was 0.039 and 0.049 mol $CO_2\cdot$mol ILs^{-1}, respectively. Equilibrium was reached within 10 minutes and $[BEIM]BF_4$ showed higher equilibrium CO_2 solubility than $[BMIM]BF_4$.

However, both RTILs revealed low physical CO_2 absorption without the amines. To improve CO_2 absorption, 30% mass fraction of MEA and MDEA was blended with the RTILs, respectively. It was found that CO_2 absorption could be greatly improved by these added amines and the CO_2 equilibrium could be achieved within 120 minutes. Because the amines dominate the absorption of CO_2, the result was expressed using the units of mol CO_2•mol amine^{-1}. Four CO_2 absorption curves are shown in **Figure 4**. The equilibrium CO_2 loadings of MEA-[BMIM]BF_4 and MEA-[BEIM]BF_4 at 298 K were 0.494 and 0.515 mol CO_2•mol amine^{-1}, respectively. Chemical reaction between MEA and CO_2 was confirmed by the detection of a new resonance at 165.08 ppm using ^{13}C NMR technology. Two MEA molecules combined with one molecule of CO_2 to form carbamate without the presence of water. The CO_2 absorption of MDEA-[BMIM]BF_4 and MDEA-[BEIM]BF_4 was accomplished within 80 minutes and the equilibrium CO_2 loadings at 298 K were 0.214 and 0.334 mol CO_2•mol amine^{-1}, respectively, which was relatively lower than the MEA-RTILs systems. It has to be pointed out that for the MDEA-RTILs systems, no additional peak could be detected after CO_2 absorption. This means that only physical absorption exists in this mixed system. The CO_2 molecule is stabilized in the absorbent by van der Waals force, which is much weaker than the force of chemical bonding. Consequently, regenerating the absorbent requires less heat compared with the system where chemical reaction happens. Furthermore, amine-[BEIM]BF_4 performs better than amine-[BMIM]BF_4 on CO_2 absorption. It can be concluded that RTILs with higher CO_2 capacity are able to enhance the amine absorption of CO_2 as well.

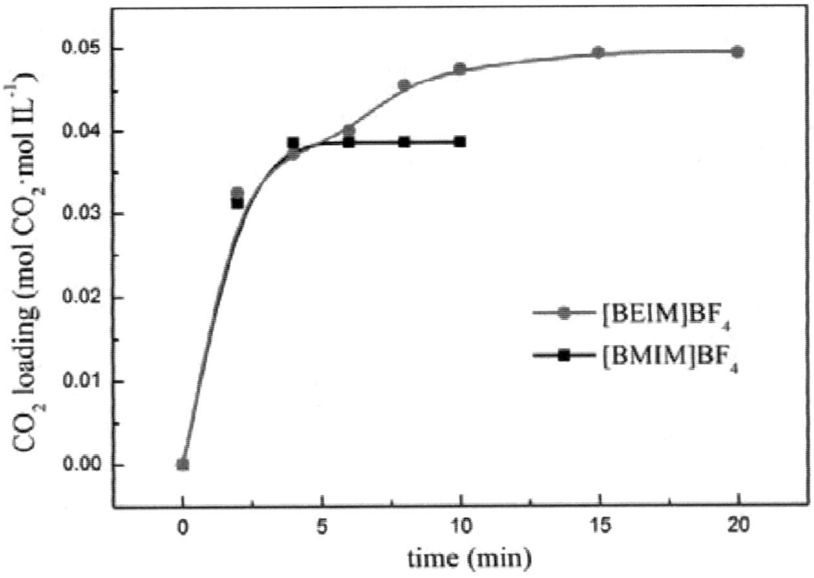

Figure 3 The CO_2 loading of pure RTILs versus time at 298 K

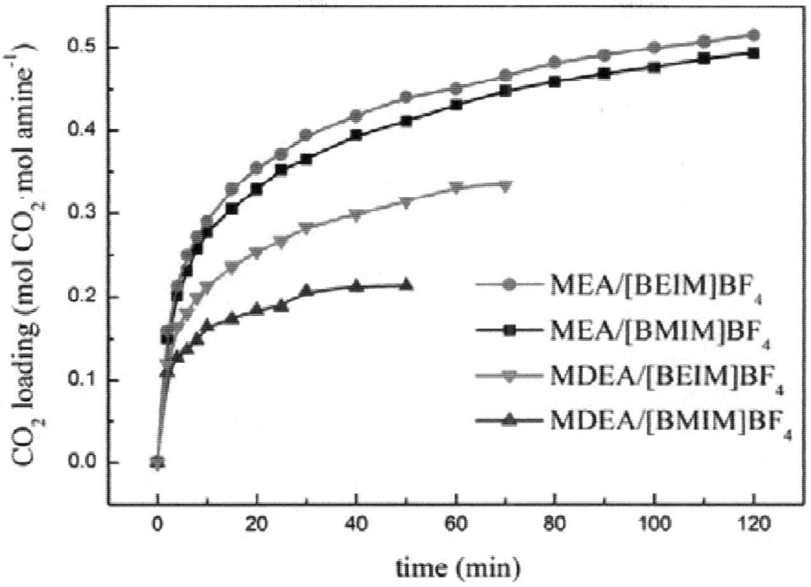

Figure 4 The CO_2 loading of mixed system with amine mass fraction of 30% at 298 K

3.4. Viscosity of mixed system before and after CO_2 absorption

To further study the viscosities of the mixed MDEA-RTILs systems, approximately 1 mL of each absorbent was exposed to a pure CO_2 atmosphere for 180 minutes. It has to be pointed out that during the CO_2 absorption using amine-RTILs mixed system, MEA was found to precipitate out in the form of carbamate after reacting with CO_2. The viscosity of MEA-RTILs system was thus unable to be measured. MDEA was fully dissolved into the RTILs before and after CO_2 absorption. Thus the viscosities of the MDEA-RTIL systems at 298 K were measured and are summarized in **Table 2**. Results show that the viscosity of MDEA-[BMIM]BF_4 increased from 49.46 cP to 63.40 cP after CO_2 absorption while for MDEA-[BEIM]BF_4, it increased from 49.79 cP to 77.77 cP. It can be concluded that the viscosity of both absorbents increased slightly after CO_2 absorption. In the case of [BEIM]BF_4, it showed higher viscosity than [BMIM]BF_4, the MDEA-[BEIM]BF_4 system revealed higher viscosity than MDEA-[BMIM]BF_4 before and after CO_2 absorption as well.

Table 2 The viscosity of mixed MDEA-RTILs system before and after CO_2 absorption at 298 K

	Before CO_2 absorption	After CO_2 absorption
	η(cP)	η(cP)
MDEA/[BMIM]BF_4	49.46 ± 1.02	63.40 ± 1.23
MDEA/[BEIM]BF_4	49.79 ± 1.02	77.77 ± 1.61

4. Conclusion

In this work, the thermal decomposition temperature and viscosity of ionic liquids [BMIM]BF_4 and [BEIM]BF_4 was measured. Both ionic liquids showed good thermal stability. [BEIM]BF_4 was found to be more stable than [BMIM]BF_4 due to its longer carbon chain. Meanwhile, the viscosity of [BEIM]BF_4 was slightly higher than that of [BMIM]BF_4. The absorption of CO_2 by amine-RITLs mixed systems was investigated as well. The viscosity of MDEA-RTILs was measured before and after CO_2 absorption. Results show that MDEA-[BEIM]BF_4 proves to be a potential binary system for CO_2 capture. The mixed system is thermally stable, able to capture CO_2 quickly and shows relatively low viscosity even after CO_2 absorption. Although MEA-RTILs show higher CO_2 capacity than MDEA-RTILs, it has to be pointed out that only physical absorption occurs in MDEA-RTILs systems, which means the mixed system can readily be regenerated by heating to strip out CO_2. Consequently, MDEA-[BEIM]BF_4 is considered to be an energy saving absorbent.

Acknowledgements

Financial support from the National Natural Science Foundation of China (Nos. 21476064, 21376067 and U1362112), National Key Technology R&D Program (Nos. 2012BAC26B01 and2014BAC18B04), Innovative Research Team Development Plan-Ministry of Education of China (No. IRT1238), Shanxi Yanchang Petroleum (Group) Co., LTD, Specialized Research Fund for the Doctoral Program of Higher Education (No. 20130161110025),and China's State 'Project 985' in Hunan University-Novel Technology Research and Development for CO_2 Capture as well as Hunan University to the Joint International Center for CO_2 Capture and Storage (iCCS) is gratefully acknowledged.

References

1. Joan F. Brennecke, Maginn EJ. Ionic Liquids: Innovative Fluids for Chemical Processing. *AIChE Journal.* 2001;47(11):2384-2389.
2. Zhang Y, Zhang S, Lu X, Zhou Q, Fan W, Zhang X. Dual amino-functionalised phosphonium ionic liquids for CO2 capture. *Chemistry.* 2009;15(12):3003-3011.
3. Eleanor D. Bates, Rebecca D. Mayton, Ioanna Ntai, James H. Davis. CO_2 Capture by a Task-Specific Ionic Liquid. *J. Am. Chem. Soc.* 2002;124(6):926-927.
4. Seo S, Quiroz-Guzman M, DeSilva MA, et al. Chemically tunable ionic liquids with aprotic heterocyclic anion (AHA) for CO_2 capture. *The journal of physical chemistry. B.* May 29 2014;118(21):5740-5751.
5. Harris KR, Kanakubo M, Woolf LA. Temperature and Pressure Dependence of the Viscosity of the Ionic Liquid 1-Butyl-3-methylimidazolium Tetrafluoroborate: Viscosity and Density Relationships in Ionic Liquids. *J. Chem. Eng. Data.* 2007;52:2425-2430.

COMPARATIVE STUDY ON LOW AND HIGH PRESSURE CO2 ADSORPTION CAPACITY OF ORGANIC MATERIALS

Mert Atilhan and Ruh Ullah*
Chemical Engineering Department, Qatar University, Doha

Santiago Aparicio
Department of Chemistry, University of Burgos, Burgos, Spain

Hasmukh Patel
Northwestern University, Evanston, USA

*Cafer T. Yavuz**
Department of Chemistry, KAIST, Daejeon, Republic of Korea

Abstract

Application of solid sorbents for CO_2 capture and separation has been under consideration since last decades owing to the promising feature of repeatable use, low cost of regeneration and thermo physical stability. However, the capturing capacity, selectivity and the overall performance of adsorbents at different temperatures and pressures still require further investigation to make these materials capable of replacing the mono-ethanol system. Here in we report synthesis, characterization and application of covalent organic polymers. Capturing capacity of COPs at low temperature and pressure has been experimented for two COPs materials (namely COP-9, COP-10) for CO_2 and N_2 adsorption. Materials were also characterized with BET and TGA.

1. INTRODUCTION

Till date cost of mono-ethanol amine regeneration and reduction in the electricity production are still among the major challenges of the plants using fossil fuels as an ultimate source for power generation. The demand of more power production due to the technological development for the improvement of life style is increasing world-wide which in turn enhances the usage of coal and subsequently rising CO_2 emission and concentration in the environment [1]. The renewal energy sources such as solar power, wind power, nuclear power and bio fuel power are still in the research and initial stages and consequently the world will have to depend mostly on fossil fuels for next few decades. It has been estimated that on average a single power plant using fossil fuels as power production source emit at least 3.5 million tons of CO_2 per year, whereas the number of such power generation industries increases globally. Cement industries and transportation, which are the second and third major sources of CO_2 emission and environment pollution are also multiplying every year and accordingly enhancing the concentration of CO_2 in atmosphere. Research has shown that the content of CO_2 in atmosphere may exceed 500 ppm by 2020 if not handle properly [2, 3]. This alarming increase in CO_2 emission has severely threatened life on the earth and issues like global warming, severe weather conditions spread of very dangerous disease such as respiratory illnesses and asthma are all due to the excessive and uncontrollable emission of CO_2 [4-9]. The new energy production plants using fossil fuels however now changes to oxy-combustion fuel system and integrated gasification combine cycle system where CO_2 can be effectively captured using advanced technologies as well as MEA[4, 10]. Solid sorbents possess various advantages such as simple regeneration, repeated applications, low manufacturing cost, no extra-cost on electricity production, environmental and economic viability, tunable capturing capacity and changeable selectivity. Solid sorbents mainly have applications in pre-combustion system like IGCC, where, CO_2 can be separated from syngas after water shift in the process[10]. However, this requires the adsorbents to be physically strong which can hold high pressure and high temperature unlike the post-combustion process where CO_2 can be separated at atmospheric pressure. In addition to the thermal and physical stability, solid sorbent needs to be extremely selective for separation of CO_2 from the stream, which have H_2 as fuels, should have high surface area and large pore volume and better affinity for CO_2. Various solid sorbents such as activated carbon, metal organic framework, zeolites, SBA-15, nanoclays, and many organic polymers have been investigated both for low and high-pressure CO_2 adsorption and separation. Attempts have been made to increase the CO_2 capturing capacity of solid materials by inserting various amine function groups within the porous structure of adsorbents[11]. As an example SBA-15 has been impregnated with different function groups and tested for low and high pressure CO_2 adsorption. Results indicated that amine modification have very limited effect on the capturing capacity at high pressure. Here in we report synthesis and characterization of covalent organic polymers, which possess internal nitrogen functionality and can be assumed to be the more favorable compounds as long as the affinity of CO_2 [12-15]. We prepare two COPs materials using different building block monomers and linked with two different linkers to investigate the effect of different functionality on the CO_2 and N_2 adsorption capacity at low and high pressure under different temperatures.

2. EXPERIMENTAL

2.1. Materials and Synthesis procedure

N,N-Diisopropylethylamine, 1,3,5-benzene tricarbonyltrichloride, 2,5-Dimethylpiperazine, cyanuric chloride, 2-imidazolidinone, dimethylacetamide, and tetrahydrofuran (THF), purchased from Sigma-Aldric, however, dioxane and ethanol were obtained from TCI, Japan. All the chemicals were used without any further modification and/or dilution. COP-9 and COP-10 were prepared through the following procedures. In a typical procedure N,N-Diisopropylethylamine (1.9 mL, 10.8 mmol) was added to trans-2,5-Dimethylpiperazine (0.48 g, 4.2 mmol) dissolved in THF at 0-5 ^{0}C. Cyanuric chloride (0.5 g, 2.7 mmol) dissolved in THF was added dropwise to the above solution with continuous stirring at 0-5 °C in N_2 environment. The solution was stirred at 0-5 °C for 1 h. Subsequently, the temperature of the reaction mixture increased to 25-30 °C and stirred for 2 h, followed by heating at 68-70 °C for 21 h. The precipitate was washed with THF, dioxane and ethanol (three times each). Finally, the precipitate was dried at room temperature for 2 h and 80-90 ^{0}C for 5 h under vacuum and labeled as COP-9. N,N-Diisopropylethylamine (2.1 mL, 12 mmol) was added to 2-imidazolidinone (0.4 g, 4.65 mmol) dissolved in Dimethylacetamide (DMAc, 35 mL) at 25 ^{0}C. 1,3,5-Benzenetricarbonyl trichloride (0.8 g, 3 mmol) dissolved in DMAc (25 mL) was added dropwise to the above solution with continuous stirring at 25 ^{0}C. The solution was stirred at 25 °C for 2 h. Subsequently, the temperature of the reaction mixture increased to 80 ^{0}C and stirred for 16 h. The reaction mixture cooled to room temperature and 250 mL of de-ionized water added which resulted precipitate. The precipitate was washed with water and ethanol (three times each) and it was dried at room temperature for 2 h and 80-90 ^{0}C for 5 h under vacuum and labeled as COP-10.

2.2. Characterization and Adsorption Setup

Thermal stability of COP-9 and COP-10 was conducted with Perkin Elmer Pyris 6 (Thermogravimetric Analysis) TGA model 4000 system. TGA system was properly balanced and calibrated according to the purging gas (N_2), crucible and sample holder stem. TGA machine was equipped with chiller from Huber model Pilot One to automatically maintain the temperature of the system and N_2 was used as the heating environment, where the materials were heated directly from 30 °C to 800 °C at the rate of 1° per minute. Physical parameters such as pore volume and surface area of materials were estimated using BET analyzer Model micrometrics ASAP 2420. The solid materials were degased for at least 12 hours prior to the BET measurement to avoid any blockage of pore due to the water.

PCT-Pro was used for the low-pressure adsorption process, where each pressure step was 0.1 bars up to the maximum of 1 bar. Here only CO_2 and N_2 were calculated at three different temperatures of 298 K, 308 K and 323 K [16]. Two gas cylinders (high purity grade CO_2, and N_2) were connected to gas dosing system (GDS) via external pump to pressurize the required gas into the measuring cell. The maximum set pressure (1 bar) was applied stepwise by increasing the pressure gradually from 0 bar up to 1 bar using each step of 0.1 bar. After reaching the maximum set pressure points, pressure of the system was gradually reduced exactly with the same measurement points starting from the high pressure and reaching to the last evacuation point to complete the desorption process. The system is fully automated and the pressure goes to next higher point after completing the previous measurement point.

3. RESULTS AND DISCUSSIONS

Thermal gravimetric analysis of both covalent organic polymers COP-9 and COP-10 (shown in figurer 1) reveals that COP-9 is thermally more stable than COP-10, since; the former is almost stable up to 400 °C. TGA further indicates that COP-10 is thermally degraded in three steps, while COP-9 decomposes in two steps. The initial degradation of COP-10 starts at around 75 °C which can be associated to the evaporation of stored water within the structure, 2nd degradation of COP-10 starts at around 200 °C, while the third one starts at 400 °C. Contrary, COP-9 is about completely stable up to 400 °C however, beyond this point it decomposed quickly and in another 100 °C increase in temperature, almost 50 % of the material losses. TGA analysis confirmed that COP-9 is much stable than COP-10, however, it degrades gradually while a sharp degradation in COP-9 can be observed after 400 °C.

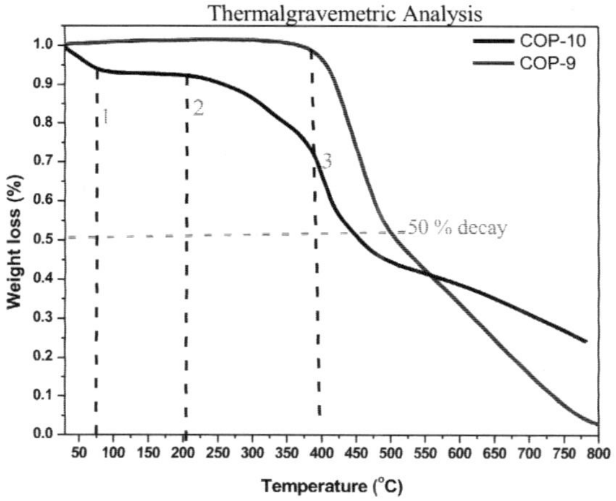

Figure 1. Thermal-gravimetric analysis of COP-9 and COP-10

Physical properties and proposed chemical structure of COPs are given in table 1 below which shows that COP-9 has smaller pore volume than COP-10, however, the former has larger surface area and lower tapped bulk density. It has been assumed that low pressure gases uptake by solid materials is mostly governed by the physical parameters such as pore volume and surface area[17]. On the other hand adsorption capacity at high pressure does not necessarily depend upon the porosity of the structure, but most often varies with affinity of the materials. Thus the material with lower porosity and lower surface area may have higher affinity at high pressure and therefore will capture more CO_2 at elevated pressure as compared to those adsorbents which have larger physical parameters [18, 19].

Table 1. Physical Parameters and structure of covalent organic polymers (COPs)

Sample	**Structure**	**Surface area (m^2/g)**		**Pore volume (cm^3/g)**	**Density (cm^3/g)**	
		BET	**Langmuir**		**Bulk**	**Tapped**
COP-9	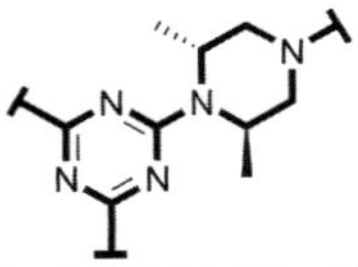	146.4	204.3	0.244	0.181	0.202
COP-10	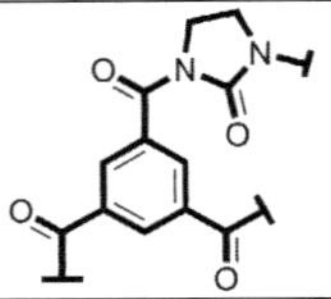	84.5	114.6	0.634	0.129	0.258

Figure 2, shows the low pressure CO_2 and N_2 adsorption of COP-9 (c, d) and COP-10 (a, b), which clearly indicates that due to the larger pore volume of COP-10 it captures comparatively more CO_2 at lower pressure of 1 bar. Xiang et al reported that COP-1 possess the highest surface area captured more N_2 and CH_4; however, COP-2 with larger pore volume captured more CO_2 and H_2 than all other organic polymers[20]. On the other hand COP-10 uptakes very low quantity of N_2 at 1 bar and three different temperatures as compared to COP-9. It must be noted that selectivity of COP-10 for CO_2:N_2 at low pressure is much higher than that of COP-9. Additionally as given in table 2 selectivity of COP-10 for CO_2:N_2:H_2 is also better than that of COP-9 at 308 K, while at all other temperatures selectivity of both polymers is similar.

Additionally, the selectivity of COP-10 at atmospheric pressure is increasing with rising temperature and becomes infinity at temperature above than 308 K. As shown in figure 2b COP-10 is totally inert for N_2 at higher temperatures, although it captures a bit smaller quantity of N_2 at standard temperature and pressure. Figure 2c and 2d indicates that COP-9 uptake less CO_2 than COP-10, but its N_2 uptake is comparatively larger than that of COP-10. Table 2 indicates that the overall performance of COP-10 at atmospheric pressure and different temperatures is comparatively better than COP-9, since the former adsorb more CO_2 and less N_2 which makes this material favorable for the pre-combustion CO_2 capture and separation.

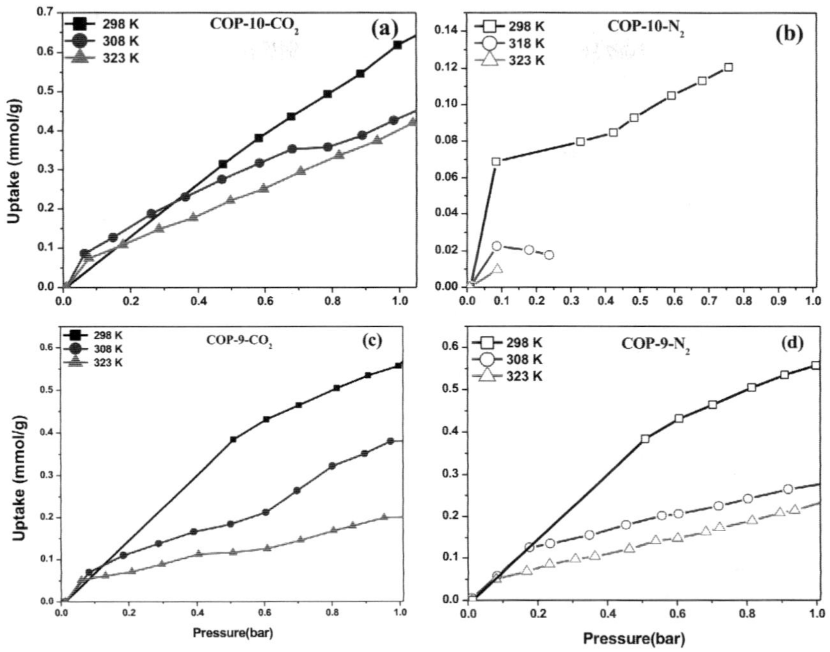

Figure 2. CO_2 adsorption of COP-9 (c & d) and COP-10 (a & b) at maximum of 1 bar and three different temperatures obtained with PCT-Pro.

It can be concluded that two covalently bonded organic polymers COP-9 and COP-10 were prepared from two different core building block monomers and linked with different linking agents. Both COPs material were tested for low and high pressure CO_2 and N_2 adsorption at three different temperatures. It was found that COP-10 has almost three times larger pore volume than COP-9 and therefore captures slightly larger CO_2 than COP-9 at atmospheric pressure and room temperature. Additionally, COP-10 uptakes significantly lower quantity of N_2 at atmospheric pressure and room temperature while it becomes almost N_2 phobic at high temperature and lower pressure, thus resulting in very good selectivity under these conditions.

In addition to the effect of pore volume and surface area on the gases uptake capacity of solid sorbents, affinity of the gas molecules toward the attractive sites also contribute to the adsorption capacity of materials. It can be assumed that, COP-9 may have more interacting sites than COP-10 since the core monomer of COP-9 has three nitrogen nodes whereas the core monomer of COP-10 has less nitrogen species. On the other hand COP-10 may have more oxygen OH functionality that may have less affinity for CO_2 than the N functionality. Thus it can be argued that, although COP-9 has lower pore volume but it has more and stronger attractive sites for CO_2 than COP-10, therefore, it captures comparatively more CO_2.

ACKNOWLEDGMENTS

This paper was made possible by the support of Qatar National Research Fund, National Priorities Research Program grant (NPRP# 5-499-1-088) and (NPRP# 6-330-2-140). The statements made herein are solely the responsibility of the authors.

REFERENCES

1. Aaron, D. and C. Tsouris, *Separation of CO2 from Flue Gas: A Review.* Separation Science and Technology, 2005. **40**(1-3): p. 321-348.
2. Boot-Handford, M.E., et al., *Carbon capture and storage update.* Energy & Environmental Science, 2014. **7**(1): p. 130-189.
3. Chu, S. and A. Majumdar, *Opportunities and challenges for a sustainable energy future.* Nature, 2012. **488**(7411): p. 294-303.
4. Ahmed, U., et al., *IGCC process intensification for simultaneous power generation and CO2 capture.* Chemical Engineering and Processing: Process Intensification, 2016. **101**: p. 72-86.
5. Awadallah-F, A. and S. Al-Muhtaseb, *Carbon dioxide sequestration and methane removal from exhaust gases using resorcinol–formaldehyde activated carbon xerogel.* Adsorption, 2013. **19**(5): p. 967-977.
6. Berger, A.H. and A.S. Bhown, *Selection of Optimal Solid Sorbents for CO2 Capture Based on Gas Phase CO2 composition.* Energy Procedia, 2014. **63**(0): p. 2092-2099.
7. Cavka, J.H., et al., *High Pressure Adsorption of CO2 and CH4 on Zr-MOFs.* Industrial & Engineering Chemistry Research, 2014. **53**(40): p. 15500-15507.
8. Chen, J., L.S. Loo, and K. Wang, *High-Pressure CO2 Adsorption on a Polymer-Derived Carbon Molecular Sieve.* Journal of Chemical & Engineering Data, 2008. **53**(1): p. 2-4.
9. Gonzalez-Salazar, M.A., et al., *Comparison of Current and Advanced Post-Combustion CO2 Capture Technologies for Power Plant Applications.* Energy Procedia, 2012. **23**: p. 3-14.
10. Ekström, C., et al., *Techno-Economic Evaluations and Benchmarking of Pre-combustion CO2 Capture and Oxy-fuel Processes Developed in the European ENCAP Project.* Energy Procedia, 2009. **1**(1): p. 4233-4240.
11. Hedin, N., et al., *Adsorbents for the post-combustion capture of CO2 using rapid temperature swing or vacuum swing adsorption.* Applied Energy, 2013. **104**(0): p. 418-433.
12. Ullah, R., et al., *Insights of CO2 adsorption performance of amine impregnated mesoporous silica (SBA-15) at wide range pressure and temperature conditions.* International Journal of Greenhouse Gas Control, 2015. **43**: p. 22-32.
13. McKeown, N.B. and P.M. Budd, *Polymers of intrinsic microporosity (PIMs): organic materials for membrane separations, heterogeneous catalysis and hydrogen storage.* Chemical Society Reviews, 2006. **35**(8): p. 675-683.
14. Modak, A. and A. Bhaumik, *Porous carbon derived via KOH activation of a hypercrosslinked porous organic polymer for efficient CO2, CH4, H2 adsorptions and high CO2/N2 selectivity.* Journal of Solid State Chemistry, 2015. **232**: p. 157-162.
15. Plaza, M.G., et al., *CO2 capture by adsorption with nitrogen enriched carbons.* Fuel, 2007. **86**(14): p. 2204-2212.
16. Deniz, E., et al., *A combined computational and experimental study of high pressure and supercritical CO2 adsorption on Basolite MOFs.* Microporous and Mesoporous Materials, 2013. **175**: p. 34-42.
17. Belmabkhout, Y., V. Guillerm, and M. Eddaoudi, *Low concentration CO2 capture using physical adsorbents: Are metal–organic frameworks becoming the new benchmark materials?* Chemical Engineering Journal, 2016. **296**: p. 386-397.
18. Veneman, R., et al., *Adsorption of CO2 and H2O on supported amine sorbents.* Energy Procedia, 2014. **63**(0): p. 2336-2345.
19. Zulfiqar, S., M.I. Sarwar, and C.T. Yavuz, *Melamine based porous organic amide polymers for CO2 capture.* RSC Advances, 2014. **4**(94): p. 52263-52269.

20. Xiang, Z., et al., *Covalent-organic polymers for carbon dioxide capture.* Journal of Materials Chemistry, 2012. **22**(42): p. 22663-22669.
21. Mason, C.R., et al., *Enhancement of CO2 Affinity in a Polymer of Intrinsic Microporosity by Amine Modification.* Macromolecules, 2014. **47**(3): p. 1021-1029.
22. McKeown, N.B. and P.M. Budd, *Exploitation of Intrinsic Microporosity in Polymer-Based Materials.* Macromolecules, 2010. **43**(12): p. 5163-5176.
23. Ullah, R., et al., *Synthesis, characterization and evaluation of porous polybenzimidazole materials for CO2 adsorption at high pressures.* Adsorption, 2016: p. 1-14.
24. Patel, H.A., et al., *Highly Stable Nanoporous Sulfur-Bridged Covalent Organic Polymers for Carbon Dioxide Removal.* Advanced Functional Materials, 2013. **23**(18): p. 2270-2276.
25. Patel, H.A., et al., *High capacity carbon dioxide adsorption by inexpensive covalent organic polymers.* Journal of Materials Chemistry, 2012. **22**(17): p. 8431-8437.

466434 Carbonic Anhydrase Enhanced Carbon Capture: Kinetic Measurements and Pilot Plant Trials

Abstract:

Carbonic Anhydrase Enhanced Carbon Capture: Kinetic Measurements and Pilot Plant Trials

Arne Gladis[1], Maria T Gundersen[2], Philip L Fosbøl[1], John M Woodley[2], Nicolas von Solms[1]*

1) Center for Energy Resources Engineering, Department of Chemical and Biochemical Engineering, Technical University Denmark, Kgs. Lynby 2800

2) CAPEC-PROCESS Center, Department of Chemical and Biochemical Engineering, Technical University Denmark, Kgs. Lyngby 2800

Keywords:

Wetted wall column, carbonic anhydrase, carbon capture, enzyme kinetics, amine, carbonate, absorption, carbon dioxide

Abstract:

In this study the effect of carbonic anhydrase addition on the absorption of CO_2 was investigated in a wetted wall column apparatus. Four different solvents: MEA (a primary amine), AMP (a sterically hindered primary amine), MDEA (a tertiary amine) and K_2CO_3 a carbonate salt solution were tested in concentrations from 5 to 50 wt%. Necessary mass transfer parameters such as liquid side mass transfer coefficient and solvent and enzyme reaction rates

were determined in a temperature range from 298 to 328 K and benchmarked to a 30 wt% MEA solution.

The study reveals that the addition of the enzyme carbonic anhydrase (CA) dramatically increases the liquid side mass transfer coefficient for 30 wt% MDEA and 15 wt% K_2CO_3. 30 wt% AMP has a moderate increase whereas 30 wt% MEA was unchanged. The results confirm that bicarbonate forming solvent which do not produce carbamate benefit from CA. The results reveal the impact of temperature in relation to CA. A temperature increase resulted in lower liquid side mass transfer rate for 30 wt% MDEA and 15 wt% K_2CO_3 but in higher rate for 30 wt% AMP. The overall first order enzyme reaction rate (s^{-1}) was linearly dependent on enzyme concentration for 30 wt% MDEA and 15 wt% K_2CO_3 at 313 K. The derived enzymatic reaction rate constant k_{enz} ($m^3\ kg^{-1}\ s^{-1}$) for 15 wt% K_2CO_3 at 313 K was about 9 times higher than for 30 wt% MDEA and 10 times higher than for 30 wt% AMP. Temperature and concentration did not observably influence the enzymatic rate constant in the concentration range of 5 to 15 wt% K_2CO_3. The higher solvent concentration only led to a slightly higher reaction rate. A solution with 20 wt% K_2CO_3 had almost 3 times higher enzyme reaction rate compared to 15 wt% at 298 K and increased with temperature to almost 5 times faster at 328 K. The enzymatic reaction rate for MDEA decreased with both temperature and solvent concentration from 15 to 30 wt%. An increase to 50 wt% resulted in a decrease in reaction rate due to less water present.

Pilot plant campaigns were carried out for different solvents and conditions and the results were successfully modelled using intrinsic data obtained from the wetted-wall column experiments

458151 In-Vitro Dosimetry Model for Toxicity Ranking of Metal Oxide Nanoparticles

Abstract:

There is a debate in the nanosafety community regarding whether cellular toxicity of engineered nanoparticles (NPs) should be ranked on the basis of delivered dose (i.e., NP mass settled per suspension volume) or administered dose (i.e., initial NP mass concentration). Calculation of delivered dose with sufficient confidence for in-vitro toxicity ranking of NPs requires quantification of NP sedimentation considering fundamental suspension properties of NPs in cell culture media. Accordingly, we developed a first principles â€œparticles in a boxâ€ sedimentation model, which accounts for the particle size distribution, fractal dimension and permeability of agglomerated NPs. The developed sedimentation model was evaluated against both external and our own experimental sedimentation data for metal-oxide NPs, demonstrating significantly improved estimates of settled NP mass over other simplified sedimentation calculations with the assumption of impermeable and/or monodispersed NP agglomerates. The model was then applied to calculate the delivered NP doses in different cell culture media for the dose-response analyses of seven metal oxide NPs that were identified toxic from a library of twenty-four NPs in our previous studies. Hierarchical hazard ranking of the identified seven toxic NPs, according to the their EC_{50}, was not measurably different when determined on the basis of either delivered or administered doses. In contrast, simplified sedimentation calculations based on the assumption of monodispersed impermeable NP agglomerates led to significant underestimation of the settled NPs mass and thus distorted toxicity ranking.

472232 Amorphous Silicon Dioxide Nanoparticle Interactions with Pulmonary Epithelial Cells with and without a Pre-Existing Protein Corona

Abstract:

Silicon dioxide nanoparticles are increasingly abundant in consumer products and industrial applications, which increases their potential exposure to humans. In addition, these nanoparticles are under investigation for biomedical applications within the body such as targeted drug delivery. In either case, upon entering the body, nanoparticles interact with biological fluids and are immediately coated with proteins. Ultimately, it is this protein layer that interacts with the bodyâ€™s cells and tissues, and thus dictates the subsequent fate of the nanoparticles in the body. In this work, we investigated the adsorption of an abundant protein in biological fluids, albumin, to silicon dioxide nanoparticle surfaces and the impact of this protein layer on subsequent nanoparticle uptake within human cell cultures.

A previous mechanistic study conducted in our group has indicated that pH plays a strong role in the protein affinity for the nanoparticle surface (*publication in preparation*). Using attenuated total reflectance Fourier transform infrared (ATR-FTIR) spectroscopy, we obtained a real-time adsorption profile for BSA on the nanoparticle surface. Protein coverage on the nanoparticle surface was quantified using thermogravimetric analysis. Results from these studies indicated that at the isoelectric point of the BSA-silicon dioxide nanoparticle complex, pH 3.7, protein coverage was highest. Whereas at physiological pH, pH 7.4, much lower adsorption was observed over the same time period.

Cellular responses to uncoated particles and particles coated at these two pHs, to model a low-coating and a high-coating of protein on the nanoparticle surface, were conducted .Two pulmonary epithelial cell lines, 16HBE14o- and A549, were chosen for these studies as models of the different regions of the lung (bronchioles and alveoli) where nanoparticles deposit. Cell viability was measured via the MTT assay, particle uptake through TEM, and inflammatory markers through ELISA. Results from these studies indicated that coated particles behaved differently than uncoated particles in every parameter examined and that uptake depended on the cell system studied.

Overall, exposure to amorphous silicon dioxide nanoparticles induced a cellular response in pulmonary epithelial cells. Since these particles impacted the two cell lines differently, the deposition location of particle in the lungs after inhalation is important for understanding their toxicity. Surface functionalization prior to cellular exposure also played a key role in the cellular response, further supporting the idea that the protein corona affects cellular recognition and responses.

466807 Physiochemical Properties of Nanoparticles Determine Their *in Vitro* Cytotoxicity

Abstract:

Solid nanoparticles have proven useful in overcoming many of the barriers to drug delivery. Additional development has seen the attachment of a myriad of surface coatings such as polyethylene glycol (PEG). This commonly used PEG layer masks particles from innate host defence mechanisms within the body and increases circulation half-life of the therapeutic upon injection. Although this approach is widely accepted, the underlying mechanism is not well understood. This study examined the effect of hydrophilic surface coatings on solid silica nanoparticles with their interactions with Chinese Hamster Ovary (CHO) cells. Specifically, we examined the effect of differing lengths of PEG on intracellular uptake and viability.

In order to elucidate the effects of particle surface characteristics, studies were conducted with PEG 2k, 5k and 20k, as well as aminated silica nanoparticles. Two nanoparticle sizes, 60 and 120 nm, were chosen to draw size effect comparisons between the particles. These particles are generally recognized as a safe material by the Food and Drug Administration and provided a uniform foundation via size distribution, surface characteristics and mild reaction chemistry that precluded any nanoparticle-mediated effects.

In vitro distribution of the silica nanoparticles was found to directly correlate with their surface coating. Pegylated particles were found to remain interspersed among the CHO cells, whereas amine coated particles were found to produce punctate regions of fluorescence on and around the cells. Confocal microscopy confirmed uptake over simple surface association of the nanoparticles. Similar trends were correlated with flow cytometry. It was found that all PEG coatings decreased particle uptake by the CHO cells in comparison to those with an amine coating. The aminated particles had the greatest effect on cell growth. The 60 nm particles yielded greater half maximal inhibitory concentrations (IC_{50}) values over the 120 nm particles. The IC_{50} values for 60 nm aminated, PEG 2k, 5k and 20k particles were found to be 0.42 ± 0.18, 2.58 ± 0.09, 1.75 ± 0.11 and 3.535 ± 0.15 mg/mL, respectively. The IC_{50} values for the 120 nm aminated, PEG 2k, 5k and 20k particles were found to be 0.59 ± 0.05, 1.435 ± 0.13, 1.57 ± 0.125 and 1.333 ± 0.17 mg/mL. Growth inhibition directly correlated with particle concentration regardless of surface character. Pegylated 60 nm particles showed the greatest differences among their PEG coatings whereas the 120 nm pegylated particles showed similar results among all PEG sizes.

This study found that size and surface coatings have a dramatic influence on the cell viability and uptake *in vitro*. In addition, these studies show there may be an interplay between both of these variable sets. This work has provided a basis of comparison that could be extended to various nanoparticle systems. Understanding the impact of widely used solid nanoparticle compositions, sizes, coatings and their combinations on mammalian systems is an important step in developing robust drug delivery vehicles.

471757 The Impact of Titanium Dioxide Nanoparticles on the Lysosome-Autophagy System and Cellular Clearance

Abstract:

Titanium dioxide (TiO_2) is "generally regarded as safe" by the US Food and Drug administration and is found in a wide variety of consumer products, from cosmetics and pharmaceuticals to food additives. Recent studies suggest that the cellular uptake of nano-sized TiO_2 may cause oxidative stress, lysosomal impairment, and inflammation. However, the impact of nano-sized TiO_2 on cell physiology remains unclear. To address this knowledge gap, we evaluated the cellular response to the uptake of anatase TiO_2 nanoparticles (15, 50, and 100 nm in diameter), with particular emphasis on the lysosome-autophagy system, which is the cell's first response to the uptake of a material perceived as foreign and potentially harmful. Following characterization of the TiO_2 nanoparticles, cytotoxicity analyses were conducted and each of the three TiO_2 nanoparticles were found to induce minimal toxicity at concentrations in the cell culture medium up to 1000 µg/mL. Induction of the lysosome-autophagy system was then tested by monitoring the subcellular location of the transcription factor EB, a master regulator of lysosome- and autophagy-related genes. We found TiO_2 nanoparticles to induce activation of the transcription factor EB at concentrations as low as 10 µg/mL and to induce autophagic vesicle formation. However, uptake of high concentrations of TiO_2 nanoparticles (100 µg/mL and above) was also found to prevent vesicle turnover. We hypothesized that this observed effect on vesicle turnover was due to lysosomal impairment. Therefore, we analyzed lysosomal hydrolase function and membrane stability following TiO_2 nanoparticle uptake and observed a correlation between lysosomal function and integrity and vesicle turnover. In conclusion, we established that cellular uptake of TiO_2 nanoparticles induces activation of the lysosome-autophagy system, and thus upregulation of cellular clearance, but results in a blockage of autophagic flux and degradation *via* impairment of the lysosome.

458173 Potential Impact of Sublethal Levels of Nanomaterials on Interactive Behavior of Environmental Bacteria

Abstract:

Although bactericidal effects of a wide range of nanomaterials on environmental bacteria have been extensively reported, little is known about the impact of nanomaterials at sublethal levels on environmental bacteria and their interactive behaviors, which is a critical gap toward a comprehensive understanding of the impacts of nanomaterials on microbial ecosystems. In this study, using model environmental bacteria, we report the impact of single-wall carbon nanotubes (SWCNTs) and nanoSilver (nAg) on cell-to-cell interaction through influencing the production of â €œpublic goodsâ€ or the production of quorum sensing (QS) signals. Sublethal level of SWCNTs (40 or 80 Âµg/ml) inhibited the production of pyoverdine (PVD) (an important metabolite that involves in interactive behavior of microbial communities, especially in iron-limiting environments) in planktonic cultures of *Pseudomonas aeruginosa*. The inhibitory effect has been confirmed at the mRNA level by transcriptional assay and quantitative polymerase chain reaction (qPCR) analysis. We further explored the effect of SWCNTs on mature biofilms of *P. aeruginosa*. Our results showed that the exposure to SWCNTs for up to 48 h did not significantly influence PVD production in cells encased in mature biofilms; however, the exposure to SWCNTs promoted cell detachment from the biofilms and decreased PVD production of the detached cells. To explore the influence of nanomaterials on quorum sensing, we used *P. syringae* and *Pantoea stewartii* as model organisms and quantified the influence of SWCNTs and nAg on the production of different acylated homoserine lactones (AHL) signals. Our results showed for the first time that differential effects could be exhibited by nanomaterials on different QS signals. Our findings advance current knowledge toward a comprehensive understanding of the ecotoxicity of nanomaterials.

459515 Review on Rheological Studies of Sludge from Different Sections of Municipal Wastewater Treatment Plants for Improved Process Performance

Abstract:

Wastewater treatment process generates large quantities of sludge from different operational steps. The rheological behaviour of sewage sludge at different steps of wastewater treatment plant (WWTP) is useful for the design, operation, optimization of polymer dosing, hydrodynamics and mixing in anaerobic digesters, and for efficient sludge management. Rheological characterization of sludge is also known to be an essential tool to optimize flow, mixing and other process parameters in WWTPs. In this review article, the impacts of different parameters like solid concentration, temperature, pH, floc particle size, mixing ratio, ageing and dose of sludge conditioning agents on the rheological behaviour of sludge from different parts of WWTPs were analysed. Raw primary sludge, excess activated sludge, thickened excess activated sludge, mixture of raw primary and thickened excess activated sludge (mixed sludge), digested sludge, and biosolid rheological and flow behaviour under the influence of different operating parameters and the impacts on process performance were reviewed in this study. Solid concentration, temperature, primary to thickened excess activated sludge mixing ratio, and sludge age were found to affect the viscosity, yield stress, flow index and flow consistency of mixed sludge., viscosity also showed a decreasing trend with decreasing total solid concentration and percentage of thickened excess activated sludge in the mixture. In contrast, yield stress and viscosity generally showed reduction with increasing temperature. The effect of dose of conditioning agents on the rheological behaviour of thickening of excess activated sludge and digested sludge was also addressed in this review. The relationship between activation energy and viscosity for varying temperature has also been discussed. The rheology of digested sludge affects the flow hydrodynamics, dewaterability and the polymer consumption in WWTPs. The rheological characteristics of digested sludge are highly dependent on changes in total solid concentration, temperature and polymer dose. The rheological characteristics of digested sludge during polymer conditioning and

flocculation process was significantly affected by temperature and solid concentration; hence, polymer dose can be reduced by operating the dewatering process at optimum temperature condition and varying the polymer dose as a function of the total solid concentration and viscosity of the digested sludge. Volatile solid content of digested sludge fed to centrifuge has inverse relationship with total solid content of biosolid or biosolid capture rate. Hence, improving the volatile solid and COD removal/reduction by enhancing the performance of anaerobic digesters or by using other digested sludge conditioning methods will significantly improve the flow behaviour of digested sludge and dewatering performance. Investigating the relationship between rheological and physico-chemical characteristics of sludge can also serve as a tool to optimize essential process parameters. The applicability and practical significance of various rheological models such as Bingham, Power Law (Ostwald), Herschel-Bulkley, Casson, Sisko, Careau, and Cross models to represent experimental rheological characteristic of different sludge types was also investigated in this review.

460652 Carbon-Based Rods with Anomalous Water Expulsion at High Humidity

Abstract: Three water adsorption-desorption mechanisms are common in inorganic materials: chemisorption, which can lead to modification of the first coordination sphere; simple adsorption, which is reversible; and condensation, which is irreversible. Regardless of the sorption mechanism, all known materials exhibit an isotherm in which the quantity of water adsorbed increases with an increase in relative humidity. Here, we show a unique water adsorption phenomenon of carbon-based rods (PNNL-Carbon Rods; PCR). The rods adsorb water at low humidity and spontaneously expel about half the adsorbed water at high relative humidity. The water expulsion is attributed to the interfacial forces between confined rod surfaces. At wide rod spacing, a monolayer of water can form on the surface of the rods, which subsequently leads to condensation in the confined space between adjacent rods. As the relative humidity increases, adjacent rods (confining surfaces) in the bundles are drawn closer together via capillary forces. At high relative humidity, the size of the confining surfaces decreas to a critical distance, and water condensed inside the confined space is released as a vapour. The process is reversible, as indicated by over three cycles of water sorption-desorption. The rods synthesised under different reaction conditions (eg., reaction temperature, amount of $FeCl_3$ catalyst used, and $FeCl_3$ source) have different morphology, size and exhibited different RH for the onset of evaporation and amount of water expelled (Figure 1). PCR were synthesized at relatively low temperature compared to temperatures used for carbon nanotube synthesis (200 °C vs 600 °C, respectively). We will present the physical, thermodynamic and spectroscopic evidence for this behavior and discuss its potential applications.

Figure 1: (A) Water sorption isotherms of carbon-based rods synthesized using 0.1 M (PCR-1; synthesized at 220 °C) and 0.2 M $FeCl_3$ (PCR-2; at 220 °C, PCR-3; at 230 °C), (B-D) SEM image of rods. (E) Schematic of proposed water expulsion mechanism.

- Abstract Text

465679 An MINLP Model for Sustainable Water Management in Macroscopic Systems Under Economic Penalty Scenarios

Abstract:

In a sustainable system, the amount of water available from the natural sources in a watershed would be sufficient to meet the demands of the users without compromising their future availability. Furthermore, the amount and type of pollutants discharged over the water bodies would be completely degraded by the natural flora and fauna present in the aquatic ecosystems. However, this is not a common scenario.

Optimal resources distribution and wastewater treatment in a system (which involves a complex network of usage and disposal of water) cannot always be achieved just by imposing environmental regulations. In most cases, a compromise between the economic and environmental aspects has to be taken into account. This is not a simple task, because the needed investment for water treatment sometimes becomes prohibitive, especially when the users (e.g. domestic, industrial, agricultural) have many economic concerns regarding budgets, profitability of their products, market share, etc. An efficient strategy is still needed to mitigate harmful consequences on the environment.

Literature reports several efforts to deal with pollution over water bodies from a trading and penalty perspectives. These approaches are based on an ideal scenario, where charging environmental violators with fines and exchanging pollution credits among various sources achieve lower global costs while environmental constraints are met. However, none of those approaches has considered a macroscopic system which also takes into account the sustainability of water resources within a watershed and its surroundings.

This paper proposes a model to account for the sustainability of a macroscopic system through resources management, including the evaluation of alternative technologies for fresh water consumption reduction, and a distributed treatment system to treat wastewater from different sources; however, our approach also considers the possibility of penalty scenarios due to the violations of environmental regulations. The aim is to assess the implications of the penalties over the costs and the environmental goals of the problem.

Our approach results in an MINLP multi-period model which has been solved through the GAMS® modeling environment. A case study with different scenarios shows the scope of the proposed approach and the results show the significance of the results; the solution of the optimization problem allows finding the minimum investment to meet the desired environmental goals with respect to freshwater consumption and pollution over water bodies. This investment consists of the number, size and location of rainwater storage devices, the number, size, type and location of treatment plants in the distributed wastewater treatment system, the total amount of recycled wastewater and the total amount of fines charged to the users for violation of environmental constraints regarding pollution over water bodies.

469744 Nitrogen Doped Magnetic Carbon Nanocomposits Synthesized from Melamine with Enhanced Cr(VI) Removal

Abstract:

With the growth in various industries, such as metal cleaning, electroplating, pigment, mining and leather processing, heavy metals (Cr, Cd, Hg, Pd and As) have become a widespread pollutant in the surface water and groundwater. Cr(VI), a typical heavy metal ions, is a commonly identified contaminant because of its high toxicity and mobility. Recently, magnetic carbon nanocomposites with large specific surface area were introduced to remove the heavy metal in the wastewater and displayed the excellent efficiency. In this work, we successfully synthesized the novel nitrogen doped carbon composites with higher magnetism and specific surface areas using melamine as carbon and nitrogen source through a simple pyrolysis method. After that, we utilized the as-synthesized materials to treat the Cr(VI) solution, and found that the excellent performance were observed (removal capacity: 29.46 mg/g, pH at 7; 2001 mg/g, pH at 1), which can be attributed to the nitrogen dopant and the iron nanoparticles on the surface. The nitrogen dopant on the carbon surface can increase its negative charge density and then enhance the ability for the adsorption of Cr(VI) ions. These results demonstrated that nitrogen doped magnetic carbon nanocomposites are the promising materials for the environmental remediation.

470707 Techno-Economic Assessment of Membrane Distillation for Shale Gas Produced Water Treatment

Abstract:

Unconventional shale gas is a promising energy resource with major economic benefits but is accompanied by a host of environmental challenges. In particular, the rapidly developing shale gas industry faces a critical challenge of managing vast quantities of high salinity wastewater generated in the process of hydraulic fracturing. Membrane distillation (MD) is an emerging desalination technology for treating high salinity wastewaters including the produced water from shale gas activities. MD has lower operating temperature (30-90^{0}C) and pressure compared to conventional desalination technologies. Moreover, MD can treat wastewaters with up to 350,000 mg/Liter total dissolved solids (TDS) with up to 99% salt rejection where most pure thermal processes or pressure driven membrane processes are not applicable. While MD offers several advantages, techno-economic assessment (TEA) is necessary for evaluating the economic feasibility of MD for shale gas produced water treatment. To date, there has been little emphasis on evaluating the performance of MD technology for treating produced water and to best of our knowledge TEA of MD technology for produced water is not available in the existing literature.

We present a detailed TEA for a hypothetical MD plant for treating shale gas produced water based on a combination of experimental results, ASPEN Plus process model, and best available engineering knowledge. The analysis is carried out for a hypothetical 0.5 million gallons per day (MGD) MD plant that concentrates produced water from 10% (100,000 mg/Liter) TDS to 30% salinity. Preliminary results reveal that thermal energy cost for MD operation contributes the most to total cost of treating produced water in a MD plant. We also perform sensitivity analysis to identify input parameters that have the highest impact on total cost of purified water using MD technology. The results of sensitivity analysis reveal that plant capacity and feed TDS level have a significant impact on the cost of treating produced water. Subsequently, we evaluate the economic potential of MD for treating shale gas produced water under the scenario of utilizing waste heat for meeting the thermal energy requirements of the MD process. Specifically, we focus on utilizing available waste heat from existing natural gas (NG) compressor stations (CS) to meet the heat requirements of the MD process. Our prior work has developed quantitative estimates of waste heat from existing NG CS in the U.S. This information is coupled with experimental results on MD performance to propose a systems-level integration of MD with available waste heat at NG CS. Subsequently, we quantify the amount of produced water that can be treated using MD driven by available waste heat using ASPEN Plus simulation and experimentally measured flux rates. The results of this study reveal that utilizing waste heat as the energy source for MD reduces the total water cost significantly. A comparison of MD with existing desalination technologies including reverse osmosis (RO), multi-stage flash (MSF), and multi-effect distillation (MED) will be discussed. We also compare our findings with the business-as-usual scenario where produced water is transported and disposed in class II injection wells to highlight the potential and limitations of the MD technology for produced water treatment.

470025 Perchlorate Destruction from Ion-Exchange Regenerants By Catalytic Reduction Process Increases Brine Recovery

Abstract:

Perchlorate salts are generated as a byproduct from rocket fuels and other explosives. Ingestion of perchlorate inhibits normal function of the thyroid gland and contributes to hormone imbalance. Recently, the USEPA determined that perchlorate must be regulated as a water contaminant under the Safe Drinking Water Act (SDWA). One method of removing perchlorate from drinking water and potable water sources is through selective ion-exchange resin. Because perchlorate ion binds very tightly to the strong base anion exchange resin, a solution of having extremely high salt concentration, typically between 7 â€“ 12%, is required for regeneration of the resin. Furthermore, disposal of high concentration of perchlorate containing regenerant is problematic. Accordingly, there is a need for an improved method of removing perchlorate from water, including brine recovery.

Using a nanofiltration (NF) process, perchlorate is separated from the brine. More than 95% perchlorate was separated and retained in the reject. Permeate primarily containing brine was recycled back to the Ion exchange process to be used in the next stage of regeneration. Low volume of the NF reject containing high concentration of perchlorate ions is mixed with a reducing agent in a reactor where the perchlorate ions and the reducing agents undergo an oxidation-reduction (Redox) reaction. During the Redox reaction, perchlorate ions are reduced to chloride ions and the reducing agent is oxidized. The oxidized reducing agent is separated from the water which is regenerated by means of a second reducing agent, and reused in the Redox process. Redox reaction kinetic was significantly improved in presence of catalyst. Various reducing agents including Titanium (III), Manganese (II), and Iron (II) at different temperatures were investigated. The impacts of pH on the reduction kinetic were examined. All reactions were conducted with and without the catalyst. The efficiency of TiO_2 and MnO_2as the catalyst was compared.

Using an ion Chromatography (IC), perchlorate was measured. Results reveal that within the first two hours of

reaction, more than 98% perchlorate was converted to chloride ion in presence of a catalyst. Increasing temperature increased prechlorate destruction efficiency. The process, and the impacts of various process parameters on the perchlorate destruction efficiency will be presented and discussed.

456861 Low Temperature Conversion of Biomass into Activated Biochar for Water Treatment in Resource-Constrained Settings

Abstract:

Access to clean water is a major challenge in many parts of the world, especially in remote rural areas such as rain forest communities in Africa. New approaches are needed to ensure adequate water supply and quality, as they are inextricably intertwined with food security and public health. In this presentation, we will focus on the synthesis, characterization, and application of activated biochar for water purification. Unlike conventional gasification reactor systems designed to operate at large scales, not much has been investigated in the past about small-scale batch pyrolysis. Nevertheless, batch systems offer advantages such as increased char yield and net exothermicity. Various types of cellulosic biomass (eg: sawdust, coconut shells) can be used as carbon precursors via pyrolysis. We will describe a process where catalytic activating agents are used in batch setups to lower the pyrolysis temperature and simultaneously activate the carbon produced to BET surface areas in excess of 500 m2/g . A systematic study of process parameters and characterization of the resulting activated carbons is used to optimize the chemistry and minimize contaminants from the produced filter-carbon.

468039 Small-Scale Multiple Effect Distillation (MED) for Brackish Groundwater Desalination

Abstract:

Multiple effect distillation (MED) is the oldest thermal desalination process and has a typical plant capacity of 600 to 300,000 m^3/day. Because MED units do not use membranes, the capital cost, system complexity, and maintenance costs are usually lower than those for reverse osmosis (RO) systems. The purpose of this study was to engineer a small unit that could be operated in the field using renewable energy-sourced heat, namely, heat from the pyrolysis of waste biomass.

A small-scale MED unit was designed and fabricated at NMSU to produce approximately 0.5 m^3/day of very pure water (<10 ppm total dissolved solids (TDS)) from brackish groundwater. The unit is being used to study scaling behavior as a function of water chemistry and identify optimum operating conditions for each water. The fabricated MED unit consists of two falling film evaporators with horizontal smooth copper tubes in a parallel feed arrangement; the flow rates are approximately 60 kg/hr. brackish feed water and 20 kg/hr. of distilled water based on brine concentration factor of 1.5. The feedwater is sourced from the Brackish Groundwater National Desalination Research Facility (BGNDRF) in Alamogordo, NM and has a TDS concentration of 1,000-5,000 ppm. The MED operates under partial vacuum (0.2 bar absolute) and temperatures of 60-70°C. The MED performance was improved by adding a heat exchanger, water heater, and pump to bring the feedwater temperature closer to its boiling point prior to be sprayed into the effects. The aluminum parts in the MED effects were anodized to prevent corrosion. Remote thermocouples were used to monitor the temperature at the surface of copper tubes to study the effects of dry spots. The temperatures and pressures selected in this design are meant to be compatible with local brackish groundwater chemistry in terms of salt solubility to limit scaling on the heat exchanger surfaces.

472093 Bench Scale Investigation of the Biosand Filter Media As the Fluoride, Turbidity, and *E. coli* Removal

Abstract:

Biosand filter (BSF) is a simple and inexpensive facility that can be built as a community-based water treatment technique using locally-available materials. It has been proven that BSF can remove turbidity, *E. coli*, arsenic in water efficiently by modifying the filtration media. Drinking water fluoridation is prevalent in the world. High fluoride intake would cause dental fluorosis and other toxic health effects, especially for children. Therefore, drinking water de-fluoridation is necessary to safeguard public health. There is little information on the use of BSF for fluoride removal. For instance, some previous works reported that aluminum oxide-coated pumice (AOCP) was effective for fluoride removal and had better removal efficiency compared to activated alumina. Other research works found that pumice media BSF can provide similar water quality as conventional sand BSF but 30% more filtered water. Based on these findings, we have investigated the optimization of using AOCP surfaces to serve as media for BSF to clean contaminant water. Preliminary results have shown that AOCP can remove 93% of fluoride in surface water. In our setting, the traditional sand layer has been replaced by AOCP surfaces. We have found that filtration rate, contaminants removal efficacy are prime parameters that enhance BSF performance. It is expected that our findings will be able to predict the behavior of BSFs and their implementation in low-income communities or as point of use (PoU) systems.

463624 Capacitive Deionization of Brackish Waters Using Biomass-Derived Carbons

Abstract:

Irrigated agriculture water represents the bulk of demand for water in most of the countries suffering from acute water scarcity in the Near East and North Africa, as well as Mexico, Pakistan, South Africa, and large parts of China and India. Greenhouses are emerging solution. Current testing in Mozambique and Sierra Leone has demonstrated significantly reduced water use. Yet even a few gallons per day per greenhouse is challenging because many local water sources are brackish. Capacitive deionization (CDI) is a sustainable, energy efficient, and cost effective technology for removal of low or moderate salinity typical of brackish water as compared to technologies that remove the water instead. Indigenous production of the carbon for the electrodes holds the key to such technology being adopted as a source of irrigation water, simultaneously facilitating the use of such using renewable energy sources. Biochars have recently been explored as electrode materials in CD, and can be easily produced from local plant-based agricultural waste materials.

The goal of this study is thus to identify biomass process conditions to enable the best performance as electrode material in CDI. Studies on CDI so far have been targeted towards achieving the best adsorption capacity, which directs towards the use of highly processed and tailored carbons. This is mostly achieved through chemical activation processes using toxic or corrosive and expensive materials which require deionized water for subsequent washing of the carbon, for achieving better CDI performance. Hence, with the effort directed towards making the process viable for use locally in regions with acute water scarcity, a change in the approach to tailor the electrode material for use in CDI is proposed. This approach is to determine the best possible material utilization by ensuring maximum possible contact between the liquid and the biochar surface (wettability) and by generating the least Ohmic loss thus resulting in highest electrical efficiency.

472455 The Use of Electrodeionization for High Water Recovery in Brackish Water Treatment

Abstract:

Currently reverse osmosis dominates desalination because of low cost, reliability, and choices in the market place. However, there are many new desalination opportunities, especially in brackish water applications, where the size and scale make it possible to consider other technology. One of these technologies is electrodeionization. Electrodeionization is a technique in which we use wafers to remove ions to low concentration through an electricially driven process. Because the cost is related to the amount of ions being moved this technology can be cost competitive at low to medium ion concentration and low to medium scale. The purpose of this presentation is to explore places where electrodeionization can be economically feasible and show the fundamental scientific challenges with the technology. This work will specifically focus on the high water recovery possible with electrodeionization with experimental results to show high current efficiencies at concentration concentrations of greater than 10% salt.

466903 Total Phosphorus Removal from Sewage in Microbial Electrochemically Assisted Septic Tanks

Abstract:

Background

Subsurface sewage treatment systems (SSTS), commonly known as septic systems, typically consist of a septic tank and an absorption field (or drainfield or leachfield), to treat sewage from households. The septic tank as pretreatment step decreases total suspended solids (TSS), biochemical/chemical oxygen demands (BOD/COD), and pathogen levels from sewage influent, while the drainfield or soil absorption field captures nutrients (nitrogen and phosphorus) from the tank effluent so that the water leaving SSTS would have minimal environmental impact. By modifying the conventional septic tank through incorporating electrode materials to make use of microbial electrochemical process, this study assessed this novel design in terms of sewage treatment performance. Since phosphorus becomes increasingly threatening to water environment, it was the goal of this novel design to remove it from liquid phase and potentially capture it at electrode surface.

Methods

The experiments were conducted with 1-L scale simulated septic tanks at four different applied voltage levels (0.50 V, 0.63 V, 0.75 V, and 0.88 V, selection based on previous preliminary results) with control of without applied voltage. Tanks were set up and operated at two temperatures (15 and 25 °C), a hydraulic retention time of 8.3 days, two cycles of feeding and withdrawal during each HRT. The treatment conditions of the reactors are listed in REF _Ref450506281 \h Table 1 08D0C9EA79F9BACE118C8200AA004BA90B02000000080000000E0000005F0052006500660034003500300035003000360032003800310000000 . ANOVA and pairwise multiple comparisons based on Bonferroni test were conducted for water characteristics with OriginPro Version 8 at significance level of α=0.05.

Table SEQ Table * ARABIC 1. Treatments of simulated novel septic tanks for 171-day operation

	Treatment*		
Reactor #	Temp, oC	Voltage, V	Chlortetracycline (CTC)
Reactor #1	25	0	No
Reactor #2	25	0.50	No

Reactor #3	25	0.63	No
Reactor #4	25	0.75	No
Reactor #5	25	0.88	No
Reactor #6	15	0	No
Reactor #7	15	0	1 mg-CTC/L
Reactor #8	15	0.50	No
Reactor #9	15	0.63	No
Reactor #10	15	0.75	No
Reactor #11	15	0.88	No
Reactor #12	15	0.75	1 mg-CTC/L
Reactor #13	15	0.88	1 mg-CTC/L

*Water characteristics data of reactors #6 and #7 were pooled for data analysis, and the same procedure was applied to reactors #10 and #12, and reactors #11 and #13, for the reason that CTC addition was not detected to affect other water characteristics.

Results and Discussion

For total phosphorus, results suggested that septic tanks partially removed it from influent (REF _Ref450507954 \h Table 2 08D0C9EA79F9BACE118C8200AA004BA90B02000000080000000E0000005F0052006500660034003500300035003000370039003500340000000), and the novel design of microbial electrochemical further decreased its concentration (p=0.0001) with an influence from temperature (p=0.0003), while the interaction of voltage and temperature was not significant (p=0.2723). Those performance differences in total phosphorus removal were further elucidated by pairwise multiple comparisons given in REF _Ref450508201 \h Table 3 08D0C9EA79F9BACE118C8200AA004BA90B02000000080000000E0000005F0052006500660034003500300035003000380032003000310000000 . The results concluded that at 25 °C, 0.63 V or higher voltage at the selected electrode material is practically effective to remove most total phosphorus (>89.7%) from sewage, resulting in an effluent TP concentration less than 0.82 mg/L over 171-day operation; and that at 15 °C, phosphorus removal efficiency was smaller and it required 0.88 V to remove over 90% of total phosphorus from sewage.

Table SEQ Table * ARABIC 2. Total phosphorus concentrations in influent and effluents of different reactors

	Ave	SD	TP removal
	mg/L	mg/L	%
Influent	7.95	1.39	-
Control, 25 °C	6.98	1.72	12.2
0.50 V, 25 °C	1.81	3.67	77.2
0.63 V, 25 °C	0.82	2.92	89.7
0.75 V, 25 °C	0.14	2.30	98.3

0.88 V, 25 °C	0.11	2.31	98.7
Control, 15 °C	7.36	1.60	7.45
0.50 V, 15 °C	6.30	3.84	20.7
0.63 V, 15 °C	4.31	3.90	45.8
0.75 V, 15 °C	2.26	2.48	71.6
0.88 V, 15 °C	0.49	2.36	93.9

Table SEQ Table * ARABIC 3. Pairwise multiple comparisons results for total phosphorus levels in effluents

Comparisons (Bonferroni Test)	Significance at alpha=0.05 or lower
0.50 V, 25 °C vs. Control, 25 °C	*
0.63 V, 25 °C vs. Control, 25 °C	*
0.63 V, 25 °C vs. 0.50 V, 25 °C	NS
0.75 V, 25 °C vs. Control, 25 °C	*
0.75 V, 25 °C vs. 0.50 V, 25 °C	*
0.75 V, 25 °C vs. 0.63 V, 25 °C	NS
0.88 V, 25 °C vs. Control, 25 °C	*
0.88 V, 25 °C vs. 0.50 V, 25 °C	*
0.88 V, 25 °C vs. 0.63 V, 25 °C	NS
0.88 V, 25 °C vs. 0.75 V, 25 °C	NS
Control, 15 °C vs. Control, 25 °C	NS
Control, 15 °C vs. 0.50 V, 25 °C	*
Control, 15 °C vs. 0.63 V, 25 °C	*
Control, 15 °C vs. 0.75 V, 25 °C	*
Control, 15 °C vs. 0.88 V, 25 °C	*

0.50 V, 15 ºC vs. Control, 25 ºC	NS
0.50 V, 15 ºC vs. 0.50 V, 25 ºC	*
0.50 V, 15 ºC vs. 0.63 V, 25 ºC	*
0.50 V, 15 ºC vs. 0.75 V, 25 ºC	*
0.50 V, 15 ºC vs. 0.88 V, 25 ºC	*
0.50 V, 15 ºC vs. Control, 15 ºC	NS
0.63 V, 15 ºC vs. Control, 25 ºC	*
0.63 V, 15 ºC vs. 0.50 V, 25 ºC	*
0.63 V, 15 ºC vs. 0.63 V, 25 ºC	*
0.63 V, 15 ºC vs. 0.75 V, 25 ºC	*
0.63 V, 15 ºC vs. 0.88 V, 25 ºC	*
0.63 V, 15 ºC vs. Control, 15 ºC	*
0.63 V, 15 ºC vs. 0.50 V, 15 ºC	*
0.75 V, 15 ºC vs. Control, 25 ºC	*
0.75 V, 15 ºC vs. 0.50 V, 25 ºC	NS
0.75 V, 15 ºC vs. 0.63 V, 25 ºC	NS
0.75 V, 15 ºC vs. 0.75 V, 25 ºC	*
0.75 V, 15 ºC vs. 0.88 V, 25 ºC	*
0.75 V, 15 ºC vs. Control, 15 ºC	*
0.75 V, 15 ºC vs. 0.50 V, 15 ºC	*
0.75 V, 15 ºC vs. 0.63 V, 15 ºC	*
0.88 V, 15 ºC vs. Control, 25 ºC	*
0.88 V, 15 ºC vs. 0.50 V, 25 ºC	NS

0.88 V, 15 °C vs. 0.63 V, 25 °C	NS
0.88 V, 15 °C vs. 0.75 V, 25 °C	NS
0.88 V, 15 °C vs. 0.88 V, 25 °C	NS
0.88 V, 15 °C vs. Control, 15 °C	*
0.88 V, 15 °C vs. 0.50 V, 15 °C	*
0.88 V, 15 °C vs. 0.63 V, 15 °C	*
0.88 V, 15 °C vs. 0.75 V, 15 °C	*

* indicates significant difference, p-values less than 0.05; NS indicates non-significant difference, p-values equal or larger than 0.05.

Sludge solids were collected from simulated septic tanks (reactors at 25 °C, REF _Ref450512244 \h Figure 1 08D0C9EA79F9BACE118C8200AA004BA90B02000000080000000E0000005F005200650066003400350030003500310032003200340034000000) and quantified for mixed-liquor (ML) total suspended solids (TSS) and total solids (TS), sludge volume index (SVI), and composition. After log transformation, both MLTS and MLTSS displayed a linear increase along with voltage level. At 0.63 V where TP removal achieved close to 90%, the MLTSS and MLTS increased by 2.53 and and 1.17 folds, respectively. However, the corresponding SVI of the sludge was only 52.5% of the control reactor, indicating the formation of denser sludge, which suggested a smaller increase of sludge volume (85.3%) than the weight increase. The composition analysis based on energy dispersive X-ray spectroscopy (EDS) revealed the concentration and amount of phosphorus in sludge samples (REF _Ref450512615 \h Table 4 08D0C9EA79F9BACE118C8200AA004BA90B02000000080000000E0000005F005200650066003400350030003500310032003600310035000000). More phosphorus was captured in sludge sediment when voltage was increased, but the phosphorus balance, assumed to be captured on electrode surface, was higher at voltages levels of 0.50 and 0.63 V than 0.75 and 0.88 V.

Figure SEQ Figure * ARABIC 1. Mixed-liquor total suspended solids (A), mixed-liquor total solids (B), and sludge volume index (C) of the sludge from reactors with different voltage treatments at 25 °C.

Table SEQ Table * ARABIC 4. Phosphorus balance in different reactors.

	Amt of eff P	Amt P removed	Amt P in Sed	P of balance (on electrodes)
	mg	mg	mg	mg
Inf	178	-	-	-
Control, 25 °C	143	35	13	21
0.50 V, 25 °C	37	141	26	114
0.63 V, 25 °C	17	161	37	125
0.75 V, 25 °C	3	175	92	83
0.88 V, 25 °C	2	176	92	83

Conclusions

These results confirmed the phosphorus retention capability of the novel septic tank, suggesting a lower burden on the soil absorption for nutrient removal. The amount of sludge solids was also increased, which may require a more frequent tank cleaning.

464838 Microwell Arrays for Screening Interactions Between Root-Associated Microbes

Abstract:

Plant growth promoting rhizobacteria (PGPR) associate with root surfaces in order to provide benefits to the plant host, some of which include improved nutrient uptake, pathogen protection, and plant hormone regulation. An improved understanding of the relationships between PGPR and root hosts has spurred recent interest in developing constructed, artificial communities of PGPR for use as a low-energy, sustainable alternative to chemical fertilizers. However, wide-spread application of constructed communities has been limited due to the fact that the vast majority of microbe-microbe interactions occurring between PGPR in the rhizosphere are unknown or poorly characterized. This limitation stems from the fact that traditional microbiological assays designed to characterize microbial interactions do so using co-culturing methods that are qualitative and low-throughput. Additionally, these assays fail to re-create the physical and chemical complexities of natural ecosystems, such as soil or rhizospere, which are spatially confined microenvironments. To address these types of limitations, we have developed a microfabricated cell-screening platform termed the â€œmicrobe arrayâ€ which uses a novel biofabrication method to randomly assemble thousands of independent bacterial meta-populations into microscale wells with controlled physicochemical properties. It was recently shown that by tuning well dimensions, interacting bacteria can be combined into wells at low dispersity, allowing for the assembly of replicate initial populations; or at high dispersity, allowing for thousands of unique meta-populations to be assembled across a single array of wells. After this, communities can be simultaneously screened for growth or decay to identify combinations of bacteria and environmental parameters that promote or inhibit colonization. In development of this platform, we investigated the effect of spatial confinement on the growth of *Pseudomonas* bacteria. Here, populations were assembled into wells at high dispersity, and growth screening experiments revealed that successful colonization of microbial communities in a spatially confined, diffusion limited environment is highly dependent on inoculum levels. Our current work focuses on screening rhizobacterium populations in this platform to identify pathogenic, competitive or symbiotic interactions that inform the development of constructed communities for biofertilizer and biocontrol applications.

EFFECT OF SIZE AND CHARGE OF METAL IONS ON HYDROGEN PEROXIDE STABILITY IN SILICA HYDROGELS

Ezgi Melis Dogan, Fulya Sudur Zalluhoglu, Nese Orbey, University of Massachusetts Lowell, Lowell, MA, USA

Abstract

Currently, the most efficient method to increase the stability of hydrogen peroxide (H_2O_2) is microencapsulation in silica hydrogels and xerogels via environmentally-friendly sol-gel chemistry using sodium silicate as the starting precursor. The high stability of H_2O_2 in these systems arises from strong hydrogen bonding between silanol groups (Si-OH) and peroxide molecules. Our previous study showed that the amount of Na^+ ions have a strong effect on the structure and morphology of hydrogels. Therefore in the present work, the effect of the size and charge of the ions other than Na^+ on the properties of the hydrogel and stability of peroxide was studied. Monovalent (K^+) and divalent (Mg^{+2}) ions were introduced into the sodium silicate precursor in the form of their respective salts. The pH of the sol was changed by using an ion exchange resin to replace a portion of the ions in the solution with H^+ ions. Gelation time and structure, pH of the sol and stability of H_2O_2 in hydrogels were investigated. Results showed that the stability and gel time can be adjusted to the desired level by changing the ion content and pH of the sol. The increase in total ion concentration in the sol resulted in a decrease in gel time and the best H_2O_2 stability was obtained with the addition of the magnesium salt: seventy percent of the peroxide was retained in hydrogels containing Mg^{2+} ions, as compared to 61 % retention with hydrogels containing Na^+ ions at the end of 70 days. Hydrogels were characterized by using scanning electron microscopy (SEM) and Brunauer-Emmett-Teller (BET) analysis.

Introduction

Hydrogen peroxide (H_2O_2) is an environmentally friendly oxidant which has a limited stability and tend to decompose into oxygen and water according to following reaction.[1]

$$H_2O_{2(aq)} \rightarrow H_2O_{(l)} + \frac{1}{2}\,O_{2(g)} \qquad (1)$$

The concentration of the peroxide, temperature, pH of the aqueous solution, and presence of impurities or stabilizers significantly affects the decomposition rate of H_2O_2. Storage at low temperatures is required for H_2O_2 solutions due to this instability.[2] However, H_2O_2 stability at ambient temperatures is desired by many industrial applications such as sanitation,[3] disinfection,[4] and tooth whitening.[5]

Various methods have been experimented to decrease the decomposition rate and increase the stability of H_2O_2.[6-10] Among these methods, microencapsulation of H_2O_2 in silica hydrogels and xerogels has been the most efficient methods to enhance peroxide stability because of formation strong hydrogen bonds of H_2O_2 molecules with the silica surface.[11]

The synthesis of hydrogels and xerogels is performed using sol-gel chemistry which is a simple and low cost method.[12] The most common precursor of the gels is sodium silicate solution. A colloidal suspension known as sol is first formed by introducing silanol groups into aqueous sodium silicate and removing sodium ions from the solution. For this purpose, a cation exchange resin is used to adjust the sodium content via ion exchange in the solution.[11] This exchange occurs according to following reaction where Rz is the styrene-divinylbenzene copolymer and SO_3H is the sulfonic acid group.

$$Na^+ + RzSO_3^-H^+ \rightarrow H^+ + (RzSO_3)Na \qquad (2)$$

A hydrogel is observed in the sol-gel transition point where the viscosity of the solution significantly increase. At this point, Si-O-Si bonds are formed via polymerization of silanol groups in the solution resulting a three-dimensional interconnected network (hydrogel). Xerogels are more contracted gel networks obtained by drying the hydrogels. Silica xerogels containing H_2O_2 has been studied by Żegliński et al.,[6] Bednarz et al.[7] and Wolanow et al..[8] Hydrogel and xerogel properties depend on the drying and storage temperatures, pH of silica sols, and initial precursors. Sudur et al.[9, 10] studied the transition conditions of hydrogel to xerogel and observed the effects of sodium ion on the pH of the sol and H_2O_2 (3.3 to 33.2wt %) stability at room temperature (RT). Significant effects of sodium ion content on the structure, the release rate of entrapped H_2O_2, and the thermal properties of hydrogels and xerogels were found.

The effects of size and charge of ions present in the sol on the structure of hydrogels and stability of entrapped H_2O_2 were questioned because of the observations in our previous work.[9] In the literature, the effect of ion type and concentration on the binding power of sodium silicate was studied for applications in petroleum drilling and construction.[13-15] However, H_2O_2 was not incorporated into the formulations used in these studies. To best of our knowledge, the effect of size and charge of ions on the hydrogel formation or hydrogen peroxide stability has not been studied.

In the present work, H_2O_2 was encapsulated in silica hydrogels using K^+ ions into the sodium silicate precursor in the form of potassium chloride (KCl) at three different molar ratios of K^+: Na^+ (0.14:1, 1:1, and 1.76:1). The effects of K^+ on the gelation time, pH of the sol, stability of H_2O_2 and gel structure were studied. Similar to Na^+, K^+ is monovalent but it has a larger atomic size. On the other hand, silica hydrogels were obtained using Mg^{2+} in the precursor solution to study the effect of charge of the ion on H_2O_2 stability.

Experimental Procedure

Materials

Sodium silicate solution (reagent grade, 10.6% Na_2O, 26.5% SiO_2), hydrogen peroxide (30wt % in H_2O, ACS reagent), Amberlite IR120 (hydrogen form), orthophosphoric acid (85% wt in H_2O); potassium permanganate (ACS reagent >99%) and sulfuric acid (99.999%) were purchased from Sigma Aldrich Chemical Co. Potassium chloride, magnesium chloride and calcium chloride (ACS reagent, >99%) were purchased from J.T. Baker.

Hydrogel synthesis

The procedure of hydrogel synthesis, previously available in the literature[9] was modified. Hydrogels containing Mg^{2+} and K^+ ions along with Na^+ ions were synthesized dissolving magnesium chloride ($MgCl_2$) and potassium chloride (KCl) in deionized water before the addition of sodium silicate. Sodium silicate solution was added dropwise to salt solutions at a volume ratio of 1:5. For hydrogels containing K^+, three different K^+: Na^+ molar ratios of 0.14:1 (low), 1:1 (equimolar), and 1.76:1 (excess) were formulated. In hydrogels containing Mg^{2+}, the molar ratio of Mg^{2+}: Na^+ was adjusted as 0.2:1. Then, all formulations are combined with 30wt % H_2O_2 solution and 7 μL of phosphoric acid (H_3PO_4). The pH of the sols were controlled by adding different amounts (0-3.2 g) of ion exchange resin (Amberlite IR 120, hydrogen form) to replace a portion of the ions in the solution with H^+ ions. The mixture was stirred with an overhead mixer for 4 min. The sol was transferred to a glass Petri dish (D = 100 mm, h = 15 mm) after observing the settlement of the resin. The petri dish was placed in a BlueM gravity oven at 40 °C until observing the gelation.

Characterization

The pH of the sol was changed using different amounts of ion exchange resin at a constant K^+: Na^+ ratio and an initial H_2O_2 concentration (19.9wt %). Gelation time was observed with respect to pH value of the solution for each K^+ formulations. H_2O_2 stability was observed in hydrogels containing K^+ as a function of pH and molar ratio of K^+: Na^+ ions. Stability of H_2O_2 in hydrogels containing Mg^+ was compared based on the presence of ion exchange resin in the starting sol. Hydrogels were stored under ambient conditions at room temperature, and H_2O_2 content was determined using titration method with potassium permanganate ($KMnO_4$) as the titrant. Measurements were made daily for the first 10 days and, then periodically until the concentration of H_2O_2 dropped to 2wt %. Two repeated titrations were performed for each sample and the reproducibility was found to be within 1%.

Quantachrome Autosorb-3B nitrogen sorption porosimeter and JEOL JSM 7401F scanning electron microscope were used to study the morphology and structure of the hydrogels. The surface area and pore size were measured through BET (Brunauer-Emmett-Teller) analysis in the porosimeter. Before these measurements and SEM imaging, aerogels were obtained from hydrogels using a critical-point dryer (Tousimis SAMDRI-795) to preserve the solid network and retain the pore structure. Aerogel samples were coated with gold/palladium target by using a vacuum sputter coater. SEM images of samples were obtained at an acceleration voltage of 10 to 15 kV.

Results and Discussion

Gel time as a function of pH and K^+: Na^+ ratio

Gelation time of low and equimolar of K^+: Na^+ ratio in the sol was studied for different pH values of the sol. Results are plotted in Figure 1. Gelation was most rapid between pH=4 and 6 because the sol passed through the isoelectric point, at which it did not carry a net electrical charge and rapid aggregation of the particles was observed.

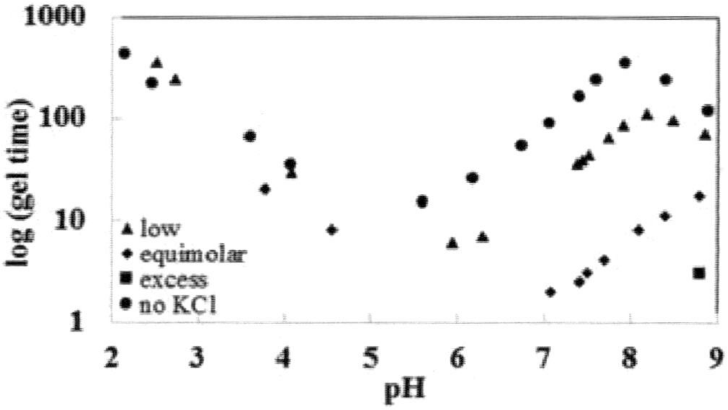

Figure 1: Relationship between gel time and pH for different molar ratios of K^+: Na^+ ions ▲: low (0.14) ◆: equimolar (1) ■: excess (1.76) ●: previous results obtained with unmodified sodium silicate precursor[9]

For all the pH values studied, addition of the ions resulted in faster gelation compared to the previous results obtained in the presence of Na^+ ions only.[6] At the pH range of 2-4, a slow gel formation was resulted because of low concentration of hydroxyl ions. However, longer gel times were obtained after the isoelectric point (pH=6-8) because the gel become negatively charged by high concentration of hydroxyl ions. In the presence of KCl, a shift toward the basic side was observed in Figure 1. This is because the K^+ and Cl^- ions have higher affinity for silicate ions rather than OH^- ions and, therefore more OH^- ions remain in the sol.

Stability of H_2O_2 in hydrogels containing K^+

H_2O_2 content was measured during 70 storage days to study the stability of H_2O_2 as a function of the K^+: Na^+ ratio and the pH of the sol. Stability results of hydrogels containing different K^+: Na^+ ratio at a constant pH of 8.8 are shown in Figure 2. H_2O_2 stability was observed to increase with the addition of K^+ ions as seen more clearly in short term stability (first 10 days during storage).

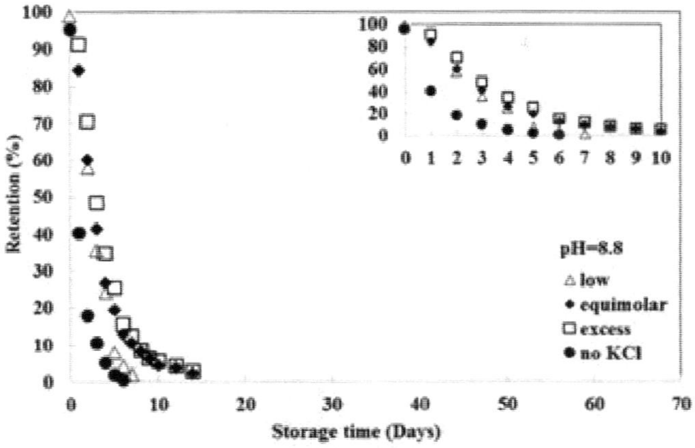

Figure 2: Room temperature stability of H_2O_2 as a function of storage time at constant pH=8.8 (for different molar ratios of K^+: Na^+ Δ: low ♦: equimolar □: excess ●: previous results obtained with unmodified sodium silicate precursor[9])

For example, 18.1wt % peroxide retention was obtained at the end of 2 days of storage at room temperature in our previous work when no K^+ ion was present in the system.[9] In this study, the peroxide retention of hydrogels containing low, equimolar and excess K^+:Na^+ molar ratios was 58.1, 60.2 and 70.6wt %, respectively, during the same period. Moreover, complete decomposition of H_2O_2 in hydrogels containing K^+ took longer (7 to 14 days) than that in previously studied hydrogels[9] (5 days).

Figure 3 a & b show the effects of pH on peroxide stability for low and equimolar amounts of K^+: Na^+, respectively. For each ratio, when the pH of the sol decreased, higher retention of H_2O_2 was observed. At lower pH values of the sol, higher amounts of H^+ ions increase the entrapment efficiency and retention by creating more available sites for peroxide in silica network.[16]

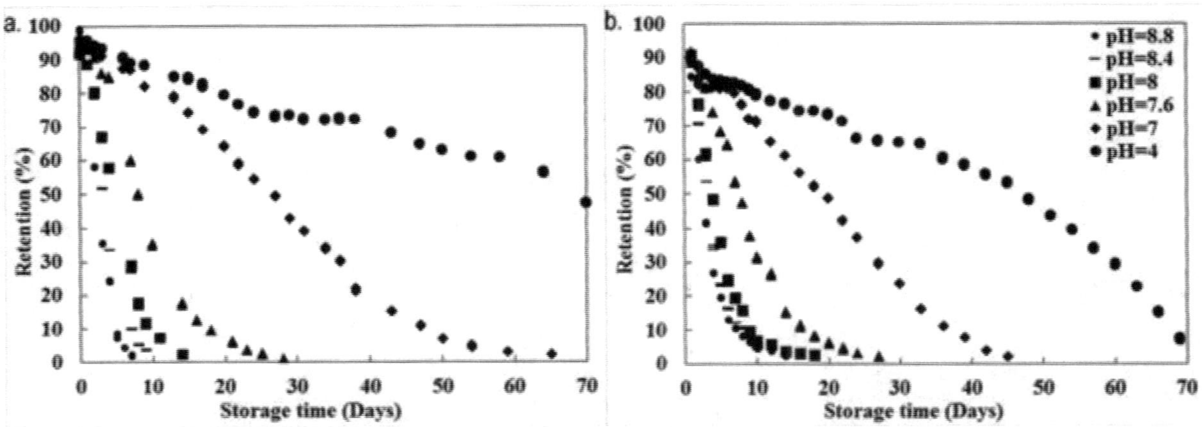

Figure 3. The comparison of H_2O_2 stability in hydrogels containing low (a) and equimolar (b) ratios of K^+: Na^+ at different pH values

Stability of H_2O_2 in hydrogels containing Mg^{2+}

H_2O_2 stability of hydrogels containing Mg^{2+} was studied to observe the effect of ion charge added into the starting silicate solution. Magnesium was selected to be partially exchanged with the sodium in the silicate at 1:0.2 molar ratio of Mg^{2+}: Na^+ because magnesium ion is divalent and smaller in size than Na^+. Magnesium ion could not be completely replaced with Na^+ because magnesium silicates are not water soluble unlike potassium silicates. The following reaction shows the precipitation reaction of magnesium silicates.

$$2(\equiv SiO^{-}Na^{+}) + MgCl_2 \rightarrow 2 \equiv SiO^{-}Mg^{2+} + 2NaCl \qquad (3)$$

The pH of the sol containing Mg^{2+} was decreased using ion exchange resin (Amberlite IR120) which replaces ions with H^+. Figure 4 shows H_2O_2 retention in hydrogels synthesized in the presence (pH=4.6) and absence (pH=8.8) of ion exchange. Previous stability results of hydrogels containing only Na^+ are also included in Figure 4 for comparison. The presence of Mg^{+2} showed improved peroxide stability regardless of the presence of ion exchange. We believe that the reason of this result is more sites available for hydrogen bonding between silanol groups and peroxide molecules due to the smaller size and higher charge of magnesium ions than Na^+.

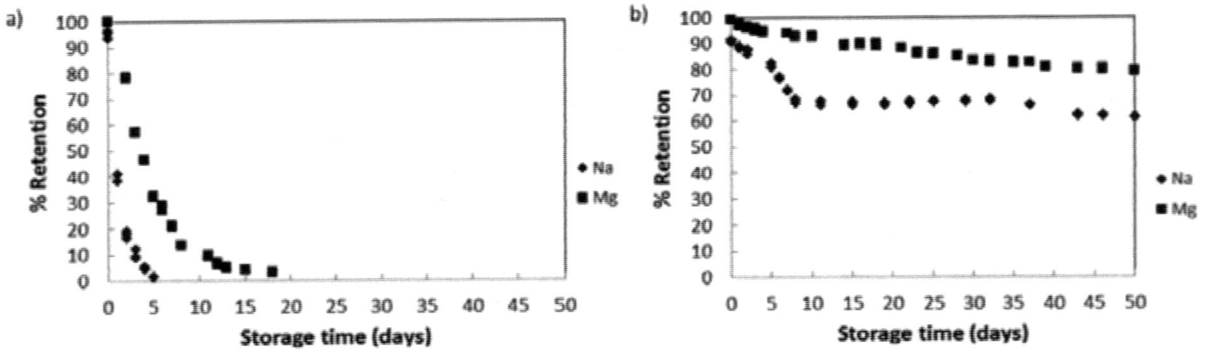

Figure 4. Room temperature stability of H_2O_2 as a function of storage time (a) without ion exchange (pH=8.8) and (b) with ion exchange (pH=4.6)

Morphology and structure of hydrogel

BET and SEM analyses showed that the presence of potassium ion resulted less compact gel structure at constant pH. This is because K^+ has larger atomic size than Na^+ and results larger pores in the hydrogel structure. Figure 5 shows SEM images of samples having different molar ratios of K^+: Na^+ at pH=8.8.

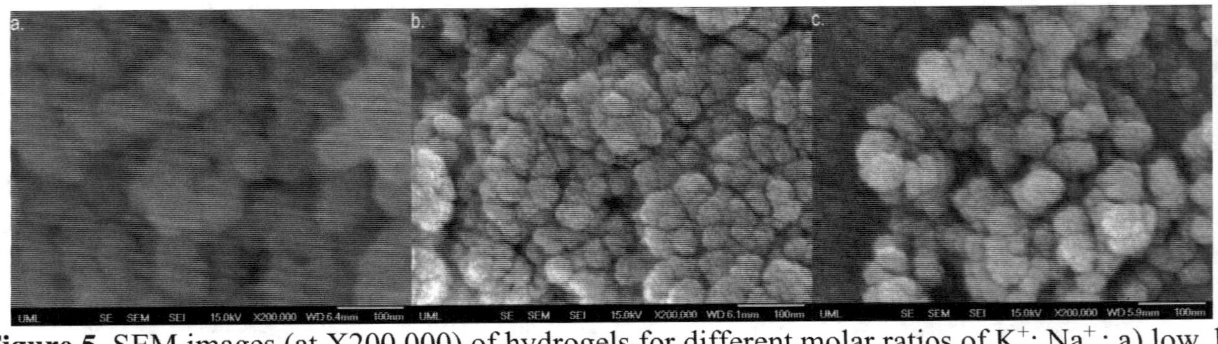

Figure 5. SEM images (at X200,000) of hydrogels for different molar ratios of K^+: Na^+ : a) low, b) equimolar, c) excess (at constant pH=8.8)

Nitrogen sorption studies were carried out to obtain quantitative results of the effect of potassium ion on pore size and surface area of hydrogels. Results are shown in Figure 6. When K^+: Na^+ molar ratio was increased, larger pore sizes were obtained because of the larger atomic size of K^+ (Figure 6a). For each case, small pore sizes were observed as pH decreased because hydrogen ions have capability to bring silica particles together by charge screening effect. In Figure 6b, a decrease in surface area was resulted in the presence of K^+. However, surface area of hydrogels increased at lower pH values as expected due to the decrease in pore size.

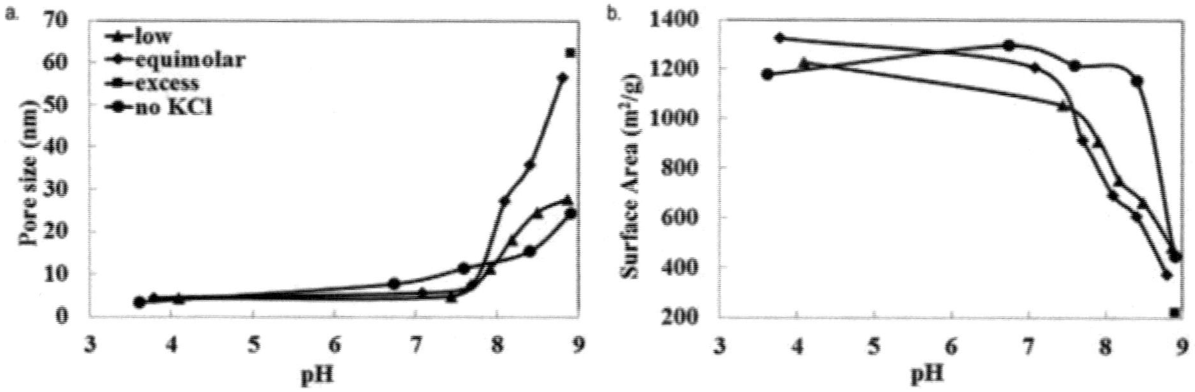

Figure 6. Structural properties of silica hydrogels with changing pH and K^+: Na^+ molar ratio, ▲: low ♦: equimolar ■: excess ●: previous results obtained with unmodified sodium silicate precursor[9]

Conclusions

A significant effect of the size and charge of ions on H_2O_2 stability and the hydrogel structure was observed. Gelation time decreased as K^+: Na^+ ratio were increased in the starting precursor. The stability of encapsulated H_2O_2 was improved with the addition of Mg^{2+} and K^+ ions and decreasing pH of the sol. With the increase in K^+: Na^+ ratio, average pore diameter was found to increase and surface area decreased at constant pH (8.8), therefore the gel became less compact. The higher retention and better stability of H_2O_2 at room temperature was resulted by controlling the ion content (size and charge) and pH. The properties of hydrogels can be tailored to control the gel time and H_2O_2 stability. A detailed study might be required for better understanding of the effect of size and charge of ions on peroxide stability and the hydrogel structure. For example, calcium ions can be included in hydrogel synthesis because the sizes of Ca^{2+} and Na^+ ions are similar whereas it has higher charge like Mg^{2+}.

References

1. Schumb WC, Satterfield CN, Wentworth RL. Hydrogen peroxide. ACS Monograph Series. New York: Reinhold Publishing Corporation, 1955.
2. Sigma Aldrich Product Information Sheet: Hydrogen Peroxide Solution. http://www.sigmaaldrich.com/
3. Colodette JL, Rothenberg S, Dence CW. Factors Effecting Hydrogen Peroxide Stability in the Brightening of Mechanical and Chemimechanical Pulps. J. Pulp. Pap. Sci. 1989; 15: 3−10.
4. Sulieman M, Addy M, MacDonald E, Rees JS. The Effect of Hydrogen Peroxide Concentration on the Outcome of Tooth Whitening: An In Vitro Study. J. Dent. 2004; 32: 295−299.
5. Kim JC, Lee KU, Shin WC, Lee HY, Kim JD, Kim YC, Tae G, Lee KY, Lee SJ, Kim JD. Monoolein Cubic Phases Containing Hydrogen Peroxide. Colloids and Surfaces B: Biointerfaces. 2004; 36: 161−166.

6. Żegliński J, Cabaj A, Strankowski, M, Czerniak J, Haponiuk JT. Silica Xerogel-Hydrogen Peroxide Composites: Their Morphology, Stability, and Antimicrobial Activity. Colloids and Surfaces B: Biointerfaces. 2007; 54: 165−172.
7. Bednarz S, Rys B, Bogdal D. Application of Hydrogen Peroxide Encapsulated in Silica Xerogels to Oxidation Reactions. Molecules. 2012; 17: 8068-8078.
8. Wolanov Y, Shurki A, Prikhodchenko PV, Tripolskaya TA, Novotortsev VM, Pedahzur R, Lev O. Aqueous Stability of Alumina and Silica Perhydrate Hydrogels: Experiments and Computations. Dalton Trans. 2014; 43: 16614.
9. Sudur F, Pleskowicz B, Orbey N. Hydrogen Peroxide Stability in Silica Hydrogels. Ind. Eng. Chem. Res. 2015; 54 (6):1930−1940.
10. Sudur F, Orbey N. Properties of Hydrogen Peroxide Encapsulated in Silica Hydrogels and Xerogels. Ind. Eng. Chem. Res. 2015; 54(44): 11251-11257.
11. Bergna HE, Roberts WO. Colloidal Silica: Fundamentals and Applications. Boca Raton, FL: CRC Press, 2005.
12. Baccile N, Babonneau F, Thomas B, Coradin T. Introducing Ecodesign in Silica Sol-Gel Materials. J. Mater. Chem. 2009; 19: 8537−8559.
13. Brady AP, Brown AG, Huff H. J. The polymerization of aqueous potassium silicate solutions. Colloid Sci.1953; 8(2):252-276.
14. Shroff AV, Shah DL. Grouting Technology in Tunneling and Dam Construction. Rotterdam, Brookfield: Oxford & IBH Publishing Company, 1992.
15. Chun-xi Z. Recent Advances in Water glass Sand Technologies. Shanghai, China: School of Material Science and Engineering, Shanghai Jiaotong University; 2007.
16. 27. Iler RK. The Chemistry of Silica: Solubility, Polymerization, Colloid and Surface Properties and Biochemistry. New York: Wiley, 1979.

472182 UV-Assisted Synthesis of Carbon Nanotube-TiO_2 Nanocomposites for Enhanced Photocatalytic Air Purification

Abstract:

Efficient photocatalysts consisting of carbon nanotubes (CNTs) decorated with TiO2 nanoparticles were synthesized by a simple two-step process that is free of aggressive acid treatments that often introduce defects to CNTs. The novel synthesis process involved UV-assisted oxidation of CNTs using H_2O_2, followed by the deposition of TiO2 via a wet chemical method using TiCl4 as a titanium source. The resulting CNT/TiO2 nanocomposites have been characterized by X-ray diffraction (XRD), transmission electron microscopy (TEM), X-ray photoelectron spectroscopy (XPS), and Raman spectroscopy. XRD results revealed that the crystalline structure of TiO2 is anatase and the TEM results showed the undamaged tubular morphology of CNTs decorated with TiO2 nanoparticles. The XPS data revealed the intimate chemical interactions between the CNTs and TiO2 in the nanocomposites. The expected synergistic effect and efficient separation of photogenerated electron-hole pairs in the synthesized CNT/TiO2 nanocomposites were verified in gas-phase photocatalytic degradation of VOCs.

470129 NOVEL Magnetic Nanocomposite Materials for RAPID Removal of Polychlorinated Biphenyls from Contaminated Water Sources

Abstract:

Magnetic nanocomposite systems able to capture polychlorinated biphenyls (PCBs) in aqueous solution were developed, providing a cost-effective water remediation technique. Two distinct methods were employed to develop these polyphenolic nanocomposite materials, where the polyphenolic moieties were incorporated to create high affinity binding sites for organic pollutants. The first method utilized a surface initiated polymerization of polyphenolic-based crosslinkers and co-monomers on the surface of iron oxide magnetic nanoparticles to create a core-shell nanocomposite. The second method utilized a bulk polymerization method to create macroscale films composed of iron oxide nanoparticles incorporated into a polyphenolic-based polymer matrix, which were then were processed into microparticles. Both methods produce nanocomposite materials that can bind chlorinated organics, can rapidly separate bound organics from contaminated water sources using magnetic decantation, and can use thermal destabilization of the polymer matrix for contaminant release and material regeneration. The polyphenol functionalities used to bind organic pollutants were quercetin multiacrylate (QMA) and curcumin multiacrylate (CMA), which are acrylated forms of the nutrient polyphenols with expected affinity for chlorinated organics. Particles were characterized using transmission electron microscopy (TEM), dynamic light scattering (DLS), Fourier transform infrared spectroscopy (FTIR), thermal gravimetric analysis (TGA) and UV-visible spectroscopy stability analysis. Pollutant binding studies were performed using PCB 126 as a model PCB and chlorinated organic pollutant to determine binding affinity and capacity, and this was quantified using gas chromatography coupled to electron capture detector (GC-ECD). It was demonstrated that the materials effectively bound PCBs in aqueous media.

472163 Influence of Silica-Based Nanoparticles Embedded in Sand Bed Filtration for Cleaning-up Industrial Wastewater

Abstract:

Sand filtration technology is widely used to remove the suspended solids and colloids from the industrial wastewater effluent. However, such standalone technology cannot remove the soluble organic compounds from the industrial wastewater. In the current application, we are targeting to improve not only the removal of suspended solids but also increase the removal efficiency of total organic carbons (TOC) by enhancing the sand surface area via integrating it with functionalized silica-based nanoparticles. These kinds of nanoparticles are environmentally friendly and naturally occurring materials at the macro-level.

In this study, different types of functionalized silica-based nanoparticles were successfully prepared, characterized, and tested for TOC adsorptive removal within batch and continuous adsorption processes. Briefly, the Brunauerâ €"Emmettâ€"Teller (BET) surface area, X-ray diffraction (XRD), Fourier transform infrared (FTIR) spectroscopy and high resolution transmition electron microscopy (HRTEM) were used to confirm the surface area, crystalline domain size, functionality, and structural topology, respectively. The characterization results showed that the prepared nanoparticles are spherical in shape with a crystalline diameter of 10 nm and BET surface area around 180 m^2/g. Adsorption experiments were conducted in a batch mode to investigate the adsorption affinity and capacity of nanoparticles toward TOC removal from a locally supplied industrial wastewater sample. The results indicated that the TOC removal efficiency was improved significantly by 50% using the functionalized sand with nanoparticles at low mass fraction (<10 wt%). Afterwards, the filter bed column, continuous mode, was conducted using the functionalized and the un-functionalized sands. Interestingly enough, the functionalized sand with in-house prepared silica-based nanoparticles showed prolonger breakthrough time (>50 times) in comparison with the un-functionalized sand. These promising results are believed to help the current conventional wastewater treatment methods not only on a lab scale but also on an industrial level.

464632 Inactivation of *E. coli.* Using a Novel TiO_2 Nanotube Electrode

Abstract:

Electrochemical technology has gained increasing interests for water/wastewater treatment in recent years. In particular, it may find applications in decentralized treatment due to its simplicity, amenability to automation, compact instrumentation, and ease of operation. The performance of electrochemical processes highly depends on the material of electrodes. However, the high cost of electrodes, formation of toxic byproducts, and short electrode lifetime have impeded its wide application. Therefore, development of cost-effective anode materials are needed to support future adoption of electrochemical technologies for environmental applications. In our study, a novel TiO_2 nanotube anode has been fabricated in the lab via simple electrochemical anodization using metallic Ti as the substrate followed by electrochemical hydrogenation/cathodic polarization. The one-dimensional nanostructure is shown in FESEM images. The TiO_2 nanotube anodes with stainless steel as the cathodes were tested for inactivation of a model microbial contaminant *E. Coli.* using a filter-press type cell. Effects of current density applied (0.625 ~ 2.5 mA/cm^2), electrolyte concentration (0.3 mM~3 mM NaCl), and different electrolyte (NaCl, Na_2SO_4, Na_2HPO_4, $NaNO_3$, and $NaHCO_3$) on *E. Coli.* inactivation were investigated. Electrochemically generated oxidizing species were measured as free chlorine. Further, production of reactive oxygen species (•OH, O_2•-, SO_4^{2-}•) were examined using probe compounds. In addition, electro-disinfection experiments were conducted in real water matrices, i.e. samples of surface water, treated surface water before disinfection, and secondary effluent before disinfection collected from local water/ wastewater treatment plants. Formation of inorganic (e.g. ClO_3^-) and organic (THMs) disinfection byproducts were also monitored. Our results have demonstrated that the novel TiO_2 nanotube anode can effectively achieve 5~6 log reduction of E. Coli. with low byproducts formation. The low energy consumption also indicates that such disinfection system may be operated using solar energy stored in a DC battery for decentralized treatment or as a point-of-use device.

451725 Metal Organic Framework Derived Nanoporous Carbon As a Novel Adsorbent for Water Treatment

Abstract:

Metal organic framework derived nanoporous carbon

as a novel adsorbent for water treatment

Zahra Abbasi[a], Ezzatollah Shamsaei[a], Soo Kwan Leong[a], Bradley Ladewig[a,b], Xiwang Zhang[a], Huanting Wang[a*]

[a]Department of Chemical Engineering, Monash University, Clayton, Victoria 3800, Australia

[b]Department of Chemical Engineering, Imperial College London, Exhibition Road, London SW7 2AZ, United Kingdom

Phone: +61 3 9905 3449, Email: Huanting.wang@monash.edu

Nanoporous carbon materials have attracted a great deal of attention because of their potential application in many areas such as adsorption and catalysis. These materials have desirable properties such as high surface area, good chemical and thermal stabilities as well as fast kinetics, which make them suitable for adsorption applications. Porous metal organic frameworks (MOFs) consist of crystalline structures that provide high internal surface areas and tunable porosities which could be considered as starting materials for developing novel carbon based nano composite

adsorbents.

Different types of adsorbents have been studied for the removal of organic pollutants from water and wastewater such as activated carbon, fly ash waste materials and different kind of MOFs. However, their adsorption capacity is low in comparison with nanoporous carbons. Therefore, developing novel carbon based materials with simple synthesis methods for adsorption of pollutants from water is of great importance.

Zeolitic immidazolate frameworks (ZIFs) are a class of MOFs that exhibit remarkable features like high thermal and chemical stability and demonstrate similar pore topologies to zeolites. Among various ZIFs reported, ZIF-8 with the formula of $Zn(2\text{-methylimidazolate})_2$ and with a sodalite-related zeolitic structure has been used for many applications such as adsorption and catalysis. The surface and bulk properties of ZIF-8 can be modified to make it a more competent adsorbent for water treatment purposes. Carbonization of ZIF-8 could be the easiest and most practical modification approach to amend this metal organic frame work's characteristics.

The adsorption capabilities and characteristics of nanoporous carbon particles derived from ZIF-8 were investigated in this study. Direct carbonization of ZIF-8 was used to synthesize carbon nano particles. SEM and TEM images revealed that the ZIF-8 derived carbon carbonized at several temperatures retained the original structure and morphology of ZIF-8. These carbon nano-particles exhibited significantly greater adsorption capacities for the removal of dye from water compared to the ZIF-8. The high adsorption capacity is attributed to the change in surface charge and pore size. The surface functionality and hydrophobicity of the materials were investigated by Raman Spectroscopy, N_2 adsorption-desorption, FTIR, TGA and contact angle measurements. The change in the surface charge of the carbon particles was confirmed by zeta potential of the samples. The dye uptake followed pseudo-second order kinetics and the equilibrium data was best fit to the Langmuir model for the carbon nano-particles. Intra-particle diffusion studies showed that the adsorption process was governed by both the diffusion step and external mass transfer. The ZIF-8 derived nano porous carbons were found to be highly efficient adsorbents for water and waste water treatment.

458430 Efficient Modification of Outer Selective P84 Nanofiltration (NF) Hollow Fiber Membranes for Cadmium Removal

Abstract:

Efficient Modification of Outer Selective P84 Nanofiltration (NF) Hollow Fiber Membranes for Cadmium Removal

Jie Gao, Shi-Peng Sun, Wen-Ping Zhu, Tai-Shung Chung*

Department of Chemical and Biomolecular Engineering

National University of Singapore, Singapore 117576

*Corresponding author

Tel: +65-65166645; fax: +65-67791936; Email: chencts@nus.edu.sg

Abstract:

Cadmium is a common heavy metal present in industrial effluents from pesticide, smelting, dye manufacturing and electroplating industries. Like other heavy metals, cadmium is non-biodegradable. Accumulation of cadmium can occur in living systems and results in harmful environmental effects. Side effects, such as nausea, bone lesions, renal disturbances and cancer, may happen if cadmium is accumulated in human beings. Thus, the removal of cadmium from wastewater before discharge has important meanings to the human society. One method to remove those heavy metals is nanofiltration (NF) because it is able to offer good rejections to multi-valent heavy metal ions without much compromising water fluxes as compared to those of reverse osmosis (RO) membranes.

There is a growing interest in thin-film composite (TFC) NF hollow fiber membranes. The TFC layer can be formed on both inner and outer surfaces of hollow fiber membranes. The outer selective TFC hollow fiber membranes may have less mass transfer resistance and lower chances being blocked by foulants compared to the inner ones. Thus, the outer selective TFC fibers can be fabricated into smaller fiber dimensions with a higher package density in modules. Despite these advantages, the molecular design of an ideal TFC layer on the outer surface of hollow fiber membranes remains a challenge, making the mass production difficult. Since most TFC layers are made from interfacial polymerization of m-phenylenediamine (MPD) and trimesoyl chloride (TMC), removing excess solution after the MPD coating is the major obstacle in the process. Vacuum assist interfacial polymerization or wipe drying are required to make a defect free polyamide layer. Another obstacle is that a large amount of alkane solvents has to be used during interfacial polymerization, which brings extra costs and environmental issues. Eliminating alkane solvents is critical to lower manufacture costs and yield a green production process.

In this study, green methods to prepare the outer selective TFC hollow fiber membranes were explored by firstly modifying the membrane substrate with polyethyleneimine (PEI) and then by water soluble small molecules such as glutaraldehyde (GA) and epichlorohydrin (ECH). Using P84 polyimide as the substrate, not only do these modifications decrease substrate's pore size, but also vary surface charge by making the membranes less positively charged. As a result, the resultant membranes have higher rejections against salts such as Na_2SO_4, NaCl and $MgSO_4$. The PEI and then GA modified membrane has the best separation performance with a NaCl rejection over 90 % and a pure water permeability of 1.74 ± 0.01 $Lm^{-2}bar^{-1}h^{-1}$. It also shows an impressive rejection of 94 % for $CdCl_2$ removal during long-term stability tests. The $CdCl_2$ rejection remains higher than 90% when testing it from 5 to 60 °C. This study may provide useful insights for green manufacturing of next-generation NF hollow fiber membranes.

Acknowledgements:

The authors would like to thank Chinese Gansu province for the research project entitled "Membrane development to removal heavy metals" with the grant number of R-279-000-360-597 and GSK-EDB Trust Fund for the project entitled "New membrane development to facilitate solvent recovery and pharmaceutical separation in pharmaceutical syntheses" with the grant number of R-706-000-019-592. The authors also acknowledge Mr. Z.L. Cheng, Dr. X. Li, Ms, Y. Zhang, Dr. Z.Z. Zhou and Dr. Y.K. Ong for all their helps and suggestions on the experimental work. Thanks are also due to Mr. K.P. Ng for his help on the NF set-up.

462463 Self-Generated RO Concentrate Pulse Backwash of Ultrafiltration Module in an Integrated Seawater Desalination UF-RO Pilot Plant

Abstract:

The effect of self-generated RO concentrate pulse backwash was evaluated in a novel integrated UF-RO seawater desalination pilot plant (with RO permeate production of up to 45.2 m^3/day at 35% feed water recovery). Direct usage of RO concentrate to backwash the UF module was further enhanced with pulse backwash (PBW) generated using bladder-type hydraulic accumulators. The accumulators were charged directly from the RO concentrate stream within a period of 30-40 s. Experimental and theoretical model analysis of the hydraulic accumulator operation (i.e., both pressure and flux time profiles), validated via a series of field experiments, demonstrated the ability to discharge the pulse over a short period along with continuous UF backwash (directly from the RO brine stream). This combined backwash strategy enabled peak UF backwash flux up to 4.2 - 4.6 times higher than the normal UF filtration flux. Along with the above backwash approach, the study successfully demonstrated automated control of the frequency and intensity of the PBW system in seawater field study. Given the rapid accumulator charging and discharge periods, UF backwash was feasible with multiple consecutive backwash pulses that was more effective than with a single pulse. Field testing of UF PBW with adaptive triggering of UF backwash and variable PBW instances per filtration/backwash cycle enabled highly effective UF operation over a wide range of water quality conditions and without the need for chemical cleaning. The results also showed that the UF system was capable of providing consistently high quality RO feed and the RO system operation was stable without deterioration in performance even during severe fouling conditions that have occurred during stormy weather.

458899 Membrane Fouling and Electrochemical Regeneration at a Sub-Stoichiometric TiO_2 Reactive Electrochemical Membrane

Abstract:

In this study, a non-invasive and non-destructive electrochemical impedance spectroscopy (EIS) characterization method was used to study membrane fouling at a novel ultrafiltration (UF) reactive electrochemical membrane (REM). The REM was fabricated from a titanium dioxide UF precursor and converted to a substoichiometric titanium dioxide using a high temperature reduction method. The EIS method was used to spatially characterize membrane fouling by interpreting the EIS data using a theoretical transmission line model. Model simulations indicate that different modes of membrane fouling (e.g., pore blocking, pore constriction, and cake layer formation) produced distinct EIS spectra. The model was used to interpret experimental EIS data collected during filtration experiments in the presence of two model membrane foulants, humic acid and polystyrene microspheres. Results indicate that the transmission line model could accurately characterize membrane fouling and was used to aid in the development of a chemical free electrochemical regeneration method. Electrochemical regeneration was able to recover the flux of a fouled REM from 3% to between 76% and 99% of the initial flux over 5 continuous fouling/regeneration cycles for humic acid fouling and full flux recovery for polystyrene microsphere fouling. The operating cost of the electrochemical regeneration method is estimated at 1.3% of that for NaOH cleaning, which indicates the great promise of this new method for separation processes.

462886 Molecular Dynamics Simulation of Water Desalination Via Functionalized Covalent Organic Framework Membranes

Abstract:

The scarcity of fresh water has become a great concern with a rapidly growing global water gap. As an energy-efficient desalination technique, reverse osmosis (RO) accounts for more than half of the existing desalination capacity. In order to screen suitable nanoporous membrane materials, and to investigate the effects of pore size and functionality, we design a series of single layer covalent organic framework (COF) membranes based on an experimentally synthesized COF structure TpPa-1. They consist of 1,3,5-triformylphloroglucinol (Tp) and *p*-phenylenediamine (Pa) with various functional groups, namely TpPa-AM2, TpPa-AMC_2NH_2, TpPa-OC_3OH, TpPa-OC_4H_9, TpPa-AMCOOH, TpPa-OBn and TpPa-AM3. The molecular dynamics (MD) simulation results indicate that water permeances in these COF membranes are affected by both pore size and functionality. Water traverse across the pore in AMC_2NH_2, TpPa-OC_3OH and TpPa-AMCOOH is enhanced, due to hydrophilic functional groups pointing inward the pores. When the polarity of functional groups is similar, the pore size becomes a dominant factor exhibiting a positive correlation with water permeance. Nevertheless, all the COF membranes show exceptionally high water permeance ranging from 1216 to 3375 kg/(m^2·h·bar), more than three orders of magnitude higher than commercial RO membranes. Furthermore, water density distribution within the membrane pores are found to be highly dependent on hydrophilicity of the pores, resulting in higher water density for hydrophilic membranes. The salt rejection is calculated to be 95.8% in TpPa-AMCOOH, and above 98.3% in other TpPa membranes. Among all the TpPa membranes designed, TpPa-OC_3OH has high water permeance (2816 kg/(m^2·h·bar)) with excellent salt rejection (99.4%). This simulation study provides a microscopic insight into the key factors governing water desalination, and suggests that TpPa-OC_3OH might be a potential candidate for high performance water desalination.

461747 Novel Amphiphilic Zwitterionic Copolymer Membranes of ~ 1nm Pore Size for Industrial Wastewater Treatment and Reuse

Abstract:

Membranes with ~1 nanometer effective pore size that allow removal of high oil/organic content (dyes, industrial chemicals, surfactants etc.) have numerous applications in the treatment of wastewater from textile plants, oil refineries, fracking and oil extraction. The small effective pore size of such membranes can potentially provide better effluent quality and prevent the environmental release of a wider range of pollutants. However, there are very few commercial membranes in this pore size range, and those that are available exhibit poor selectivity, charged surfaces which lead to complex separation mechanisms, and fouling due to the accumulation of feed components on the membrane surface. In addition to high flux, sharp selectivity and exceptional fouling resistance to common foulants in wastewaters, membranes for these applications should also have high chemical resistance and be easily fabricated into large-area membranes. Zwitterionic groups have been shown to strongly resist biomacromolecular fouling due to their high degree of hydration, which makes them promising for membrane applications. We have introduced a new class of membranes whose selective layers are made of self-assembling zwitterionic amphiphilic random copolymers. These membranes derive not only fouling resistance but also their permeability and selectivity from this self-assembled nanostructure. In this study, we evaluate the potential and performance of these new membrane materials in wastewater treatment. We synthesized random copolymers of a hydrophobic monomer (trifluoroethyl methacrylate) and two zwitterionic monomers (sulfobetaine methacrylate, phosphobetaine methacrylate) by free radical polymerization. These copolymers self-assemble into bicontinuous networks of ~1 nm nanochannels, driven by the strong electrostatic interactions between the zwitterionic groups, and documented by transmission electron microscopy (TEM). Membranes manufactured by coating these copolymers on commercially available support membranes exhibit size-based selectivity with a ~1 nm size cut-off demonstrated by filtering negatively charged dyes. We have studied how feed composition such as ionic strength and presence of foulants in varying wastewater streams affects membrane performance. Specifically, we have tested the performance of these membranes by filtering single foulants, and

simulated wastewater feeds. These membranes are extremely resistant to fouling by proteins and oil emulsions, fully retaining their permeability even during the filtration of the foulant solution in dead-end mode for up to 7 days. They are also stable in a wide range of chemical solutions: salt solutions (upto 1M NaCl), acidic and basic buffers, as well as chlorine. They exhibit a sharp, size-based cut-off for organic molecules such as dyes found in textile wastewater. Their low salt rejection allows operation at lower pressures. These results indicate the promise of this new family of membrane materials for industrial wastewater treatment, reclamation and reuse.

464906 Analysis of Air Scouring Effect in Flat Sheet Type MBR

Abstract:

MBR (Membrane Bioreactor) is a water treatment process to treat the sewage and the waste water. Membrane module is immersed in an activated sludge tank and filtrated by microfiltration membrane or ultrafiltration membrane. As compared with conventional process, activated sludge method, MBR has advantages of space saving and treated water quality, but inferior in much power consumption caused by the aeration for restraining the membrane fouling. Reduction of power consumption is the biggest challenge in MBR process. There are two methods to reduce aeration quantity, one is developing the low fouling membrane, and the other is increasing washing efficiency for membrane surface without increasing power consumption. To develop the method for increasing washing efficiency for membrane surface, it is important to understand the scouring effect for membrane surface caused by aerated bubbles. Several researches studied washing effect for membrane surface caused by aeration, but there are few researches focusing on quantifying the washing effect for membrane surface caused by aerated rising bubbles.

This study is focused on analyzing the scouring effect of rising bubbles in flat sheet type MBR. The relationship between aerated bubbles and shear stress at the membrane surface was analyzed using PIV analysis. We prepared the tank which modeled the zone between adjacent membranes in flat sheet type MBR and filmed the motion of tracer particles around a rising bubble. Velocity distribution of liquid around a rising bubble was analyzed, and then shear rate and shear stress at the membrane surface was derived. It was confirmed that the aeration condition affect the shear stress at the membrane surface by using PIV analysis. And also, relationship between shear stress at the membrane surface and washing effect for membrane surface caused by aeration was studied in model experiment. From the experiment, it was studied what kind of shear stress, such as average and maximum, has more relation with the washing efficiency. These results showed that PIV analysis was a useful method to quantify the washing effect of rising bubbles restraining membrane fouling in MBR. Also it was confirmed that the increase of shear stress at the membrane surface could improve the detachment of foulants and washing efficiency for membrane surface.

472351 Novel Composite Adsorptive Membranes for Ammonia Removal from Wastewaters

Abstract:

Natural zeolite is a superior and inexpensive adsorbent material for ammonia removal from municipal, industrial, and aquaculture wastewaters. However, zeolite used directly as an adsorbent material leads to its undesirable leaching and dissolution water. Thus, it is important to incorporate zeolite particles in a host matrix in order to improve their stability.

Novel composite fiber membranes were developed via incorporation of zeolite particles into polymeric matrices. These composite membranes demonstrate high porosity, high capacity and uniform particle distribution. It was shown that the capacity for ammonia removal increases with the increase of zeolite loading with fast ion exchange kinetics. The ammonia captured can be recovered and concentrated as fertilizers and fuels. These membranes are easy to regenerate and can be used repeatedly.

471195 Study of TL Response of Silver Nanoparticles in Borate Glasses Containing Dy^{3+} and Eu^{3+} Ions for UV and Gamma Dosimetry

Abstract:

Abstract

Dy^{3+} and Eu^{3+} doped the borate glasses samples containing Ag nanoparticles were prepared using a conventional melt-quenching method. The synthesized materials were characterized by scanning electron microscopy (SEM) and their morphology and sized distributions were determined. The photoluminescence (PL) spectra as well as UV-Vis absorbance were also analyzed and show the presence of the rare-earth in the matrix. The thermoluminescence (TL) response of Dy^{3+} doped lithium borate glasses irradiated with ultraviolet (UV) radiation was studied in detail before and after doping with silver.

471313 Effects in Morphology and Thermoluminescent Characteristics of Lif Crystals Synthesized By Using Nonionic and Cationic Surfactants

Abstract:

LiF crystals doped with Ce^{3+} and Mg were prepared using co-precipitation, using as solvent a mixture of ethanol and water in presence of Nonionic (PVP and Pluronic f-127) and cationic (CTAB) surfactants. The LiF crystals were characterized by scanning electron microscopy (SEM) obtaining average sizes of 100 nm, a single cubic crystalline phase was determined by XRD, the PL emission spectra shows two bands characteristics of Ce^{3+} and UV-Vis absorbance shows the presence of the rare-earth in the LiF crystals. The TL response to both X-ray and gamma from a Cs-137 source irradiation was studied.

470030 Thermoluminescent Dosimetric Analysis of Ag and Cu Doped LiB3O5

Abstract:

Methods

Production

Synthesis of Ag and Cu doped LiB_3O_5 crystals was carried out by precipitation, using water as a solvent. Impure crystals were produced at a synthesis temperature of 70 °C. A 5g of Ag and Cu doped LiB_3O_5 precipitate was obtained upon evaporation of water in the solution. The precipitates were calcined at a temperature of 750 °C for one hour. The calcined materials were grounded finely and subjected to severalcharacterizations.

Characterization

SEM analyses were performed for the doped and undoped samples of LiB_3O_5. The XRD properties of the samples were obtained. The UV-Vis absorbance spectra of the samples were obtained at room temperature. Spectra were recorded between 200 and 1200 nm. The photoluminescence (PL) measurements were carried out. The glow curves of the samples has been obtained using TLD reader.

Results

The XRD patterns obtained for the samples showedthat production of LiB_3O_5 with orthorhombic structure was successful since allof the main peaks of LiB_3O_5 coincide with the JCPDS data. It was seen that the dopant added did

not interfere withthe crystal structure. As it can easily be detected addition of the dopants didnot have any significant effect on the crystal structure of LiB_3O_5 produced.

SEM was carried out to check the morphologies of the samples.According to the SEM images, the samples produced by the co-precipitation method shows the agglomerated surface.

The UV-Vis absorption spectra for the samples showed that the maximum absorbance was observed at 250nm. The result showed that a clear monotonic and increasing dependence of the intensity of the absorbance as a function of dopants.

The dosimetric properties of LiB_3O_5 doped with Ag and Cu were determined by using HarshawTLD Reader Model 3500. The heating rate was chosen as 10°C/s was usedfrom 50 to 350°C. The examined dosimeters were exposed to X-rays at room temperature. The sample amount was 0.5g. The samples have shown the TL response. The pattern of the glow curves changed with the change of dopants.

Conclusions

The TL response of LiB_3O_5 has been compared with the commercial TLD100. The observation concludes that the doped LiB_3O_5 has the TL behavior. The glow curves of the LiB_3O_5 shows different shape with the change of dopants.

469662 Effect of Thermal Treatment on the Characteristics of PES/PVA Nanocomposite Membranes Modified with TiO2 Nanoparticles :a Comparitive Study Between 1-Step and 2-Step Thermal Treatment

Abstract:

PES/PVA thin film composites were modified with TiO2 nanoparticles via dip coating method. Thermal treatment of total modified membranes plays a semi-"catalytic role" in coating process. Hence, 4-temprature levels were selected and investigated, noticing that there is an important restriction due to the stability limitation of polymers under thermal treatment process. This phenomenon also leads to the concept of 2-step treatment. 2-step thermal treatment is a more moderate process which leads to a more stable structure.

Polyethersulfone (PES) ultrafiltration membrane was prepared via phase inversion induced by immersion precipitation. Poly(vinyl-alcohol) (PVA) (1.5 wt%) dissolved in distillated water was coated on the surface of PES barrier layer with dip coating technique. The PVA-coated membranes were then immersed in the cross-linking solution containing 5 wt.% of GA and 0.5 wt.% of H2SO4 for 2min at room temperature in order to reduce the membrane swelling by chemical cross-linking. H2SO4 was used as catalyst of reaction. TiO2 nanoparticles were dispersed in distillated water by sonication for 1hr and stirred vigorously for 30 min. the PVA coated membranes were immersed in 0.2 wt% TiO2 solution for 15 min. the last step is thermal treatment which in this study, four level of temperature (85,95.105.115) were investigated. All modified membranes were thermal treated for 3 min.

The performance of modified membranes was tested with filtration of yeast industrial wastewater(COD=11500 mg/L). Permeate flux and COD of four modified membranes have tested. According to the results, the PES/PVA/TiO2 membrane that was thermal treated in 105 had the best output.

469788 Synthesis, Structural and Morphological Characterization of Cu and Ag-Doped $Li_2B_4O_7$

Abstract:

Method:

In this paper the lithium tetra borate (LTB) was produced by water/solution assisted synthesis method. After the synthesis, metals (doping with Copper (Cu) and Silver (Ag)) were doped into lithium tetra borate ($Li_2B_4O_7$) to enhance its thermo luminescent properties. The heating temperature was 750°C for 2 hr as well as 150°C for 1hr the sample produced by water solution method with 0.1% Cu and 0.1% Ag. The pellets were irradiating by the X-ray.

***Result*:**

The identification and characteristics of undoped and doped $Li_2B_4O_7$ were determined by X-ray diffraction (XRD), Scanning electron microscope (SEM), Photoluminescence (PL), Ultra violet visible spectroscopy (UV Vis). The chemical compositition and its morphologies of the obtained LTB and LTB: Cu, Ag was confirmed by XRD and SEM results. The most intense peak of the XRD pattern of the lithium tetraborate sample is verified by the reference data of the lithium tetra borate with tetragonal structure. The glow curves were obtained by thermo luminescence (TL) reader. The glow curves of the pellets exposed to different doses exhibited a clear response to X-ray irradiation, which is a significant evidence that this material is a candidate for radiation dosimetry. Finally this energy response was also compared with TLD100. The comparison of the results showed that the obtained lithium tetra borate have good thermo luminescence (TL) dose response.

Enhancement of Heat Transfer Coefficient by Using Fe_2O_3 –Water Nanofluids

N.I. Zouli[1], *Muthanna Al-Dahhan*[1]
[1]*Missouri University of Science and Technology, Rolla, MO, 65409*
Email: niz8vb@mst.edu

Introduction

Research efforts have been made to increase the heat transfer rate, energy, and efficiency. Some of these efforts include the application of fins, compact heat exchangers, microchannel, and many others. Recently, nanofluids have been introduced to enhance the heat transfer rate of the heat transfer fluids by adding a certain concentration of nanoparticles into the heat transfer media. Most of the work that has been published in the literature focuses on measuring the thermal conductivity of the nanofluid or measuring the overall heat transfer coefficient. Among the efforts for enhancement of heat transfer the application of additives to liquids is noticeable (Bergles, 1973; Ahuja, 1975). Choi (1995) was the first who studied the enhancement of the nanoparticles on the heat transfer coefficient and showed that the nanoparticles could considerably affect the thermal conductivity of the nanofluid to increase. Since Choi is work (Choi,1995), various studies have been conducted on nanofluids and these studies shown that nanofluids increase heat transfer properties and stability while reducing costs, clogging, and energy required to pump the fluid. Heris et al. (2006) found that the heat transfer coefficient increased as the volume concentration of the nanofluids increased when the fluid was composed of water and metal oxides.

In the present study, effects of Fe_2O_3/water on the nanofluid's heat transfer and thermal conductivity will be discussed. The experimental investigations are done to study heat transfer performance of the nanofluid at the entrance region under laminar flow condition varying the particle size and volume concentration of nanoparticles.

Experimental setup and methodology

Schematic of the experimental setup used in the current study is shown in Figure1. It consists of a flow loop made of copper material, a heating unit, a measuring and control unit. A straight copper tube with 190.5 mm length, 25.4 ± 0.02 mm inner diameter, and 31 ± 0.05 mm outer diameter was used as the test section. The heat transfer probe comprises of a heat flux sensor (0.15 in x0.08 in) with a fast response time of about 0.02 sec and a thermal resistance of 0.003 °F/BTU/ft2hr. The sensor was placed in a slit mounted inside the test section wall flashed to the wall without disturbing the flow. The probe measures the heat flux and the surface temperature. In addition, two thermocouples (type T) were mounted inside the test section to measure the bulk temperature. The sensor was placed on the middle distance between the two thermocouples.

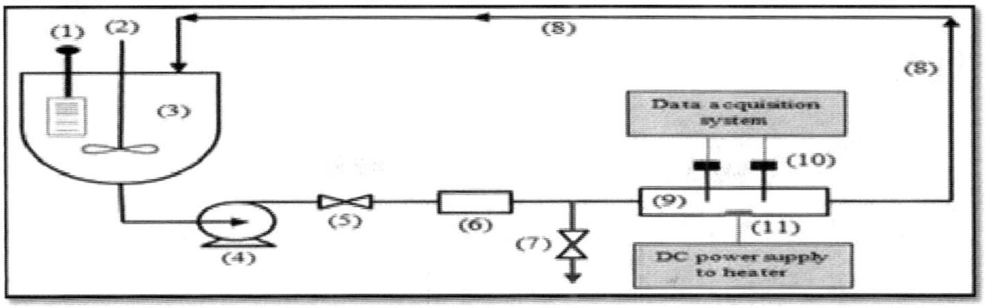

Figure 1. Schematic of experimental setup components: (1) Heater, (2) Agitator, (3) Reservoir tank, (4) Pump, (5) Needle valve, (6) Flow meter, (7) Drainage valve, (8) Copper tube, (9) Test section of copper tube, (10) Thermocouples, (11) Heat flux foil sensor.

Heat flux, temperature of the sensor surface, and the bulk media temperature are directly measured and recorded using Data acquisition system. Further they are used to estimate local instantaneous and averaged heat transfer coefficient using eqn (1) and (2):

$$h_i = \frac{q_i}{T_{si} - T_{bi}} \; ; \; \Delta T_i = T_{si} - T_{bi} \qquad (1)$$

$$h = \frac{1}{n} \sum_{i=1}^{n} \frac{q_i}{T_{si} - T_{bi}} = \frac{1}{n} \sum_{i=1}^{n} h_i \qquad (2)$$

where h_i: local instantaneous heat transfer coefficient ($kW/m^2.K$)
q_i: instantaneous heat flux measured by the sensor (kW/m^2)
T_{si}: instantaneous surface temperature of the probe (K)
T_{bi}: instantaneous bulk temperature of the media (K)

Results and discussion

The thermal conductivity of Fe_2O_3 water based nanofluids was measured at different concentrations. From Figure 2 an enhancement of 22.35% in the thermal conductivity value has been achieved with 0.09% volume concentration of Fe_2O_3 nanoparticles in water with an average diameter of 20 nm. As can be seen from Figure 2, the thermal conductivity enhancement increases linearly with increased particle volume concentration.

The concentrations for each test run were increased from 0.02-0.09% by volume and run at a constant flow rate of 6 L/min. The experimental results were obtained with 20 nm particles. And it was found that the average increase in the heat transfer coefficient was between 5.95-18.34% with the highest increase being 73.7% at 0.09% volume concentration at 5 mins into the process. It was observed that as the concentration of the nanofluid was increased so did the heat transfer coefficient. The enhancement in heat transfer coefficient was calculated by following;

$$[(h_{nf} - h_{bf}) / h_{bf}] * 100\% \qquad (3)$$

h_{nf} = heat transfer rate of nanofluid
h_{bf} = heat transfer rate of base fluid

A typical result for 20 nm particles can be seen in Figure 3. Overall the results show that the local heat transfer coefficient increases with increasing the size and the volume fractions of the Fe_2O_3 nanoparticles compared to water. The effect of Reynolds number on Nusselt number is shown in Figure 4. It is seen that the measured Nusselt numbers for nanofluids are higher than those of base fluid, and they increase remarkably with Reynolds number. The observed enhancement of the Nusselt number could be due to the suppression of the boundary layer, the viscosity of nanofluids as well as the dispersion of the nanoparticles.

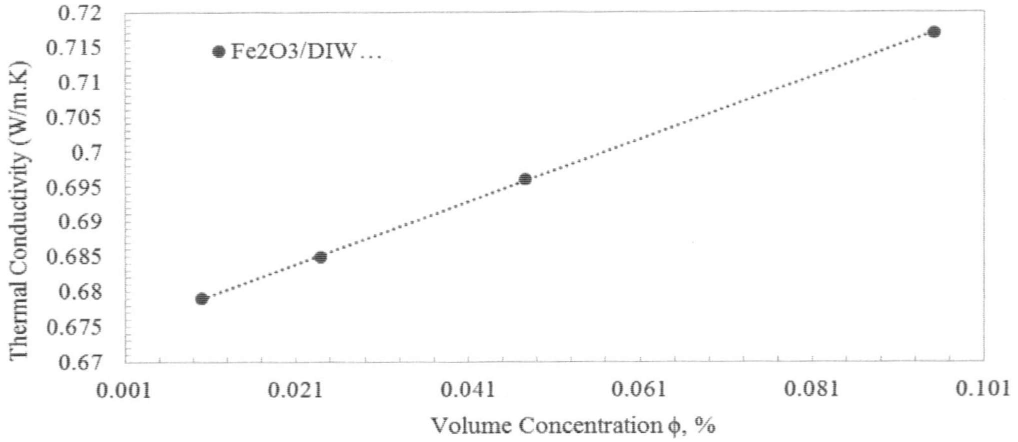

Figure 2. Thermal conductivity of Fe_2O_3 water based nanofluids at different volume fractions (0 and 0.09 %).

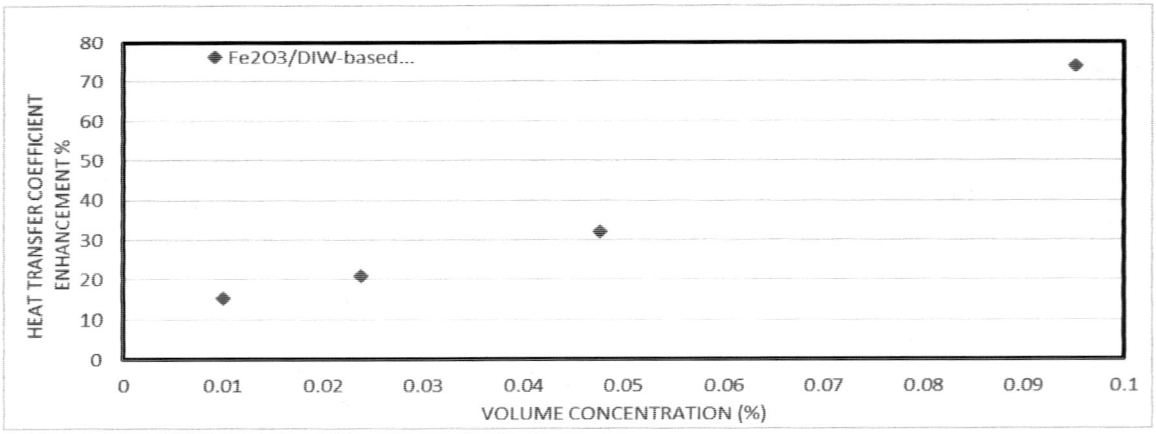

Figure 3. Variation of heat transfer coefficient with different volume concentration of Fe_2O_3 water based nanofluids.

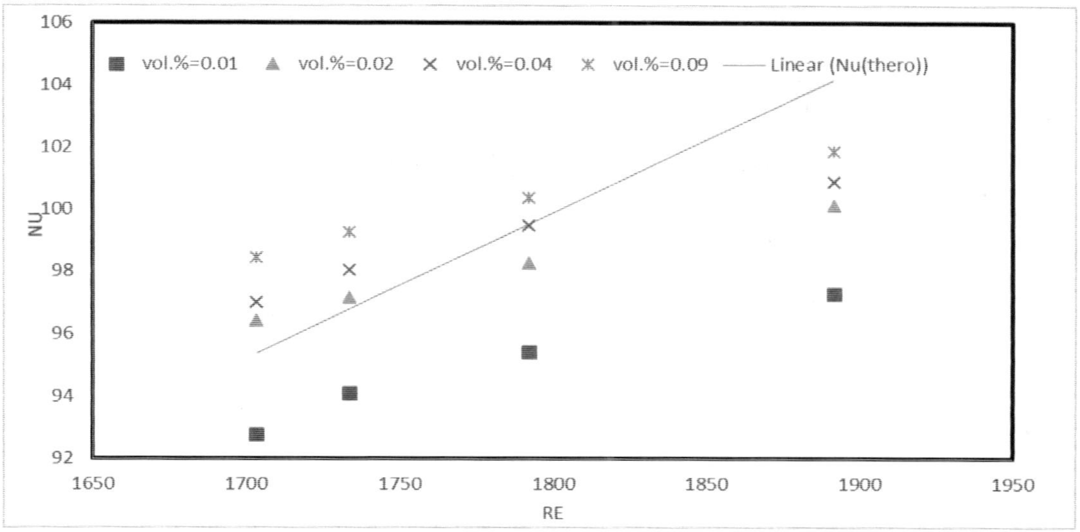

Figure 4. Effect of Reynolds number and Fe_2O_3 nanoparticle concentration on Nusselt number

Conclusions

The paper is mainly concerned with the heat transfer of nanofluids made of water and Fe_2O_3 nanoparticles in the laminar flow regime. It was found that use of aqueous Fe_2O_3 nanoparticles can significantly enhance heat transfer coefficient in laminar flow regime, and the enhancement increases with Reynolds number, as well as particle concentration under the conditions of this work. The enhancement of the heat transfer coefficient could not be only credited to the enhancement of the effective thermal conductivity. A reason for the enhancement, Particle passage which lead to reduce the thermal boundary layer thickness and a non-uniform distribution of thermal conductivity.

Acknowledgments

I would like to acknowledge the financial support provided by the government of Saudi Arabia. I would also like to thank Prof. Al-Dahhan for his supervision and support which made this work possible.

References

1. Brooke, A., Kendrick, D., Meeraus, A. A. (1992). GAMS- A User's Guide (Release 2.25). *The Scientific Press*. San Francisco, CA.

2. Gooding, W. B., Pekny, J. F., McCroskey, P. S. (1994). Enumerative Approaches to Parallel Flowshop Scheduling via Problem Transformation. *Comput. Chem. Eng.*, *18*, 909.

3. OSL. (1991). *OSL, Guide and Reference (release 2)*, IBM: Kingston, NY.

4. Zhang, X., Sargent, R. W. H. (1994). The Optimal Operation of Mixed Production Facilities - A General Formulation and some Approaches for the Solution. *In Proceedings of the 5th International Symposium on Process Systems Engineering*. Kyongju, Korea, 171.

5. A.E. Bergles, Recent development in convective heat transfer augmentation, Appl. Mech. Rev. 26 (1973) 675–682.

6. A.S. Ahuja, Augmentation of heat transport in laminar flow of polystyrene suspension: I. Experimental and results, J. Appl. Phys. 46 (8) (1975) 3408–3416.

7. Choi, S.U.S. Enhancing thermal conductivity of fluids with nanoparticles. in American Society of Mechanical Engineers, Fluids Engineering Division (Publication) FED. 1995.

8. Heris, S.Z., S.G. Etemad, and M.N. Esfahany, *Experimental investigation of oxide nanofluids laminar flow convective heat transfer.* International Communications in Heat and Mass Transfer, 2006. **33**(4): p. 529-535.

447375 Eco-Friendly Dyeing of Electrospun Cellulose Nanofibers with Reactive Dye Using Ultrasonic Energy

Abstract:

Nanofibers have created enormous opportunities and challenges for the textile and apparel industry, moreover the innovative methods for dyeing of nanofibers have recently gained more attention. Electrospun cellulose nanofibers are increasing significant interest in textile industry in view of its incomparable properties. However, the dyeing of nanofibers faces problem of lower color yield due to the higher surface area of nanofibers. In this study electrospun cellulose nanofiber dyed with using cold pad batch dyeing, a well-known and ecofriendly method of dyeing for cotton base fibers. The dyeing of cellulose nanofibers was carried out with CI Reactive Violet 5 by conventional and ultrasonic (US) method. Scanning electron microscopy and Fourier transform infrared spectroscopy were used to investigate morphology and chemical structure of electrospun cellulose nanofibers before and after dyeing and the color strength (K/S value) and color fastness of dyed nanofibers also assessed.

The results indicates that the use of ultrasonic energy not only shortens the batching time but the color strength (K/S),color fixation and color fastness also considerably increased ,the K/S value of dyed samples enhanced around 5-8% compared with conventional method. Although, washing fastness of sample increased up to 25%, but, color fastness to light was the same as conventional method.

The results proved that, the morphology of dyed cellulose nanofibers with both dyeing method were smooth with structural uniformity and remained unaffected after being dyed.

452910 Combined Quartz Crystal Microbalance with Dissipation and Generalized Ellipsometry to Characterize the Deposition of Titanium Dioxide Nanoparticles on Model Rough Surfaces

Abstract:

Titanium dioxide nanoparticles (TiO_2NPs) are the most extensively manufactured engineered nanomaterials, and understanding of their interactions with soil and sandy aquifer materials is necessary for effective natural resource management [1]. Quartz crystal microbalance with dissipation monitoring (QCM-D) has been widely used to study TiO_2NPs at water-metal oxide interfaces as model systems for soils and aquifers. However, natural sands and sediments exhibit various degrees of surface roughness, and the effects of surface roughness are not accounted for in current QCM-D studies of nanoparticle adsorption, which use the Voigt-Kelvin model [2].

A new mechanical model that considers both viscoelastic and surface roughness effects [3] is described to analyze QCM-D measurements. Obtained parameters include adsorbate areal mass density, adsorbate viscoelastic properties, the vertical roughness length, and the permeability length. Silicon slanted columnar thin films (SCTFs) were deposited onto QCM-D sensors by electron-beam glancing angle deposition and conformably coated with an alumina film via atomic layer deposition to act as model rough surfaces. The adsorption of polyacrylate-stabilized TiO_2NPs onto flat surfaces and the SCTFs was measured by QCM-D and generalized ellipsometry (GE), an optical reference technique.

We report that the surface-roughness-modified mechanical model excludes water that is retained by the SCTF surface roughness and instead considers the surface mass density of TiO_2NPs and water coupled to the TiO_2NPs [4]. GE is also sensitive to the attachment of the TiO_2NPs. When the results of both techniques are combined, new information, the hydration of the polyacrylate-stabilized TiO_2NPs, can be evaluated.

[1] B. Kim et al. J. Environ. Monit., **14**, 1128 (2012).

[2] M.V. Voinova et al. Phys. Scr., **59**, 391 (1999).

[3] L. Daikhin et al. Anal. Chem., **74**, 554 (2002).

[4] N. Kananizadeh et al. J. Hazard. Mater., In press. dx.doi.org/10.1016/j.jhazmat.2016.03.048

465346 Use of Monte Carlo Simulation Techniques Assess Remedy Complete Time Frame for First Order Contaminant Decay with Concentration Fluctuations

Abstract:

Many sites undergoing in situ treatment for dilute groundwater plumes utilize monitored natural attenuation (MNA) as a last element of a treatment train approach after a period of active treatment, and with MNA, groundwater concentrations commonly exhibit first-order concentration decay characteristics for individual monitoring wells. In some instances, attainment of a remedy complete (no further action or no monitoring required) is hindered by slow contaminant attenuation rates exhibited by the existing data. If the existing data also exhibits concentration fluctuations in addition to the first order decay trend, it can be difficult to determine a date of future cleanup objective attainment with acceptable uncertainty, as is frequently needed by the financially responsible parties. It is common practice to extend a decay trend into the future to determine the expected date of cleanup achievement, using first order attenuation rates. This approach gives only a single future date, however it is evident that, with a case of significant fluctuating concentration data, a single date will not best describe the future occurrence. A range of dates representative of both the decay trend and the concentration fluctuations is a more suitable representation, and likely much more helpful for the responsible part in their program planning. A Monte-Carlo approach to future groundwater concentration prediction with first order attenuation is presented, which provides a range of dates at which attainment of the groundwater cleanup level is anticipated.

449380 Destruction of Hazardous Heterocyclic Nitrogen Containing Hydrocarbons By Supercritical Water Oxidation

Abstract:

SCWO of Nitrogen Containing Heterocyclic Hydrocarbons Found in Hazardous Aqueous Waste Streams: Process Enhancement by Addition of Isopropyl Alcohol

B. Al-Duri[1*] and F. Alsoqyiani[1]

[1]School of Chemical Engineering, University of Birmingham, Edgbaston, Birmingham B15 2TT, United Kingdom.
Email: B.Al-Duri@Bham.ac.uk

Abstract

This work investigates the destruction nitrogen-containing heterocyclic hydrocarbons frequently encountered in hazardous wastes, by supercritical water oxidation (SCWO), considering the process enhancement by adding isopropyl alcohol (IPA) as co-fuel. DBU$ is selected for SCWO in a continuous plug flow reactor, under a temperature range of 400 to 575¡C, oxidant ratio of 0.8 to 2 times stoichiometric value, initial DBU concentration of 5mM (reactor conditions) and 25 MPa reactor pressure. IPA was premixed with DBU at various IPA/DBU ratios (0.5 to 3.5). Experimental results were presented in terms of total organic carbon (TOC) removal %, and nitrogenous products yield %. Based on GC-MS analysis exit streams, a reaction pathway from DBU to benign small compounds was proposed.

Results indicated that the destruction of N-containing hydrocarbons occurs by virtue of the vigorous free radical mechanism where HO¥ and HO_2¥ were responsible for the scission of the C-N bonds, followed by a series of C-C and C-N scissions. Temperature was found to be the main factor to influence overall destruction of DBU, while oxidant ratio predominantly affected the N speciation in the exit stream. Alcohol addition enhanced the destruction of DBU in terms of TOC removal %. Specifically, it enhanced the conversion of recalcitrant ammonia (NH_4^+) to gaseous nitrogen, which is highly advantageous in SCWO applications to nitrogenous waste destruction.

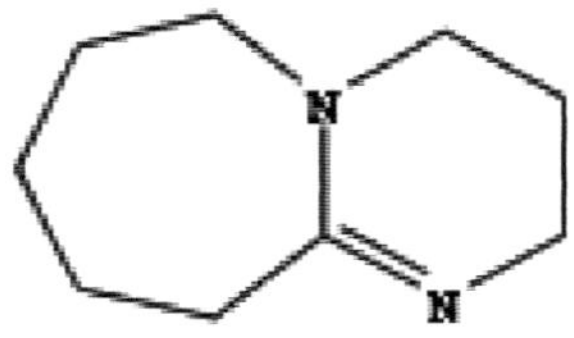

DBU

---

* Corresponding Author

$ 1,8-Diazabicyclo [5.4.0] undec-7-ene

463608 Leaching and Recovery of Vanadium from Carbon Soot

Abstract:

Carbon soot, a solid waste produced in the petrochemical industry, typically contains heavy metals that originate from crude oil. Due to the presence of these hazardous heavy metals, the unsafe disposal of carbon soot may lead to pollution that affects the environment and human health. Therefore, it is required that the carbon soot be treated before disposal to minimize any negative impacts. In this study, carbon soot containing high level of Ni and V may potentially be recovered as a valuable resource. The carbon soot could also be used as adsorbents or further upgraded to activated carbon after removal of heavy metals.

In this work, the metal of interest is V as the carbon soot contains >10,000 ppm of V. Since V is highly valuable in the market especially in the alloy industry, there is a motivation for its recovery. The leaching of V by chemical and microbial means at different conditions is being investigated. Following the leaching step, the recovery of V into solid form using selective precipitation by step-wise pH modification of the leachate is also being examined.

466285 Fundamental Study on Treatment of Osmium-Containing Wastewater

Abstract:

Osmium tetroxide (OsO_4) is often used in microscopic analysis as a dyeing reagent and in some organic reactions as an oxidizer. Most of osmium compounds including metallic osmium are easily oxidized under ambient condition into OsO_4. Although OsO_4 is highly toxic substance, wastewater containing OsO_4 (highly soluble in water) is extremely hard to treat because of OsO_4's high fugacity into gas phase. Therefore, proper treatment method of osmium-containing wastewater has not been established yet. Considering that osmium is a precious metal, recovery of OsO_4 or metallic osmium from wastewater for subsequent recycling is preferable rather than just diluting the wastewater for dumping. In this study, various treatment methods were tested and the behavior of osmium compounds was studied.

Firstly, reduction of OsO_4 by simple reductants such as ethanol and acetaldehyde was investigated in various conditions of aqueous solutions. By using ethanol, OsO_4 was successfully reduced into water-insoluble species. For example, when 0.1 mL of ethanol was added into 3 mL of 100 ppm OsO_4 solution adjusted at pH = 13, OsO_4 concentration decreased rapidly in several minutes due to reduction by ethanol, and by subsequently adjusting its pH to 3 the osmium compound precipitated and residual osmium concentration in water phase became as low as 0.02 ppm. Considering treatment of real wastewater, effect of co-existing material such as phosphate buffer upon the behavior of OsO_4 was also studied.

Secondly, treatment of osmium-containing wastewater using both supercritical water oxidation (SCWO) and supercritical carbon dioxide extraction was investigated. SCWO oxidizes all the osmium compounds into OsO_4 and simultaneously gets rid of co-existing organic matter completely; and subsequent supercritical CO_2 process extracts OsO_4, which has high octanol/water partition coefficient, from water phase. Behavior of osmium compounds in this oxidation/extraction system will be reported.

457350 Factors Effecting the Simultaneous and Consecutive Reduction and Adsorption of Anionic Cr(VI) and Orange II Dye Onto Chemically Modified Masau Seeds

Abstract:

Factors effecting the simultaneous and consecutive reduction and adsorption of anionic Cr(VI) and orange II dye onto chemically modified masau seeds

Ahmad B. Albadarin*, Maurice N. Collinsâ€Ž, and Gavin Walker

Bernal Institute, Dept. Chemical & Environmental Science, University of Limerick, Limerick, Ireland **Tel.: +353-(0)-61-237732, E-mail:* Ahmad.B.Albadarin@ul.ie

Novel chemically modified masau seeds (MMS) were prepared and used to investigate the solution chemistry factors influencing the reduction of Cr(VI) and adsorption of orange II (OII) dye in single, consecutively (after one another) and multicomponent systems. Epichlorohydrin and diethylenetriamine were employed as etherifying and crosslinking agents, respectively. The effect of reaction time, solution pH, initial concentration, temperature, loading order on reduction/adsorption mechanisms were examined. Additionally, characterisation techniques such as: BET, FT-IR, SEM-EDX and XPS were employed to investigate the physio-chemical properties of the modified masau before and after the adsorption of dye II orange and Cr(VI).

The maximum adsorption capacity of Cr(VI) and OII onto MMS reached approx. 50mg/g (pH = 2, C_o = 100mg and

dose = 2g/dm^3). This is due to the zeta potential changes in the surface charge of the MMS (surface is positive below pH = 6). A higher concentration of positively charged hydrogen ions were achieved at a lower pH resulting in increased levels of protonated amine groups of MMS distributed on its surface.

Park redox and pseudo second-order models were able to adequately predict the kinetics of the reduction of Cr(VI) and adsorption of OII dye, respectively. The experimental data showed that there are no competitive interactions between the dye molecules and Cr(VI) ions. However, Cr(VI) is reduced/adsorbed more easily than the dye molecule and more notably, Cr(VI) diffusivity is about four times higher than the dye diffusivity.

465053 Removal and Recycle of Boric Acid from Waste Water with Electrodalysis

Abstract:

Removal and recycle of boric acid from waste water with multi-stage electrodialysis

Yurina Nakamura, Atsushi Iizuka, Tadashi Shoji, Miyuki Noguchi, Akihiro Yamasaki

The multi-stage electrodialysis method was applied to waste water containing boric acid for recycling boric acid as solid. The electrodialysis unit was composed of two cells formed by an anion exchange membrane (AEM) sandwiched by two bipolar membranes (BPM), and five or ten units was stacked to form the electrodialysis equipment. The cell located at the higher potential side was called the concentration cell, and that at the lower potential side was called the feed cell. The sodium borate solution of which the initial concentration was set 100 ppm was circulated in a batch style through each cell for 60 or 120 min. The flow rate through each cell was changed in the range of 0.1 to 1.0 L/min, and the volume of the circulation solution through the feed cell to that through the concentration cell (the volume ratio hereafter) was changed in the range of 2 to 8. The overall electric potential applied in the electrodialysis equipment was constant at 12.5 V. The time changes of boron in the cells were monitored with ICP-AES.

The concentration of boron in the feed side decreased and that in the concentration cell increased with time, and almost leveled off at the later stage of the electrodialysis. This result indicates that the boric acid was transported through the AEM driven by the electric potential applied along the cells. The boric ions $B(OH)_4^-$, which is the dominant form of boric acid at pH higher than 9.14, in the feed cell would be transported to the concentration cell through the AEM, and the boron concentration decreased with time. The removal anions in the feed cell were compensated by hydroxyl ions supplied from the BPM. In the concentration cell, the boric ions would be coupled with protons supplied from the BPM to form non-valent boric acid, and the boron concentration increased with time. The boron concentration in the concentration cell at the later stage of the electrodialysis depended on the volume ratio; the higher volume ratio resulted in the higher boron concentration. When the ratio was 4, the boron concentration in the concentration cell increased at about 400 ppm at 60 min of the operation. The concentrated solution was then introduced to the second stage of the electrodialysis. As a result, the boron concentration in the concentration cell was 1500 ppm, while that in the feed cell was 286 ppm at 60 min. The solution in the concentration cell was further introduced to the third electrodialysis unit, and after 60 min, the boron concentration of the concentration solution was 4000 ppm, and at the fourth stage the boron concentration in the concentration cell was about 10000 ppm, which is close to the solubility of boric acid. The solid boric acid would be easily recovered from the concentrated solution, of which the concentration is close to the solubility. The sequences of the boron concentrations in the feed and concentration cell depended on the flow rate as well as the volume ratio, and the number of the stages required to be concentrated to the concentration close to the solubility of boric acid depended on both of them. On the other hand, at the initial stage of the electrodialysis, the boron concentration was decreased below the Japanese standard of the waste water for boron at 10 ppm. The present method can be applied to the treatment of waste water containing boric acid to achieve simultaneous removal and recycle of boric acid and their salts.

468153 The Top-Ref Approach to Improve the Resource Efficiency of Energy Intensive Industrial Processe

Abstract:

TOP-REF project aims at developing and validating indicators, methodologies and non-invasive tools devoted to the improvement of resource efficiency in energy intensive continuous industrial processes in the agro-chemical, chemical and petrochemical sectors. These methodologies and tools will constitute to the following targets:

- To path the way for achieving the SPIRE2030 objective of reducing non-renewable, primary raw material intensity up to 20 % and fossil energy intensity up to 30%, both compared with current levels.
- Reduce production costs up to 15 % compared with the current levels.
- A significant reduction of the environmental impacts (CO2 emissions, water footprint, pollutants, hazardous emissions, etc.).

To achieve these impacts, TOP-REF develops and demonstrates a robust, resource-efficiency-focused and cross-sectorial methodology based on novel exergy-based resource indicators. This methodology is implemented in three industrial demonstration plants: solid fertilizer production, crude distillation and fractionation unit of a refinery and a naphtha steam cracker. The tools developed are validated by demonstration under real conditions in three pilots.

The methodology developed by TOP-REF consists of the following steps:

1. Development of Key Resource Indicators (KRIs) to measure the performance in resource efficiency
2. Modeling and simulation of the three industrial pilots
3. Identification of the critical process parameters (CPPs) influencing the KRIs
4. Optimization of the CPPs to improve the resource efficiency

5. Demonstration and validation under real conditions

The approach to identify the critical process parameters for the production processes investigated is based on detailed process models validated by the industrial partners. These process models are analyzed using global sensitivity analysis to identify the critical process parameters. As the full simulation models are very complex, simplified mathematical surrogate models are derived prior to global sensitivity analysis. A variance-based global sensitivity analysis is performed for all process parameters to identify the most influential operating parameters on the key resource indicators (KRIs) developed within the TOP-REF project. The use of a variance-based sensitivity analysis enables the consideration of interaction effects among the process parameters. Finally, the identified critical process parameters are optimized using the KRIs as objective. The optimized operating points will be validated under real conditions in the three production plants.

Within the contribution we will explain the TOP-REF approach developed, show its application to the three industrial processes and highlight experiences made during the application.

Acknowledgment

The research leading to the results presented in this contribution has received funding from the European Communityâ €˜s 7th Framework Program for research, technological development and demonstration under grant agreement No 604140, TOP-REF.

468313 Novel Data Envelopment Analysis Approach for Sustainability Assessment: Application to Electricity Generation Technologies

Abstract:

Moving towards sustainable development implies implementing actions, projects, programs, plans and policies, which involve the balanced integration of economic, environmental, and social objectives. However, setting sustainability goals requires some knowledge and understanding of the current level of sustainability. To this end, sustainability assessment approaches can be applied for measuring the extent of progress towards sustainability and for defining quantitative targets that ensure a more sustainable development.

In this work we explore the use of data envelopment analysis (DEA) to quantify the level of sustainability attained by a system and identify targets for improvements. DEA is a non-parametric linear programming (LP) technique that measures the efficiency of a set of entities, termed decision-making units (DMUs), each transforming multiple inputs into multiple outputs (Charnes et al., 1978). By solving the DEA LP problem, it is possible to distinguish between 'efficient units' (i.e., efficiency score of 1) and 'inefficient units' (i.e., non-frontier units with an efficiency score strictly lower than 1), for which DEA provides in turn specific guidelines to become efficient by projecting them on the efficient frontier.

In the context of sustainability assessment, DEA presents two major limitations particularly critical when a wide range of economic, environmental and social indicators are considered in the analysis. First, the standard DEA makes no distinction between the units deemed efficient, which makes it difficult to select a final alternative in the absence of a ranking scheme. Secondly, DEA is very sensitive to the set of inputs and outputs, leading to very different results depending on the sustainability criteria considered in the analysis.

To overcome the limitations of the standard DEA as applied to sustainability studies, this work proposes an enhanced DEA methodology tailored to carrying out sustainability assessments. The methodology proposed integrates DEA with the concept of "order of efficiency", originally introduced to rank Pareto optimal solutions from a set of many Pareto points (Das, 1999). In essence, the main idea behind the new methodology consists of repeating the DEA calculations iteratively for each and every possible combination of inputs/outputs, to subsequently aggregate the results into efficiency metrics that assess each sustainability dimension separately and finally determine an overall sustainability level integrating the three scores (obtained in each dimension).

The capabilities of the methodology proposed are illustrated through its application to the assessment of the electricity-generation technologies expected to play a major role in the future electricity mix of United Kingdom (i.e., nuclear, gas, coal with and without carbon capture and storage, wind, solar photovoltaics and biomass with wood and *Miscanthus* pellets). Each technology is defined as a DMU that is characterised by a set of economic, environmental and social sustainability indicators determined using a life cycle assessment approach by Stamford and Azapagic (2014). We consider 18 sustainability indicators: three economic (capital cost, operation and maintenance cost, and fuel cost), nine environmental (freshwater ecotoxicity, marine ecotoxicity, global warming, ozone depletion, acidification, eutrophication, photochemical smog, land occupation, and land ecotoxicity) and six social (direct employment, worker injuries, human toxicity potential, radiation, depletion of elements, and depletion of fossil fuels).

The results generated with our approach show that gas, nuclear and wind technologies are efficient in each dimension of sustainability when all the indicators within each such dimension are considered. Gas is also the most economically and socially efficient (highest economic and social efficiency scores), while nuclear is the best in the environmental dimension. From an overall sustainability point of view, the gas technology presents the highest overall sustainability efficiency (mainly because of its very good performance in the social dimension), followed by nuclear and wind. On the other hand, coal is found economically inefficient; coal, solar PV and both biomass options are environmentally inefficient; and coal and biomass with *Miscanthus* are found socially inefficient. For these technologies deemed inefficient in a particular sustainability dimension, an inefficiency assessment is performed which provides targets that should be attained to reach the efficiency in each specific dimension.

The enhanced DEA proposed can facilitate the transition to a more sustainable future, since it enables ranking alternatives according to the extent to which they adhere to sustainability principles; and also pinpoints in a systematic manner the main sources of inefficiency and sets targets for improvements.

REFERENCES

Charnes, A., Cooper, W.W., Rhodes, E., 1978. Measuring the efficiency of decision making units. Eur. J. Oper. Res. 2, 429–444. doi:10.1016/0377-2217(78)90138-8

Das, I., 1999. A preference ordering among various Pareto optimal alternatives. Struct. Optim. 18, 30–35.

Stamford, L., Azapagic, A., 2014. Life cycle sustainability assessment of UK electricity scenarios to 2070. Energy Sustain. Dev. 23, 194–211. doi:10.1016/j.esd.2014.09.008

471458 Optimal Design of Sustainable Agricultural Water Networks

Abstract:

This work proposes a multi-period mixed integer non-linear programming model for the optimal design of agricultural water networks, which involves the optimal reuse of resources such as water, fertilizers and pesticides. This model is based on a new superstructure that includes attractive options for the use, reuse, recycle, regeneration and storage for the considered resources. The model incorporates simultaneously the maximization of the net annual profit and the minimization of the environmental impact. The profit is formed by the economic incomes due to the sale of crops minus the corresponding capital and operating costs associated to the fresh water, fertilizer, storage, piping, pumping and water treatment. Whereas the environmental impact is determined through the Eco-indicator 99 and the water footprint to evaluate the effect over the global warming, acidification, eutrophication, green water footprint, gray water footprint and blue water footprint. Two case studies from the Mexican state of Sinaloa were solved with the proposed model. The first one considers four crops of corn and it requires five irrigation periods; while in the second one twelve parcels are included where there are four crops of corn, four crops of bean, two crops of wheat and two crops of alfalfa, and there are needed seventeen irrigation periods. The operating, economic and design parameters are representative for the State of Sinaloa in Mexico; and the results are presented in three dimensional Pareto diagrams. From these diagrams is possible to find the optimal relationship between profit, eco-indicator and water footprint.

470200 Catalytic Depolymerization of Chitin into High-Value Platform Chemicals

Abstract:

Chitin, a major component of crustacean exoskeletons, is the second most abundant polysaccharide in nature after cellulose. Each year, billion tons of crustacean shells are discarded as waste. This is especially a critical problem in the Northeastern United States, where thousand pounds of seafood are processed each day. Since landfill of waste from seafood processing is prohibited, the current solution is to freeze the waste and ship it to a different country. Discovery of novel methods to better utilize this waste is thus important for local economy and environment protection. In our work, pyrolysis of chitin in the presence of solid acid catalysts is conducted. The experimental results show that the addition of catalysts during pyrolysis significantly reduces the yields of biochar and increases of yields of sugar derivatives. The yields of sugar derivatives are dependent on reaction temperature, reaction time, and catalyst loading. Key deploymerization reaction pathways leading to high-value products are identified.

468799 Assessing the Effect of Substrate and Catalyst on Catalytic Waste Gasification

Abstract:

This research is primarily aimed to extend proof-of-concept gasification experiments as a suitable alternative for waste management and sustainable technology. While initially intended to make more efficient use of resources in space exploration missions by producing high value reaction products from normally unused or wasted materials, the same principles can be applied to municipal waste as well.

The experiments performed examine gasification dynamics for different substrates promoted by different catalysts. It utilizes a high pressure batch scale reactor which allows examining controlled multi-phase catalytic isothermal oxidation of solid and melting substrates in the presence high purity air. The reaction time was varied between 2 and 24 hours and a preliminary overall kinetic model was formulated based on conversion, selectivity, and yield determinations. The studied substrates, Polyethylene and Cellulose, are both long chain organic polymers, and make up a substantial portion of both space and municipal waste composition. Although similar in nature, these substrates exhibit marked differences as it pertains to gasification, and were therefore selected as model substrates. Namely, cellulose contains oxygen as part of its molecular structure, while polyethylene does not. Since the technology investigated is aimed at working to relatively low temperatures (below 350 ^{o}C), cellulose remains as a solid particulate at the investigated reaction temperature while polyethylene melts.

Following the reaction phase, both the liquid and solid reaction residues, along with the gaseous products, are collected for characterization and analysis. Gas quantification is performed using a Gas Chromatograph. While thermal characterization (DSC) and Scanning Electron Microscopy (SEM) are used to characterize solid and liquid residuals.

Preliminary results suggest that catalytic gasification provides potential control for the liquid-phase oxidations. Under appropriate conditions, oxidation was found to follow the incomplete oxidation pathway which resulted in significant yields of hydrogen and methane, which are potentially high value products as they can be used for energy recovery.

Comparisons between the gasification rates for the two substrates are drawn. Similarly, comparative studies of different catalysts are also presented. The results are used in conjunction with a kinetic model for the gas-phase to formulate a methodological model for the gasification of solid waste. The results will be presented along with a preliminary extension for a continuous gasification process.

472942 Use of Cosolvents for Remediation of Hazardous Waste Sites and Their Effect on the Relative Mass Transfer and Reaction Kinetic Rates

Abstract:

Physical/chemical or biological remediation of sites contaminated with hazardous wastes with low aqueous solubility, is often limited by mass transfer of contaminants from the soil surface to bulk phase, consisting of reactive aqueous medium. To reduce the surface tension between soil and organic contaminants, alcohols, surfactants and household detergents were used as cosolvents to increase the aqueous solubility of hydrophobic chlorinated contaminants. Criteria for cosolvent selection were developed, which would lead to an inexpensive and nontoxic addition of the most efficient additive for subsurface remediation.

Relative efficiency of various cosolvents was determined under various pH and temperature conditions for spiked soil samples of a clean soil section from a trichloroethylene (TCE) contaminated site. TCE desorbed from the soil was reacted with potassium permanganate ($KMnO_4$) to determine the relative reaction and mass transfer resistances. Results of these studies will be presented and their significance to hazardous waste site remediation will be discussed.

451904 Case Study on the Application of Aspen HYSYS to Simulate and Optimise the Process of Plastics for the Production of Fuels

Abstract:

The global production of plastics was 311 million tonnes in 2014, and is expected to double over the next 20 years. Between 22-43% of the plastics is discarded in landfills resulting in wasting valuable resources, taking up valuable space and blighting communities. Recovering plastic wastes for recycling or for energy generation has the promise to decrease these problems. The pyrolysis of plastic wastes can provide a solution for these problems, where the plastics undergo a cracking process to produce lower molecular weight products. The products have similar compositions to gasoline, kerosene and diesel compounds, depending on the feedstock and process conditions.

However, the main problem facing plastics pyrolysis research is the complexity of the process with enormous numbers of reactions and hence, there are no models available to describe the complete process. Even that several models (mostly kinetic models) have been developed to describe this process, they have been unsuccessful. Overall, the lack of comprehensive understanding of this process dissuades its transformation into a profitable industry. The main aim of the project is to contribute to pollution prevention and treatment of plastic wastes by providing a valuable mechanism that will improve the research on plastics pyrolysis.

Low-density polyethylene (LDPE) is a type of polyolefin plastic, which is a common domestic plastic waste. Polyolefins account for 57% of the total amount of plastics present in household waste, of which polyethylene is the most abundant type of this group. The objectives of this project are to create hypo components that represent real LDPE, to design an effective and a complete computer-based modelled process that can simulate an actual LDPE pyrolysis process, to establish a method for predicting the pyrolysis products of LDPE with great accuracy, to determine the optimum operating conditions of LDPE pyrolysis for the production of liquid hydrocarbons for fuel generation and to predict the energy requirements for LDPE pyrolysis process.

In this study, two simulation frameworks were proposed in order to design and evaluate the pyrolysis process of low-density polyethylene to produce liquid hydrocarbons that are suitable for fuel production. The first simulation framework (Simple Model) was based on a simplified pyrolysis process where the products were combined into three groups (gas, oils and char), whilst the second one (Complex Model) was based on a simplified pyrolysis process where the detailed product distribution of the liquid fraction was obtained.

Aspen HYSYS was chosen as the modelling software to design and simulate the two frameworks of LDPE pyrolysis process. Aspen HYSYS is a market-leading process modelling tool for conceptual design and optimisation for oil and gas production, gas processing and petroleum refining industries. Since the products of LPDE pyrolysis process are mainly composed of hydrocarbons, HYSYS is suitable tool to model, predict and estimate the products obtained from this process.

LDPE was firstly developed as a number of hypo components in HYSYS. The software estimation methods were accurate for describing a complex material such as plastic. These components were classified into a number groups due to the variability of the composition of LDPE from plastic to plastic. By using HYSYS, the properties of the simulated plastic could be modified, to meet the experimental properties of the actual plastic either by changing the number groups or by changing their compositions. The created hypo components had properties very close to the real ones and exhibited an accurate behaviour when they undergone the pyrolysis process.

The results obtained from the simulation of the Simple Model agreed with the experimental data. This accuracy allowed for the simulation of the process at temperatures that were not achieved experimentally. The behaviours and properties of the three products of the Simple Model at different temperatures were accurately described. The classification of a large number of hydrocarbons into three different groups and the representation of each group with as a single hydrocarbon did not adversely affect the simulation accuracy of LDPE pyrolysis process. Although the material balances of these components were different from real reactions, the pyrolysis process was simulated with greater accuracy in comparison with published results.

The results of the Complex Model simulation showed a clear improvement compared to those of the Simple Model, whilst maintaining a great agreement with the experimental data. Moreover, the description of the pyrolysis process was more detailed, improving the quality of predicting the compositions of the products. Increasing the number of the products also resulted in a more detailed process flowsheet, providing more complete operational specifications, including the energy duties needed for the LDPE pyrolysis process. Pyrolysis products have different applications and the determination of the optimum application for each product is necessary to increase the profitability of the process. Thus, each component needs to be taken into account and managed accordingly.

It was observed, from the simulation results, that it is possible to describe the LDPE pyrolysis in HYSYS with more accuracy than other methods that have been studied previously. Both simple and complex models were proven to be accurate for simulating LDPE pyrolysis process. Overall, both models had shown their appropriateness to overcome the research challenges of plastics pyrolysis.

459529 Experimental Study on Gasification of Anaerobic Digestion Solid Residue Derived from Municipal Solid Waste: Effects of Temperature, Moisture Content and Gasifying Agent

Abstract:

In this work, gasification is used to harvest the energy remained in the solid residue produced from the anaerobic digester in treating municipal solid waste (lignocellulosic biomass waste and dewatered sewage sludge) due to its limited biodegradability. Compared with conventional disposal route of AD (Anaerobic Digestion) solid residue, such as incineration, gasification is more environmentally friendly with fewer pollutant emissions and the synthesis gas produced in this process can be directly utilized in gas engines or turbines to generate electricity or be upgraded to transportation fuels through some chemical process. The effects of operating parameters of a fixed bed reactor, like temperature and gasifying agent, and the property of feedstock, like moisture content, on the gasification performance have been experimentally studied. Both energy indicator, like CGE (Cold Gas Efficiency) and environmental pollutant indicator, like NOx emission and heavy metals concentration, are used to evaluate the gasification process. In conclusion, feasibility of gasification of AD solid residue is justified and optimized operating conditions are found based on both energy and environmental performance.

Chevron

human energy

Sustainable Development and Footprint Optimization in Shale Development

Janet Peargin, Footprint Program Manager
Hong Jin, Lead HES Engineer

Chevron Energy Technology Company

November 17, 2016

The challenge
footprint of shale/tight rock development

Many wells & moving parts

Footprint changes over time

- Jobs
- Community GDP
- Desired energy

- Land/habitat
- Emissions
- Trucks
- Noise
- Waste
- Resource competition

Chevron's environmental principles

- Include the environment in decision making
- Reduce our environmental footprint
- Operate responsibly
- Steward our sites

Footprint optimization: everybody plays a role

We can't solve problems by using the same kind of thinking we used when we created them.

– Albert Einstein

Shifting your perspective is so much more powerful than being smart.

– Astro Teller, Google "X" 2016 TED Talk

Chevron

The opportunity: sustainable development

SEE across the asset life to create value and protect people and the environment

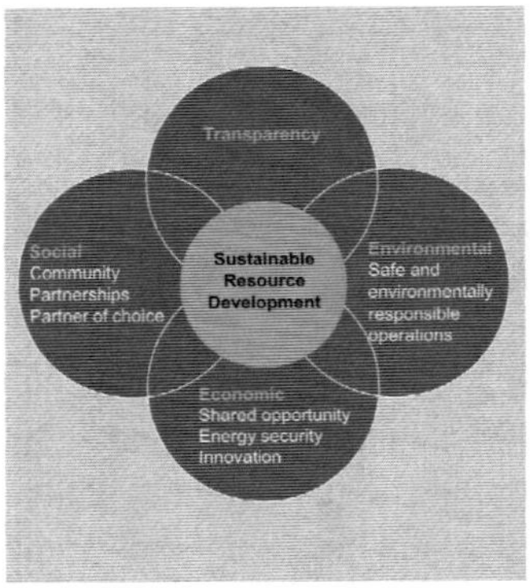

SEE triple bottom line value

SEEk win-win (all-win) ideas
Social, Environmental, Economic

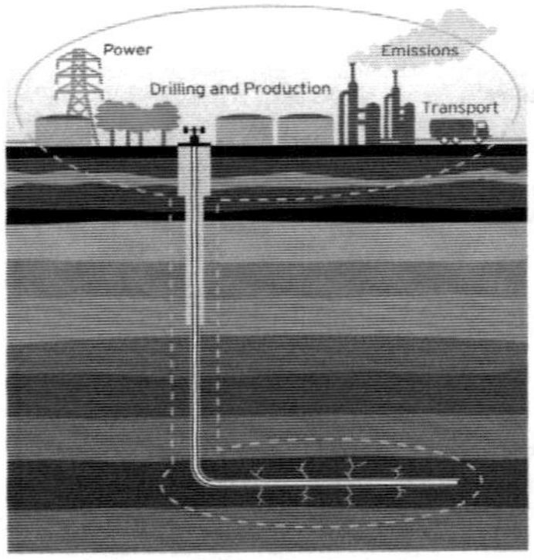

Optimize footprint

Integrated system of materials and energy that changes over time

Design with the end in mind

Decommission, restore, reuse

Chevron

The win-win (all-win): optimize footprint to enhance value and reduce cost/risk

Transparency

- Compact AND safe
- Conserve water and natural resources
- Preserve habitat and biodiversity
- Reduce waste, energy, noise, air emissions
- Optimize logistics and truck traffic
- Provide net benefit to community

Social

achieved through

Environmental

1. **Integrated, long-term plans.** Frame for SEE drivers
2. **Informed decisions.** Include SEE drivers. Use ISO LCA*
3. **Good design.** Mobile, compact, flexible, efficient

Economic

*Life Cycle Assessment Standard 14040-14044

Improve planning: SEE through the lens of sustainable development and SEE win-win (all-win) opportunities

Look at the big picture (asset, region)
Look at the long-term and change over time
Include diverse stakeholder perspectives

Take a systems view. Integrate across functions, activities, hydrocarbon/water/power
Plan and design for social, environmental, and economic (SEE) value drivers

Consider:

- Pace and scale of development
- Sequencing of events
- Degree of centralization
- Logistics and supply chain
- Ability to share infrastructure
- Power/fuel options
- Integration points

Improve decision quality: SEE more broadly and longer-term

Compliance is expected.

A traditional financial analysis is narrow and short-term.

SEE more broadly and longer term to identify risks and opportunities ... and enhance decision quality.

Improve decision quality: quantify environmental tradeoffs with ISO life cycle assessment (LCA)

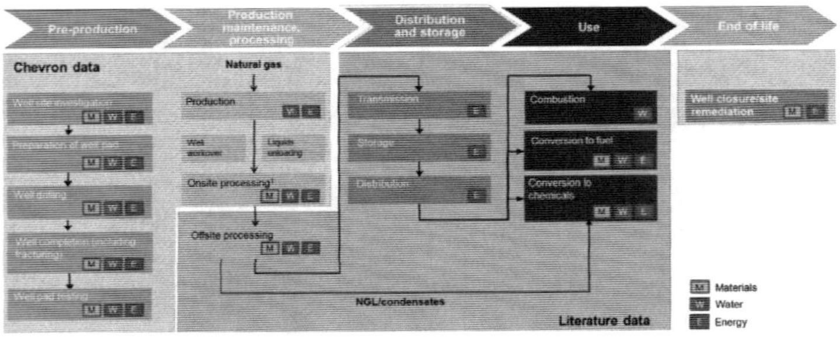

- What is our environmental footprint?
- Where are our big sources?
- What is the "best" fuel for power?
- What if I use this technology vs. that?
- What are the on and offsite impacts?

Improve design: apply win-win (all-win) principles

Plan/design for sustainable development	• Develop long-term integrated plans • Frame for SEE (social, environmental, economic)
Integrate and innovate	• Look across the asset life at hydrocarbons/water/power & different activities (drilling, operations, etc.)
Question business as usual	• What do stakeholders want? Is there a better way?
Do *more* with *less*	• Reduce – reuse – recycle – be (energy/materials) efficient • Eliminate no-value steps & equipment
Disturb less land	• Fewer pads; share infrastructure/right-of-ways; preserve habitat corridors
Conserve water	• Use less & lower quality; treat to quality needed for use; consider reuse
Reduce waste	• Flares/emissions, discharges, energy use, no-value steps/equipment
Green the process	• Consider non-toxic chemistry, power/fuel efficiencies, fewer steps
Leverage technology	• Look to downstream, offshore, suppliers, competitors, others • Think mobile, modular, compact, flexible, prefab, energy efficient, automated, standard
Leverage supply chain	• How can suppliers help? Can we hire and source locally?
Make informed decisions	• SEE the big picture. Consider short- *and* long-term costs and value. • Use life cycle assessment to quantify environmental aspects of alternatives

Chevron

Win-win example: electricity from field gas

Better emissions/safety/reliability/production/cash flow

Problem

Need cost-effective, low-emissions power source at remote site

Solution

Micro-turbines make electricity from field gas*

Win-Win Do *more* with *less*

MORE

Safety, flexibility, schedule certainty, reliability (8–10%), production/cash flow (sooner)

LESS

Equipment, permitting, O&M, trucks, emissions, material transfer/spill potential

** Great for ESPs. Rod pumps doable with more systems engineering*

Chevron

Everyone plays a role: ask questions

- What are our stakeholder concerns?
- What perspectives are we missing?
- What are our footprint boundaries?
- What are the priorities and tradeoffs?
- What does good (enough) look like?
- How can we minimize environmental and community impacts?
- How do we do more (product) with less (material, equipment, waste, etc.)?
- How do we design for change over time to operations, region, community?
- What can we learn from other processes?
  - LeanSigma®
  - Environmental, social, health impact assessment
  - Energy Efficiency and Waste Minimization assessments

LeanSigma is a federally registered service mark of TBM Consulting, Inc.

Key takeaways

- We all play a role in creating our company footprint – a system of materials and energy that changes over time
- Think beyond compliance. Include SEE (social, environmental, economic) drivers in planning, designs, and decisions.
- SEE through the lens of sustainable development to SEE opportunities. SEE the difference.
  - Cost savings from less equipment, optimized layout/network
  - Win-win ideas to save $ AND protect people & the environment
  - Decisions that are more robust
- A financial analysis is only part of the risk/uncertainty. We <u>can</u> quantify and include SEE drivers in decision-analysis.
- Sustainable development is good business. SEE it reduce cost and risk, add value, and improve performance.
- Everyone plays a role. You can contribute. Just change your perspective to SEE the win-win (all-win) opportunities to do more (product, energy) with less (energy, materials, waste, emissions, trucks, etc.)

References

- Hardisty, P.E. *Environmental and Economic Sustainability.* CRC Press, New York. ISBN 978-14200594891-56679-643-2, 378 pp. 2010.

- Sivapalan, M.; WorleyParsons. *Sustainable Development of Unconventional Oil & Gas Resources.* Prepared for presentation at *GasTech* 2012 Conference, London, 2012.

- Tanzil, D.; Dragan, M.; Bourque, B.; Seel, K.; Taylor, T.; Golder Associates. *Evaluating Social and Environmental Risk and Opportunities in Site Selection and Routing.* Copyright 2015, Society of Petroleum Engineers. SPE-173488-MS. Prepared for presentation at the SPE E&P Health, Safety, Security, & Environmental Conference – Americas held in Denver, Colorado, USA, 16–18 March 2015.

- Thomas, Charlotte. I*ndustrial Ecology: Moving Beyond Traditional Engineering Borders.* Society of Women Engineers magazine, spring edition, pp. 24-30. 2010

461852 Green Net Value Added As a Sustainability Metric Based on Life Cycle Assessment: An Application to Bounty ® Paper Towel

Abstract:

Sustainability measurement in economics involves evaluation of environmental and economic impact in an integrated manner. In this study, we combine system level economic data with environmental impact from a life cycle assessment (LCA) of a common product. We are exploring a costing approach that captures traditional costs and incorporates externality costs to provide a convenient, easily interpretable sustainability metric. Green Net Value Added (GNVA) is a type of full cost accounting that incorporates total revenue, the cost of materials and services, depreciation, and environmental externalities. It adjusts the traditional economic value added for environmental externality costs. Two, but not all, of the potential environmental impacts calculated by the standard LCIA method (TRACI) could be converted to externality cost values. We compute externality costs disaggregated by upstream sectors in order to accurately reflect the externality costs associated with different industries along the supply chain. We preset full cost, green profit and GNVA to evaluate the relative sustainability of Bounty® paper towels manufactured at two production facilities. We found that the longer running, more established line had a higher GNVA than the newer line. The dominant factors contributing to externality costs are to come from the stationary sources in the supply chain: electricity generation (27-35%), refineries (20-21%), pulp, and paper making (15-23%). Health related externalities from Particulate Matter (PM2.5) and Carbon Dioxide equivalent (CO_2e) emissions appear largely driven by electricity usage and emissions by the facilities, followed by pulp processing and transport. Supply chains for other facility purchases, product distribution, and end-of-life management (landfilling/incineration) contribute the least to the environmental health-related costs.

474653 Sustainability Metrics for Solar Powered Charging Infrastructure for Electric Vehicles

Abstract:

This presentation will include an analysis of sustainability metrics associated with the installation of solar panels in parking lots designed to provide electricity for commercial use and shade for those who park under the solar panels. The installation will also include electric vehicle supply equipment to charge batteries in electric vehicles. The system is connected to the electrical grid so that the electricity generated can be used efficiently and electric vehicle batteries can be charged at all times. Important sustainability metrics are cost of electricity generated, cost of vehicle charging, value of shaded parking, convenience of vehicle charging, land area, water, air quality, greenhouse gas emissions, and materials for the installation and equipment. Social metrics include impact on health status of the community, psychological well being, harmony, values, agency, and environmental justice.

458893 Evaluation of Low-Energy Recovery of N-Methyl-2-Pyrrolidone Solvent from Waste Streams in Specialty Resin Manufacture

Abstract:

N-methyl-2-pyrrolidone (NMP) is a common solvent used as a reaction medium in the manufacture of specialty polymeric resins and their precursors. NMP is used primarily because of its thermal stability, dipolar aprotic properties, and ability to produce a high molecular weight resin precursor. Since NMP is not consumed by the reaction, an aqueous hazardous waste stream is generated. A previous case study at a DuPont specialty chemical plant by the authors, shows that a distillation recovery system that can effectively purify NMP from the waste stream. The distillation recovery process reduces operating cost by 83%, but only reduces the life cycle carbon footprint by 40%. This is due to the high energy demand of the distillation column. Additionally, high NMP purity levels in the resin manufacture process and the complex multicomponent nature of the waste stream, required a more extensive design for the recovery operation.

Alternative distillation designs were explored, but none reduced energy consumption and life cycle carbon emissions as significantly as the economic life cycle impact. More sustainable low energy separations were investigated. Various membrane, adsorption, and extraction separation strategies have the potential to reduce energy demand, but each has its own limitations. An extraction operation was selected as the best candidate for separation, and the proposed process considers alternative uses for the recovered NMP waste in other manufacturing scenarios.

Results with a chloroform extraction operation to produce NMP of suitable quality for alternative use, provides the best option for solvent recovery. The new process can not only yield environmental life cycle improvements, based on various impact factors, but provide economic benefits as well. Yearly projections of waste reductions, life cycle improvements, energy utilization, and cost savings are presented. Life cycle carbon emissions are reduced by over 60% compared to the base case. Since NMP is a prominent solvent for several commercial sectors, methods aimed at reducing waste help promote broader sustainability goals. This study shows how sustainability metrics can be used to guide decision making when evaluating other processing alternatives and uses of the final product. (This work is partially supported by a grant from the EPA Pollution Prevention program).

468272 Thermoelectric Power Technology Choices Based on Water Availability

Abstract:

Thermoelectric Power Technology Choices Based on Water Availability National Energy Technology Laboratory (NETL) has developed a prototype water-energy model to investigate issues surrounding the energy-water nexus as it relates to the effect of power generation plant design choices on water availability and the overall costs of supplying water for thermoelectric power generation.Â The model utilizes the database developed by a team of researchers with the Water Security Program at Sandia National Laboratory (SNL). The database includes water availability, cost, and water use for electric power, agriculture, municipal and industrial for 38 conterminous states representing various Interconnects within the U.S. Data on current and future water use for thermoelectric power are derived from NETL databases on water use by technology and projections from the Energy Information Administrationâ€™s National Energy Modeling System (NEMS). Current and forecasted water availability data from the World Resources Institute are used to develop three water availability scenarios. The prototype water-energy model is used to project the amount of water available for power generation, to project water demand for future power generation based on a NEMS forecast, to identify regions that experience potential water shortfalls, and to estimate the cost of meeting water demand where water shortfalls exist for each water availability scenario. A case study of the Brazos River Basin in Texas is used to illustrate a specific application and to highlight the capabilities of the prototype water-energy model.

AUTHOR INDEX

AUTHOR INDEX

AUTHOR INDEX

AUTHOR INDEX

AUTHOR INDEX

AUTHOR INDEX

AUTHOR INDEX

AUTHOR INDEX